INSIGHTS INTO THE REACH TO GRASP MOVEMENT

ADVANCES IN PSYCHOLOGY

105

Editors:

G. E. STELMACH

P. A. VROON

NORTH-HOLLAND
AMSTERDAM • LONDON • NEW YORK • TOKYO

INSIGHTS INTO THE REACH TO GRASP MOVEMENT

Edited by

Keree M. B. BENNETT
General Psychology Department
University of Padova
Padova, Italy

Umberto CASTIELLO
Department of Psychology
University of Bologna
Bologna, Italy

1994

NORTH-HOLLAND
AMSTERDAM • LONDON • NEW YORK • TOKYO

ELSEVIER SCIENCE B.V.
Sara Burgerhartstraat 25
P.O. Box 211, 1000 AE Amsterdam, The Netherlands

Library of Congress Cataloging-in-Publication Data

Insights into the reach to grasp movement / edited by Keree M.B.
Bennett, Umberto Castiello
p. cm. -- (Advances in psychology)
Includes bibliographical references and indexes.
ISBN 0-444-89931-6
1. Hand--Movements. I. Bennett, Keree M. B., 1960- .
II. Castiello, Umberto, 1959- . III. Series: Advances in
psychology (Amsterdam, Netherlands) ; 105.
QP334.I565 1994
152.3--dc20 93-47507
CIP

ISBN: 0 444 89931 6

Transferred to digital printing 2005

To Eliah

Preface

The number and diversity of authors who have agreed to contribute to this publication is sufficient evidence in itself to indicate that the study of the reach to grasp movement is not a trivial venture. The professions covered in this book range from those interested in the basic sciences to those more interested in practical application. Neurophysiologists and biomechanists join with therapists and neural modelers to present an extensive overview of current developments in this field. Evolutionary and developmental aspects are included together with descriptions of how this movement is affected by central nervous system damage. Purely theoretical aspects of the motor control of this movement are interspersed with treatment applications and robotics.

Interest in the reach to grasp movement was largely stimulated by the works of Marc Jeannerod who has kindly agreed to write the first chapter. In an elegant series of experiments he demonstrated that the reach to grasp movement is composed of two major components. One is the transport component which acts to bring the arm and hand to the target object to be grasped. The other is the manipulation component which enables appropriate preshaping of the hand prior to enclosure and successful grasping of the target object. Each component was found to have a distinct kinematic arrangement, prompting Jeannerod to propose a central nervous system channel for transport which was relatively independent of that for manipulation. He also found that kinematic landmarks of one component were coupled in time with kinematic landmarks of the other. This led him to the conclusion that this temporal coordination allowed for both successful hand positioning and object grasping.

There are a number of advantages in using this movement as an experimental model. It is a natural task and thus the results from pure neurophysiological studies or from the more black box behavioural studies are likely to reveal normal rather than experimentally induced central mechanisms. In the words of Marr and Nishihara ([1] p. 21) it is an operation " .. that we as human beings perform well, fluently, reliably, and hence unconsciously .. ". Despite the simplicity of its execution, detailed assessments of its kinematic and biomechanical organization can assist in our understanding of the processing within these neural channels. As a result, this movement has been well-characterized for subjects who have no neurological damage. For subjects with central nervous system damage the subtle effects upon movement kinematics enable a more complete description of dysfunction and consequently assist with diagnosis and with the design of appropriate therapeutic regimes. Similarly, studies of how the human child develops the ability to reach and grasp not only show the learning of a goal-directed skill but ultimately assist in the identification of abnormalities of motor control at each developmental stage. The utilization of the reach to grasp movement across a variety of species allows comparisons which illustrate the unique features of human movement. Finally, the results from the basic behavioural studies can be utilized for the development of neural and biomechanical models and for valid application to the design of functional robotic arms.

The first section of this book deals with the evolutionary and developmental aspects of the reach to grasp movement. Mary Marzke takes us from the characteristics of our early prosimian-like ancestors to those of our ape-like and then our hominid ancestors. This evolution has granted us the ability to reach in several directions within extracorporeal space and to firmly grasp and to manipulate objects, particularly tools, within our hands. Andrew Gordon concentrates primarily upon the human ontogeny of grasping function. He describes the excessive force generation of young children and their mode of sequentially activating the different forces required for the grasp of an object. With adulthood the grasping of an object is simplified, grip and load forces being synergically and functionally coupled. It is proposed that much of this development is dependent, not only upon function driven experiential behaviour, but upon maturation of such central nervous system structures as the corticospinal tract and somatosensory systems.

The second section presents some of the neural control mechanisms which subserve the reach to grasp movement. Andrew Schwartz provides a historical review of the neurophysiological and neuroanatomical mechanisms subserving the reach to grasp movement. Until quite recently, the brain has been thought to consist of almost independent sections, each with their own specific function. He demonstrates that this line of reasoning has been largely refuted. Rather, the central nervous system demonstrates distributed processing - different areas operating in a cooperative and parallel manner. Alan Gibson, Khris Horn and Peter van Kan discuss their recent findings with regards to the cerebellar substrates of the monkey reach to grasp movement. After testing several motor variations, such as reaches without a grasp, or vice versa, they conclude that the intermediate cerebellum is specialized more for the control of particular grasp, rather than reach, functions. The contribution of positron emission tomography (PET) in the anatomical identification of central mechanisms is presented by Jean Decety. To date, few to no PET studies have focussed on the reach to grasp movement. Nevertheless, he presents an overview of the research on the more general topic of upper limb function. Again, the evidence favours distributed processing between widely separate areas of the cerebral cortex such as the supplementary motor area, the posterior parietal cortex and the motor cortex, and between subcortical structures such as the basal ganglia and cerebellum.

The third section consists of three chapters which describe some of the latest kinematic behavioural human studies. Patricia Weir reviews the influence of object properties and task-related factors upon the organization of the reach to grasp movement. The transport and manipulation components appear to be coupled not only in a temporal sense but also in spatial and functional manners. She describes the influence of sensory information, such as visual cues, upon the control of a coordinated movement. Patrick Haggard speaks of methods which have been used to perturb the reach to grasp movement and how the resulting patterns of the perturbed motor behaviour have assisted in our understanding of the coordination between the transport and manipulation components. Consistent findings that perturbation of one component influences not only its organization but that of the other component support the idea of coupling, be it temporal, spatial or functional, between transport and manipulation. The contribution of vision for patterning and coordination is addressed by Keree Bennett, Carl Waterman, Carla Mucignat and Umberto Castiello. In studies with blind, blindfolded and full vision subjects, they demonstrate that blind subjects develop a reach to grasp movement which is kinematically similar to those of full vision subjects. They discuss how these findings give clues as to the function of vision in full vision subjects.

The fourth section gives descriptions of the abnormalities which can occur with specific central nervous system damage and outlines new approaches to physical treatment. Lorna Jakobson and Melvyn Goodale describe the motor dysfunctions resulting from focal lesions to the dorsal visual pathway connecting the primary visual cortex with the posterior parietal cortex and frontal lobe regions. Based on these clinical findings, they argue that the function of this pathway for visuomotor transformations is distinct from mechanisms which underlie visual representations of objects in extrapersonal space. Umberto Castiello, Keree Bennett and Marina Scarpa outline their kinematic studies, including perturbation, of the reach to grasp movement of

Parkinson subjects. They demonstrate that these subjects show delays of movement activation and discuss this finding in relation to the growing body of evidence suggesting deficits in the performance of sequential or simultaneous movement components. Louise Ada, Colleen Canning, Janet Carr, Sharon Kilbreath and Roberta Shepherd outline both the negative and positive dyscontrol features following upper motor neuron lesions. They propose that the nervous system utilizes what remains after such lesions to form compensatory motor patterns and propose task-specific training techniques to potentiate optimal functional reorganization of the brain.

The fifth section deals with the biomechanical and robotic aspects of this movement. Kai-Nan An and Marjorie Johnson demonstrate that shoulder function, in moving the arm within the range of a sphere, is mutually exclusive from elbow function which acts to change the position of the hand relative to the body or to the external environment. With regards to the manipulation component, they show that subjects have a self-selected wrist position which facilitates grasp strength. Andrew Fagg introduces a reinforcement learning technique for the training of a neural system to control the reach of a robot with two prismatic joints. He discusses how such work contributes to our understanding of how the various functions provided by different neural regions work together for learning computation and task performance.

In the final section Michael Arbib and Bruce Hoff provide an extensive review of neural network modeling. They propose that the many approaches to modeling nervous system and behavioural function reflect the diverse backgrounds of the researchers. Even though, it appears that optimization operations, such as the minimization of a certain motor parameter, are gaining support, they suggest that further understanding of the underlying neurophysiological, neuroanatomical and computational abilities and limitations within a single accepted paradigm should assist in furthering our understanding of motor control. Mitsuo Kawato and Yoji Uno, in the last chapter of this volume, consider the reach to grasp movement as an ill-posed motor problem; that is, the solution for movement performance is not unique. They describe, in particular, different neural modeling optimization principles which have been attempted to solve this problem and conclude that the origin of smoothness constraints may lie more at the motor command level within the central nervous system than at the level of task space, of joint-torque coordinates or of muscle tension.

In conclusion, we feel that this volume provides a necessary and valuable contribution to the field of motor control. The study of reach to grasp movement is already classical, yet its use for further investigations and for application will continue for many years to come. New approaches to this topic are already evident. For example, this book has touched little upon the cognitive, perceptual and attentional factors related to this movement. In addition, the current and exciting developments in virtual reality promise an alternative means for studying the reach to grasp movement.

REFERENCES

1 D. Marr and H.K. Nishihara, Technology Review, 81 (1978) 1

Acknowledgements

Our farming neighbours, in this hilly and scenic region just south of Parma, did not really appreciate why we spent so much time on such fascinating topics as prehension components and neural modeling, but, all the same, gave us much encouragement and fed us many times with the world famous Parma cuisine. Perhaps they were a little upset that we did not include a chapter entitled "Hands in the Kitchen", however our warm thanks are extended to all the inhabitants of Cafragna: Maria, Aldo, Vittorio, Flavia and Angelo Coruzzi, Theresa, Claudio, Costantino and Marco Berni. Angiolina, Blandina, Dario, Giorgio, Pietro and Simone Cavazzini, Fausta, Roberto and Giacomo Pozzi-Ganazzoli.

All the authors are gratefully acknowledged. They have put in a tremendous amount of time and effort and have been very patient with us. Even though this is an invitation volume, each chapter was reviewed by two and sometimes three referees. All authors have promptly and conscientiously addressed the resulting suggestions and we feel that, as a result, this has significantly contributed to the quality of the book. We also thank the referees.

We began this editorial process in America before returning to Italy. Our gratitude is thus extended to many people at Arizona State University: to Professor James Skinner, Laboratory Director of the Department of Exercise Science and Physical Education, for allowing us to use the facilities, and to Dolores, Susan Broussard, Janelle Ederer, Barbara Hamilton and Barbara Nyman for their extremely efficient secretarial assistance.

In Italy we have many people to thank. Claudia Cavatorta, Giuseppe Ferrari, Liliana Gastaldi, Fabrizio Orzi, Renzo and Liliana Tebaldi, Dr. Antonio Caretta and Dr. Roberto Tirendelli, all of the Human Physiology Institute, University of Parma, assisted in ensuring that the first round correspondence which was addressed to Umberto Castiello in Parma, found him in Bologna. Professors Gianni Brighetti, Corrado Cavallero and Paolo Neyroz are thanked for allowing Umberto Castiello to utilize their facilities at Bologna University when he first arrived there. Professor Pio Ricci Bitti director of the Department of Psychology, University of Bologna, is also acknowledged. Professor Carlo Umiltá, General Psychology Department,

University of Padova, is thanked for his continuing support and encouragement.

Finally, our gratitude is extended to our friends and families.

K.M.B. Bennett and U. Castiello
Cafragna, Italy

Permissions

We would like to acknowledge the following authors and publishers for allowing us to reproduce formerly published figures:

Ablex Publishing Corporation for Fig. 2 (Jordan, copyright 1992) of Chapter 15.

Churchill Livingstone Incorporation for Fig. 9 (Shepherd, copyright 1991) of Chapter 12.

I.E.E.E. for Fig. 5 (Kawato et al., copyright 1987) of Chapter 15.

J.B. Lippincott Company for Fig. 3 (O'Neill, copyright 1992) of Chapter 13.

Mayo Foundation for Fig. 1 (An et al., copyright 1991) of Chapter 13.

M.I.T. Press for Fig. 4 (Houk et al., copyright 1990) of Chapter 15.

Orthopaedic Research Society and the authors for Fig. 2 (An et al., copyright 1991) of Chapter 13.

Pergamon Press for Fig. 5 (Wada and Kawato, in press) of Chapter 16, for Figs. 1 and 2 (Castiello et al., copyright 1993) of Chapter 11, and for Fig. 2 (Jakobson et al., copyright 1991) of Chapter 10.

Reza Shadmehr for Fig. 1 (Shadmehr, copyright 1991) of Chapter 15.

Springer Verlag for Fig. 8 (Iberall et al., copyright 1986) of Chapter 15, for Figs. 2 and 3 (Uno et al., copyright 1989) of Chapter 16, for Figs. 1 and 4 (Kawato et al., copyright 1990) of Chapter 16, and for Fig. 1 (Castiello et al., copyright 1993) of Chapter 9.

The American Physiological Society for Figs. 4, 5 and 6 (van Kan et al., copyright 1993) of Chapter 5.

Table of Contents

List of Contributors

LOUISE ADA - Cumberland College of Health Sciences, University of Sydney, East Street, Lidcombe, N.S.W., Australia, 2141.

KAI-NAN AN - Orthopedic Biomechanics Laboratory, Mayo Medical School, Rochester, Minnesota 55905, U.S.A.

MICHAEL A. ARBIB - Center for Neural Engineering, University of Southern California, Los Angeles, CA 90089-2520, U.S.A.

KEREE M.B. BENNETT - Dipartimento di Psicologia Generale, Universitá di Padova, Piazza Capitaniato 3, 30139, Padova, Italy.

COLLEEN G. CANNING - Cumberland College of Health Sciences, University of Sydney, East Street, Lidcombe, N.S.W., Australia, 2141.

JANET H. CARR - Cumberland College of Health Sciences, University of Sydney, East Street, Lidcombe, N.S.W., Australia, 2141.

UMBERTO CASTIELLO - Dipartimento di Psicologia, Universitá di Bologna, Viale Berti Pichat, 5, 40127, Bologna, Italy.

JEAN DECETY - INSERM Unité 94, Vision et Motricite, 16 Avenue du Doyen Lépine, F-69500 Bron, France.

ANDREW H. FAGG - Center for Neural Engineering, University of Southern California, Los Angeles, CA 90089-2520, U.S.A.

ALAN R. GIBSON - Division of Neurobiology, Barrow Neurological Institute, 350 West Thomas Road, Phoenix, AZ 85013-4496, U.S.A.

MELVYN A. GOODALE - Department of Psychology, University of Western Ontario, London, Ontario, N6A 5C2, Canada.

ANDREW GORDON - Department of Physiology, 6-225 Millard Hall, University of Minnesota, Minneapolis, MN 55455, U.S.A.

PATRICK HAGGARD - University Laboratory of Physiology, Parks Road, University of Oxford, Oxford OXI 3PT, England.

BRUCE HOFF - Center for Neural Engineering, University of Southern California, Los Angeles, CA 90089-2520, U.S.A.

KHRIS M. HORN - Division of Neurobiology, Barrow Neurological Institute, 350 West Thomas Road, Phoenix, AZ 85013-4496, U.S.A.

LORNA S. JAKOBSON - Department of Psychology, University of Western Ontario, London, Ontario, K7L 3N6, Canada.

MARC JEANNEROD - INSERM Unité 94, Vision et Motricite, 16 Avenue du Doyen Lépine, F-69500 Bron, France.

MARJORIE E. JOHNSON - Orthopedic Biomechanics Laboratory, 128 Guggenheim Building, Mayo Clinic, Rochester, Minnesota 55905, U.S.A.

PETER L.E. VAN KAN - Division of Neurobiology, Barrow Neurological Institute, 350 West Thomas Road, Phoenix, AZ 85013-4496, U.S.A.

MITSUO KAWATO - A.T.R. Human Information Processing Research Laboratories, 2-2, Hikaridai, Seika-cho, Soraku-gun, Kyoto 619-02, Japan.

SHARON KILBREATH - Cumberland College of Health Sciences, University of Sydney, East Street, Lidcombe, N.S.W., Australia, 2141.

MARY MARZKE - Department of Anthropology, Arizona State University, Tempe, Arizona 85287-2402, U.S.A.

CARLA MUCIGNAT - Istituto di Fisiologia Umana, Universitá di Parma, Via Gramsci 14, 43100 Parma, Italy.

MARINA SCARPA - Clinica Neurologica, Universitá di Modena University, Largo del Pozzo, 71, 41100 Modena, Italy.

ANDREW B. SCHWARTZ - Division of Neurobiology, Barrow Neurological Institute, 350 West Thomas Road, Phoenix, AZ 85013-4496, U.S.A.

ROBERTA B. SHEPHERD - Cumberland College of Health Sciences, University of Sydney, East Street, Lidcombe, N.S.W., Australia, 2141.

YOJI UNO - A.T.R. Human Information Processing Research Laboratories, 2-2, Hikaridai, Seika-cho, Soraku-gun, Kyoto 619-02, Japan.

CARL WATERMAN - Department of Exercise Science and Physical Education, Tempe, Arizona, 85287-0404, U.S.A.

PATRICIA L. WEIR - Department of Kinesiology, University of Windsor, Windsor, Ontario, N9B 3P4, Canada.

Overture

Insights into the Reach to Grasp Movement
K.M.B. Bennett and U. Castiello (Editors)

CHAPTER 1

OBJECT ORIENTED ACTION

M. JEANNEROD

Vision et Motricité, INSERM U94, 16 Avenue du Doyen Lépine, 69500, Bron, France.

SUMMARY

This chapter reviews some of the aspects of object oriented behaviour during the action of grasping and describes the kinematic aspects of grip formation during object acquisition. It is postulated that hand shaping during grasping is largely based on a ***pragmatic*** *representation of the object attributes which are relevant to action. This mode of representation is contrasted with another,* ***semantic****, mode for object recognition and categorization. A possible cortical mechanism underlying the* ***pragmatic*** *representation is outlined.*

INTRODUCTION

Hand movements have been the focus of an increasing number of studies in the last ten years. The advent of computerized motion analysis in man, together with the use of neurophysiological techniques in behaving monkeys, have made it possible to describe the pattern of hand movements during object oriented actions, and to identify some of the brain mechanisms involved in the control of these movements. Grasping, however, cannot be reduced to its motor aspects - it is the motor counterpart of a broader function. During handling and manipulation, for which grasping is a pre-condition, signals for object identification, arising from sight and touch, are co-processed. The fingerpads have been considered by some authors (e.g., Sherrington) as the somatosensory "macula". Thus, the hand brings objects to be manipulated within the central field of vision, so that "the finest movements of the fingers must be under simultaneous control from the very centers of the visual and tactual maculae" [1]. Berkeley in his famous treatise on **A New Theory of Vision** [2], emphasized the fact that objects can only be known by touch, which he considered as the ultimate means of exploration and knowledge of the world. Vision is subject to illusions which arise from the distance-size problem (size must be extracted from apparent distance) or from the 3-D reconstruction problem (the visual third dimension is extracted from a two-dimensional map using indirect cues from perspective). Touch, and particularly active touch, is not subject to these constraints, as it involves direct assessment of size and volume. In addition, touch is critical for perceiving object properties like hardness, compliance, texture, temperature, weight, etc., which can hardly be accessed by sight alone.

One of the main aims of this paper is to defend the idea that grasping is driven by internal representations about objects. Thus, the initial stage of the grasp (the preshape that occurs during the reach), is a transitional stage where the hand posture and movements express the representation that the subject has formed about the object. Fingers shape in anticipation to object size and shape, and the wrist rotates in anticipation of object orientation to give optimal stability to the grasp. In addition, during this stage, forces are generated which will be applied immediately at the time of contact, in order to grasp (grip force) and lift (load force) the object. These forces are largely adjusted on the basis of visual cues about whether the object is deformable, slippery, heavy, etc. (see [3]).

THE HAND, A PRIMATE ATTRIBUTE

In the motor domain, the hand (together with the speech apparatus) is one of the highest achievements of man. Napier [4] considered that human prehensile movements can be described along two main motor patterns. "If", Napier stated, "prehensile activities are to be regarded as the application of a system of forces in a given direction then the nature of prehensile activity can be resolved into two concepts - that of precision and that of power" (p. 906). The precision and the power grip patterns can be used alternatively or in combination for almost every object. In other words, the pattern of the grip is not determined solely by the shape or the size of the object (e.g., a rod can be held with a precision grip, as in writing, or a power grip, as in hammering); it is the intended action which is the main determinant of the type of grip for each given action.

The two grips differ anatomically by the relative postures of the thumb and the other fingers. Precision grip is mostly characterized by opposition of the thumb to one or more fingers. Opposition means that the thumb is abducted at the metacarpo-phalangeal and at the carpo-metacarpal joints, so that its pulp surface is diametrically opposed to the pulpar surface of the other fingers. In power grip, the fingers are flexed to form a clamp against the palm, the thumb is abducted at the two joints, and there is no opposition between the thumb and the other fingers. The two types of grip have clearly different degrees of involvement in manipulative actions, only the precision handling allows movements of the object relative to the hand and movements of the object within the hand [5]. The patterns of neuromuscular innervation for the two types of grip seem to be very different. In power grip, fingers are flexed by powerful bi-articular muscles, while during precision handling, there is an interplay between flexion and extension.

Precision grip with true opposition of the pulpar surfaces of the thumb and the index finger is considered as the top attribute of dexterous hands. The problem of whether this attribute is specific to the human hand or not, is a matter of discussion. The Heffner and Masterton scale for ranking digital dexterity [6], based on anatomy of the hand, includes only man in the topmost category, that with opposable thumb and precision grip (see also [7]). It remains that the classification based on anatomy of the hand is probably under inclusive, and should be counterbalanced by a classification based on behavioural observation with emphasis on the capability for independent finger movements and for the use of tools. Tool use behaviour is well known in apes like Chimpanzee and Orangutan. Gibbons, due to the small size of their thumb relative to their other fingers, perform handling of small objects by means of the pulp of the thumb and the side of the middle phalanx of the index finger. It should not be concluded from this difference that these animals are not capable of opposition: they can adduct their thumb and rotate it in front of their other fingers, but the shortness of the thumb prevents contact between the pulpar surfaces [8]. More primitive monkeys like rhesus monkeys or baboons where thumb adduction is not possible, are also capable of accurate precision grips which resemble the human grip (see below). Other still more primitive animals use whole hand prehension with a non-opposable thumb, a good example of which is given by behavioural observation of the squirrel monkey. In this animal, objects are reached with all the fingers in a slightly curved convergent position; in the later stage of the reach, the fingers diverge and straighten, then close in a scooping motion at contact with the object. The fingers frequently close to the palm with the distal and medial phalanges parallel to the palm, rather than curled around the object as human fingers do [9]. Interestingly, this description of a primitive prehensile behaviour can be applied almost without change to pathological prehension in man following cortical lesions [10, 11]. This observation suggests that the various types of grasps displayed by infra-human species, as well as the progressive maturation of human grasp during development, relate to different degrees of cortical control of the hand. The high degree of development of the hand in primates is paralleled by the development of a remarkable neural apparatus. The amount of cortical surface devoted to innervation of the hand testifies to its functional importance. This includes not only the large hand areas in primary somatosensory and motor cortices, but also areas more recently discovered in monkeys, in the posterior parietal cortex [12] and in the premotor cortex [13].

HOW HANDS HANDLE OBJECTS - THE PATTERN OF GRIP FORMATION

The type of grip that is formed by the hand in contact with the object represents the end result of a motor sequence which starts well ahead of the action of grasping itself. The fingers begin to shape during transportation of the hand to the object location. This process of grip formation is therefore important to consider, because it shows dynamically how the static posture of the hand is finally achieved. No systematic investigation of this aspect of grasping (preshaping) seems to have been made until the film study by Jeannerod [14].

Preshaping first involves a progressive opening of the grip with straightening of the fingers, followed by a closure of the grip until it matches object size. The point in time where grip size is the largest (maximum grip size) is a clearly identifiable landmark which occurs within about 60% to 70% of the duration of the reach, that is, well before the fingers come in contact with the object ([14, 15]; see also [16, 17]). This biphasic opening-closure motor pattern might not be unique to man: a few observations based on films during prehension in rhesus monkeys have revealed a very similar opening of the grip followed by closure before contact with the object (unpublished results, courtesy of S. Faugier-Grimaud).

The amplitude of grip aperture during grip formation covaries with object size [14-20]. Marteniuk et al. [20] found that for an increase of one centimeter in object size, the maximum grip size increases by 0.77 cm. One possible explanation for the biphasic opening-closure pattern of grip formation relates to the thumb-index finger geometry. Because the index finger is longer than the thumb, the finger grip has to open wider than required by object size, in order for the index finger to turn around the object and to achieve the proper orientation of the grip. Indeed, the movement of the index finger contributes the most to grip formation, whereas the position of the thumb with respect to the wrist tends to remain invariant [21]. The extra opening of the grip during preshaping might also represent a safety margin which compensates for the effects of the variability of the reach. Indeed, maximum grip size increases in a number of conditions (e.g., lack of visual control) where the variability of the reach is likely to be increased (e.g., [22]).

As already stated, the pattern of finger movements that arises prior to and during grasping, reflects the activity of higher order visuomotor mechanisms for detecting the shape of the object and for generating appropriate motor commands. The problem is for the motor system of the hand to build an "opposition space" which would take into account both the shape of the object and the biomechanics of the hand [23-25]. Experimental data suggest that there are preferred orientations for the hand opposition space. For example, Paulignan et al. [26] showed that the same orientation of the hand was retained during prehension of the same object (a vertical cylindrical dowel) placed at different positions in the working space. This implies different degrees of rotation of the wrist or of the elbow. In other words, the kinematic redundancy of the whole limb, and not only its distal segments, would be exploited in building the appropriate hand configuration for a given object. This type of observation (see also [33]) strongly suggests the existence of a higher order coordination mechanism which couples the different components of prehension. Description of this mechanism is one of the major challenges in this field of research (see [27]).

A more global approach to the problem of grip formation has been developed by several authors, showing that preshaping, manipulation and tactile exploration of objects are "knowledge-driven". Klatzky et al. [28] showed that subjects tend to classify usual objects into broad categories, the boundaries of which are determined by the pattern of hand movements these objects elicit when they are to be grasped, used and manipulated. Four main prototypical hand shapes (e.g., poke, pinch, clench, palm) seem to be sufficient for defining the interaction between the hand and most everyday objects. In addition, this differentiation of hand shapes according to the forms of objects is retained in preshaping during actual reaching (see also [26]). The basic idea is that characteristics of reach and grasp movements depend on prior knowledge gained from previous interactions with objects.

These findings suggest that hand movements and postures during object oriented behaviour are largely determined at the representational level. Indeed, visual feedback signals seem of very little importance during the movement itself, as both the pattern of grip formation and the coordination of the reaching and grasping components are correctly achieved in situations where the hand remains invisible to the subject. Similarly, the size of the maximum grip aperture correlates with the size of the object in the absence of visual feedback from the hand [14, 15]. Representation of object affordances, however, must be reactivated by visual input, in order to steer the movement until its completion. The visual signals involved in this process must arise from the central retina. Prehension movements directed at objects presented within the peripheral visual field are not only slower and less accurate, but show an incomplete grip formation and improper shaping of the fingers [29].

OBJECT ATTRIBUTES FOR PERCEPTUAL AND MOTOR FUNCTIONS

An essential aspect of object oriented behaviour is that the same object has to be represented in multiple ways, simply because the environment asks different questions to the nervous system and because the answer to each of these questions requires the accessing of different types of representations. Concerning visual objects, the three main questions arising from the environment are: ***Where*** is it?, ***What*** is it? and ***How*** to deal with it?.

Let us consider first the questions of ***What*** and ***How***. In order to answer the question of ***What***, the function of the representation will be to form a perceptual image of the object, to access its meaning and to reach an overt identification. Although, according to our common experience, this seems an easy operation, it requires a number of steps, most of which are still largely hypothetical. One of the essential steps must be to **bind** the many elementary attributes of the object into a single identifiable and meaningful entity. These are likely to be processed separately at the lower levels of vision, where each visual module solves its own particular problem. Accordingly, volumetric primitives have to be extracted from stereoscopic depth cues or coherent motion cues, form has to be extracted from contours and contrast cues, etc. Another important step is to code the relevant information in such a way as to preserve the invariance of the object in all conditions, including when the relative positions of the object and the perceiver change. Shapes should be recognizable from all vantage points. As stated by Marr [30] and by Biederman and Cooper [31], this can only be achieved by an object-centered coordinate system, independent from the position of the viewer with respect to the object. A viewer-centered description would impose a redescription of the shape each time the position of the viewer changes.

At variance with the above ***semantic*** representation, the representation for answering the question of ***How*** involves a sensorimotor transformation. It has a purely ***pragmatic*** function, in that it relates to the object as a goal for an action, not as a member of a perceptual category. The object attributes are represented therein as affordances, that is, to the extent that they trigger specific motor patterns for the hand to achieve the proper grasp. In addition, this function does not seem to imply binding of object attributes into a single entity. Instead, each attribute contributes to the motor configuration of the hand by selecting the relevant degrees of freedom. This point was emphasized by Arbib in his motor schema theory [32]. Motor schemas are functional units which can be assembled into a limited number of postures: the posture selected during the preshape defines the optimal opposition space for applying the required forces to the object [25].

The study of hand patterning during the preshape and the grasp itself strongly suggests that the object representation used by the motor system also operates in object-centered coordinates. The main argument is that the hand configuration for grasping an object is not affected by its location with respect to the body and that, conversely, the opposition space tends to remain invariant with respect to the object. There are limitations to this assertion, however, because the arm geometry creates constraints that may require changes in hand configurations for extreme positions of the object. One example of such a constraint is shown in Stelmach et al.'s experiment where subjects grasped the same object placed at different orientations with respect to the body. At a given orientation, the type of grasp (with an unpronated hand) which was spontaneously used by the subjects suddenly shifted to a different type with a pronated hand [33]. This indicates that the representation for object oriented movements must include, not only the commands for building the opposition space, but also constraints imposed by arm geometry.

The representation for answering the question of ***Where*** is built in a completely different way from those for ***What*** and ***How***. This mechanism deals with the object as a locus in space, not as a set of attributes. For this reason, the representation has to be built in a viewer-centered system of coordinates, that is, a system with the body as a reference (see [34]). Its function is to carry the hand to the appropriate location, an action which is mostly achieved by the proximal joints of the arm. Although this may obscure the issue, one has to remember from the previous paragraph that proximal joints may also contribute to the formation of opposition space. To solve this apparent discrepancy with our hypothesis of separate representations, one would need a better definition of the object properties which are relevant to the questions of ***Where*** and ***How***, respectively. If the criterion for classifying one property into the ***Where*** category is that it should involve a proximal movement, then object orientation should be included in this category.

These considerations thus imply that, during the action of grasping, several systems of coordinates coexist for describing the same object. A viewer-centered system would be used for generating movements at the proximal joints and performing the reach. An object-centered system would be used for generating movements with the distal motor apparatus and performing the grasp. The fact that the representations for ***What*** and ***How*** both operate in object-centered coordinates, in spite of their very different behavioural implications, suggests that the coordinate system is defined early in the processing of the object attributes, that is, before the object information is distributed to the ***semantic*** and ***pragmatic*** representations, respectively. An argument as to the coexistence of these two modes of representations can be drawn from the study of cases of spatial neglect. Whereas in most cases neglect affects one half of space (whereby the subject will tend to neglect all objects located to the left of his/her body midline), in some cases, it affects one half of all objects whatever their location in space [35].

DISSOCIATING THE REPRESENTATIONS FROM EACH OTHER

Dissociations can be observed between the different modalities of object oriented behaviour, whether they are created experimentally or by pathological conditions. I will briefly consider two examples, one in normal subjects, the other in patients. Furthermore, these two examples will focus on one particular dissociation, that between the iconic and the ***pragmatic*** aspects of hand function in relation to objects. Other dissociations, between grasping and reaching, for example, have also been observed (see [34]).

In normal subjects, an interesting dissociation can be obtained using a paradigm based on the temporal organization of object oriented behaviour. One has to assume first that the iconic representation, because it achieves the ***semantic*** processing of objects, can be accessed consciously, whereas the ***pragmatic*** representation is largely automatic. Situations in everyday life, like car driving, reveal that actions in response to visual events are often dissociated from conscious experience of the same events. Castiello et al. [36] designed a series of experiments to measure the time difference between the two. Subjects were instructed to reach by hand a target object as soon as it became illuminated. It took them approximately 330 ms on average to start moving. On a different set of trials, the instructions to the subjects were to signal (by a vocal utterance: "tah!") at what time they became aware of the illumination of the object. The vocal response took 380 ms to appear. Finally, on still another set of trials, the instructions were to perform the two tasks at the same time. Not unexpectedly, the values for both the motor and the vocal reaction times were very similar to those measured in the previous sets of trials, that is, the onset of hand movement aimed at the object preceded the vocal response signalling the awareness of its change in visual appearance by about 50 ms. This difference was not noticeable to the subjects, who felt that their hand movement coincided with their perception of the illumination of the object.

This temporal pattern of the two responses changed when perturbations occurred in the visual presentation of the targets, so that the motor response had to be modified during the movement. In this situation, three identical objects, separated by about 10 cm from each other, were used as targets. On most trials, the central object alone was illuminated. On some occasions, however, the light which illuminated the central object was suddenly shifted to one of the two others, exactly at the time of onset of the hand movement. In this event, subjects had to correct the direction of their hand movement in order to reach towards the second illuminated object, and to emit a vocal signal to indicate the time at which they became aware of the shift in illumination. The first sign of correction of the hand trajectory appeared early (about 100 ms) following the shift in illumination. By contrast, the vocal utterance corresponding to this same event came much later, in the order of 300 ms after the beginning of the change in movement trajectory. The subjective report was in accordance with this temporal dissociation between the two responses: subjects reported that they saw the light jumping from the first to the second object near the end of their movement, just at the time that they were about to take the object (sometimes even after they had taken it!).

The clearest effect observed in this series of experiments was that the time to awareness of a visual event, as inferred from the vocal response, kept a relatively constant value across different conditions (for details and control experiments, see [36]). Under **normal** circumstances this value was roughly compatible with the duration of motor reaction times: when the subjects moved, they became aware of the object near the time when their movement started, or shortly after. This consistency broke down when their motor reaction time shortened for generating corrections. This dissociation between motor responses and subjective experience, when it happens, as well as the more usual synchrony between the two, both reflect the constraints imposed by brain circuitry on the processing of neural information. Different aspects of the same event are processed in different pathways. The fact that the ***semantic*** processing is more complex and that it requires binding object attributes from different sources, may explain why it is delayed with respect to the ***pragmatic*** processing (see [37]).

The existence of several distinct neural pathways for processing object-related visual information was particularly emphasized by the anatomical work of Ungerleider and Mishkin [38] in the monkey. According to these authors, information arising in the visual cortex is channeled into two main pathways, one through the posterior parietal areas (the dorsal pathway), and the other through the infero-temporal areas (the ventral pathway). These pathways have complementary functions, which Ungerleider and Mishkin limited to visual localization and object identification. Early reports about the effects of posterior parietal lesions in man appeared to be consistent with this hypothesis. Visually-directed reaching movements made by these patients are inaccurate, and often systematically err in one direction (usually to the side of the lesion). In addition, these movements are kinematically altered: their duration is increased, their peak velocity is lower, and their deceleration phase is longer (see [34]). More recently, however, it was noticed that grasping and manipulation are also altered by posterior parietal lesions. Patients misplace their fingers when they have to visually guide their hand to a slit [39]. During prehension of objects, they open their finger grip too wide with no or poor preshaping, and they close their finger grip when they are in contact with the object [10, 40].

These results have prompted a reappraisal of the respective functions of the two cortical pathways. The role of the posterior parietal cortex would be to organize object oriented action, whether movements are executed by the proximal or the distal channels. This role must be dissociated from the role of other cortical structures specialized for object identification and recognition. A recent observation by Goodale et al. [41] goes in this direction. These authors reported the case of a patient who, following a bilateral lesion of occipito-temporal cortex, was unable to recognize objects. The patient was also unable to purposively size her fingers according to the size of visually inspected target objects (an easy task for normal subjects; see [42]). In contrast, when she became involved in a grasping task and was instructed to take objects by performing prehension movements, the patient was quite accurate and her maximum grip size correlated normally with object size. If one compares these results with those of posterior parietal lesions, there appears to be a clear double dissociation between impairments in perceptual recognition of objects and impairments in object oriented action.

It would be tempting to consider only central representation for answering the questions of ***What*** and ***How***. In this case, the question of ***Where*** would become "How to get there?", and would be treated by the same areas as those for ***How***. This unique representation for ***How*** would be incompatible with the foregoing description, as it would have to include two different sets of coordinates. In addition, it is likely that the representations of ***How*** and ***Where*** are separate systems: the fact that they are packed within a restricted area of cortex would, however, ensure that they are usually affected by the same lesion. The possibility remains that a dissociation could be observed under certain conditions (a restricted lesion, for example). The observation by Jeannerod, Decety and Michel (in preparation) of a case of posterior parietal lesion with an isolated deficit in grip formation is in accordance with this prediction.

The above hypothesis implies that the cortical mechanisms for object recognition or for object oriented action are selectively activated by the task in which the subject is involved. If the task involves recognizing, memorizing or forming a visual image of an object, only the ventral visual pathway should be activated. If, on the other hand, the task involves preparation or mental representation of fingers movements for grasping or manipulating an object, the dorsal pathway should be activated. Recent experiments in man using brain activation techniques indicate that this prediction might be correct. Grafton et al. [43] showed that selection of visually guided finger movements for tracking a moving object activated the same area in the posterior parietal cortex as that described by Perenin and Vighetto [39] as the site of lesions producing optic ataxia. Further experiments are needed to fully document this point.

ACKNOWLEDGEMENTS

This work was supported by a Human Frontier Science Program grant awarded to M. Jeannerod.

REFERENCES

1 C.G. Phillips, Movements of the Hand, Liverpool University Press, Liverpool, 1985.
2 G. Berkeley, A New Theory of Vision, 1709, New Edition, Dent, London, 1910.
3 A.M. Gordon, H. Forssberg, R.S. Johansson and G. Westling, Exp. Brain Res., 83 (1991) 477.
4 J.R. Napier, J. Bone Jt. Surg., 38B (1956) 902.
5 J.M. Elliott and K.J. Connolly, Devel. Med. Child Neurol., 26 (1984) 283.
6 R. Heffner and B. Masterton, Brain Behav. Evol., 12 (1975) 161.
7 J.R. Napier, Symp. Zool. Soc. Lond., 5 (1961) 115.

8 J.R. Napier, Proc. Zool. Soc. Lond., 134 (1960) 647.
9 D.M. Fragaszy, Brain Behav. Evol., 23 (1983) 81.
10 M. Jeannerod, Behav. Brain Res., 19 (1986) 99.
11 T.E. Twitchell, Brain, 74 (1951) 443.
12 M. Taira, S. Mine, A.P. Georgopoulos, A. Murata, H. Sakata, Exp. Brain Res., 83 (1990) 29.
13 G. Rizzolatti, R. Camarda, L. Fogassi, M. Gentilucci, G. Luppino and M. Matelli, Exp. Brain Res., 71 (1988) 491.
14 M. Jeannerod, in: J. Long and A. Baddeley (eds.), Attention and Performance IX, Erlbaum, Hillsdale, 1981, p. 153.
15 M. Jeannerod, J. Mot. Behav., 16 (1984) 235.
16 S.A. Wallace and D.L. Weeks, J. Mot. Behav., 20 (1988) 81.
17 A.M. Wing, A. Turton and C. Fraser, J. Mot. Behav., 18 (1986) 245.
18 M. Gentilucci, U. Castiello, M.L. Corradini, M. Scarpa, C. Umiltá and G. Rizzolatti, Neuropsychologia, 29 (1991) 361.
19 R.G. Marteniuk, J.L. Leavitt, C.L. MacKenzie and S. Athenes, Hum. Mvt. Sci., 9 (1990) 149.
20 R.G. Marteniuk, C.L. MacKenzie, M. Jeannerod and S. Athenes, Can. J. Psychol., 41 (1987) 365.
21 A.M. Wing and C. Fraser, Quart. J. Exp. Psychol., 35A (1983) 297.
22 S. Chieffi and M. Gentilucci, Exp. Brain Res., 94 (1993) 471.
23 M.A. Arbib, Hum. Neurobiol., 4 (1985) 63.
24 T. Iberall and M.A. Arbib, in: M.A. Goodale (ed.), Vision and Action. The Control of Grasping. Ablex, Norwood, 1990, p. 204.
25 T. Iberall, G. Bingham and M.A. Arbib, in: H. Heuer and C. Fromm (eds.), Generation and Modulation of Action Pattern, Exp. Brain Res. Series, 15 (1986) 158.
26 Y. Paulignan, C.L. MacKenzie, R.G. Marteniuk and M. Jeannerod, Exp. Brain Res., 83 (1991) 502.
27 B. Hoff and M.A. Arbib, in: R. Caminiti, P.B. Johnson and Y. Burnod (eds.), Control of Arm Movement in Space. Neurophysiological and Computational Approaches, Exp. Brain Res. Series, 22 (1992) 285.
28 R.L. Klatzky, B. McCloskey, S. Doherty, J. Pellegrino and T. Smith, J. Mot. Behav., 19 (1987) 187.
29 B. Sivak and C.L. MacKenzie, in: L. Proteau and D. Elliott (eds.), Vision and Motor Control, Elsevier, Amsterdam, 1992, p. 233.
30 D. Marr, Vision, Freeman, San Francisco, 1982.
31 I. Biederman and E.R. Cooper, J. Exp. Psychol., Hum. Percep. Perf., 18 (1992) 121.
32 M. Arbib, in: V.B. Brooks (ed.), Handbook of Physiology, The Nervous System II, Amer. Physiol. Soc., Bethesda, 1981, p. 1449.

33 G.E. Stelmach, U. Castiello and M. Jeannerod, J. Mot. Behav., in press.
34 M. Jeannerod, The Neural and Behavioural Organization of Goal-Directed Movements, Oxford University Press, Oxford, 1988.
35 J. Driver and P.W. Halligan, Cog. Neuropsychol., 8 (1991) 475.
36 U. Castiello, Y. Paulignan and M. Jeannerod, Brain, 114 (1991) 2639.
37 U. Castiello and M. Jeannerod, Neuroreport, 2 (1991) 797.
38 I. Ungerleider and M. Mishkin, in: D.J. Ingle, M.A. Goodale and R.J.W. Mansfield (eds.), Analysis of Visual Behavior, MIT Press, Cambridge, p. 549.
39 M.T. Perenin and A. Vighetto, Brain, 111 (1988) 643.
40 L.S. Jakobson, Y.M. Archibald, D.P. Carey and M.A. Goodale, Neuropsychologia, 29 (1991) 803.
41 M.A. Goodale, A.D. Milner, L.S. Jakobson and D.P. Carey, Nature, 349 (1991) 154.
42 M. Jeannerod and J. Decety, in: M.A. Goodale (ed.), Vision and Action. The Control of Grasping, Ablex, Norwood, 1990, p. 33.
43 S.T. Grafton, J.C. Mazziotta, R.P. Woods and M.E. Phelps, Brain, 115 (1992) 565.

Section One: Evolution and Development

Insights into the Reach to Grasp Movement
K.M.B. Bennett and U. Castiello (Editors)

CHAPTER 2

EVOLUTION

M. MARZKE

Department of Anthropology, Arizona State University, Box 872402, Tempe, Arizona 85287-2402, U.S.A.

SUMMARY

Morphological correlates of human reach and grasp movements emerged with the origin of our earliest primate ancestors. Pentadactyl hands with a divergent thumb, and with nails supporting friction pads, allowed foraging among slender branches for fruits and insects. Binocular vision evolved concurrently. The reaching aspect of this adaptation became exaggerated in our more recent ape-like ancestors, whose long, mobile arms and fingers allowed them to hang by one hand on stable branches while the other hand reached to the periphery of trees for fruits. Effective thumb/fingertip grips were sacrificed with elongation of the fingers, but were compensated for by elaboration of independent, controlled movements by the index finger and opposable thumb. With the evolution of bipedality, human reach to grasp movements were applied increasingly to the use of tools. Fossil bones and prehistoric tools reveal stages in the evolution of distinctively human grips and manipulative behavior.

INTRODUCTION

The ability of our readers to retrieve this book from a shelf has its source in the behavior of our earliest primate ancestors, whose survival depended upon their success in retrieving insects and fruits from the slender branches of trees. With grasping feet and hands they climbed through this three-dimensional substrate, reaching for hand-holds and retrieving food with one or two hands, while the remaining extremities resisted the effects of gravity. Both food collection and travel to the food sources were dependent upon a capacity to reach for objects and to enclose them with the thumb and fingers. Morphological features facilitating these behaviors were supplemented in divergent ways as primates underwent evolutionary radiation into a variety of arboreal, and later terrestrial niches. The lineage leading to humans underwent an increase in body size, requiring suspensory behavior to maintain access to the concentration of fruits in the periphery of the arboreal canopy. This behavior was permitted by enhanced mobility of the shoulder, elbow and wrist, which we exploit today in many of our activities of daily living, such as replacing overhead light bulbs.

The evolution of morphological features facilitating reach and grasp movements is traced below. There are many such features, morphological correlates to reach and grasp in the hands of animals, which vary with the varying demands of locomotor and manipulative behavior. These include the relative proportions of the thumb and fingers, the topography of joint surfaces (which affects the range of potential movement at each joint), the locations and orientations of ligaments (which constrain the directions of bone movements), the shapes and relative robusticity of the bones, the relative potential of muscles to generate movements of bones, and the arrangement and distribution of ridges and tactile receptors in the skin. We begin with a description of these features in modern humans, and then proceed to the reconstruction of evolutionary stages at which they appeared, considered in the context of living and fossil primate locomotion and manipulative behavior. In this reconstruction we are especially interested in evaluating the evidence from prehistoric tools, as we come to the stage at which our family, the Hominidae, appeared in the fossil record.

MORPHOLOGICAL CORRELATES TO MODERN HUMAN REACH AND GRASP

Reach

The human forelimb occupies a lateral position relative to the trunk, in contrast to a more medial position in most other animals. It is thus able to extend the reach of the hand over a larger radius. A shallow ball and socket joint between the humerus and scapula at the shoulder, together with joints permitting sliding and rotation of the clavicle on the scapula and sternum, facilitate reach by the hand to the front, side, and back of the trunk, and vertically over the head. This range of mobility is considerably greater than we see in most other animals, even including the majority of primates.

Similarly in the elbow region, we are capable of an unusual amount of rotation, which can bring the hand from a pronated, palm-down position to a fully supinated, palm-up position. This rotatory ability is due to joint surfaces permitting a wide range of movement between the two bones of the forearm (the radius and ulna), and to the articulation of the wrist bones with only one of the forearm bones (the one that rotates, the radius), rather than with both. Movement at joints between three wrist bones and the forearm, supplemented by movements among seven bones of the wrist, bring the hand into precise orientation for grasp.

Grasp

Human grips. Objects of many sizes and shapes are grasped by the human hand, in such a way that they may be applied forcefully or delicately to tasks and maneuvered into varying positions by the thumb and fingers. Grips used for these purposes have been classified in a number of different ways, relating to the different types of analysis requiring the classification. Much of the medical and anthropological literature referring to manipulative behavior and its morphological correlates employs a simple pair of terms, power grips and precision grips, introduced by Napier [1] to describe the difference between grips that stabilize objects against the palm and those that permit maneuvering of objects by the thumb and fingers, away from the palm. An example of a power grip is the squeeze grip of a hammer or axe handle, in which the fingers squeeze the handle against the palm and stabilize it with pressure by the thumb (Fig. 1**A**). A very common precision grip for humans is one in which an object, such as a needle, is pinched and maneuvered by the opposed pads of the thumb and index fingertips (Fig. 1**B**). Another commonly used precision grip is the ***key pinch***, in which a key or other object is held and maneuvered between the thumb and the side of the index finger (Fig. 1**C**). A third precision grip, which is perhaps best characterized as a three-jawed chuck ***stone-throwing*** grip, involves the holding and maneuvering of approximately spherical objects by the palmar aspects of the thumb, index and third fingers (Fig. 1**D**). These three elements of the hand form a functional unit, which is apparent when the hand hangs at rest. The grip rarely receives notice in the literature on manipulative behavior, yet it is a fundamental element of our gripping repertoire with unique morphological correlates [2]. It is, for example, the grip used universally by pitchers in the American game of baseball. The ***squeeze*** form of power grip and the ***stone-throwing*** precision grip seem to be used exclusively by humans, whereas the thumb/fingertip and ***key*** precision grips are in the repertoire of many nonhuman primates as well as humans.

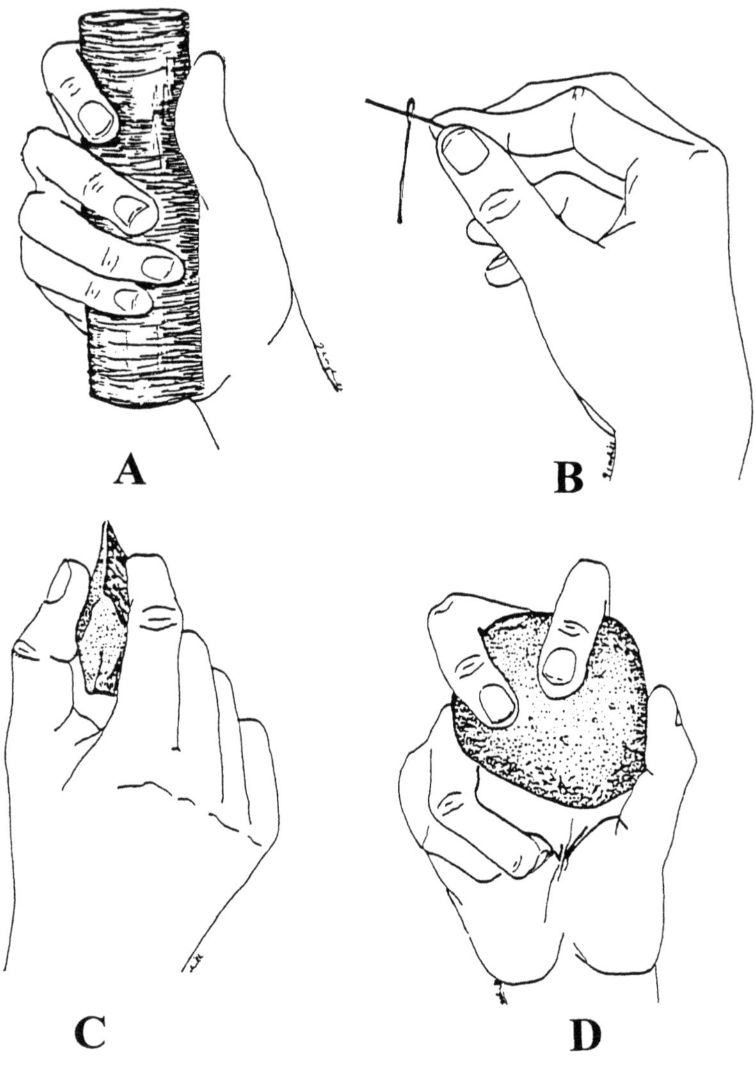

Figure 1. Four grips of tools commonly used by humans: ***A****. Squeeze* ***B****. Pad-to-pad* ***C****. Key pinch* ***D****. 3-jawed chuck.*

Morphological correlates of human grips. Demands on the human forelimb are unique in their virtually exclusive association with manipulative behavior and in the kinds of stresses associated with the tool-using aspects of this behavior. The distinctiveness of the human pattern of morphological features is thus related to our manipulative behavior. One of these demands is for the ability to accommodate the hand to objects of many different shapes and sizes. This is met by our possession of a relatively long thumb and short fingers, and by joint contours which allow opposition of the thumb to each of the fingers and alignment of the sensitive palmar friction surfaces of the thumb and all the fingers with the surfaces of the object. A second requirement is for the ability to grasp objects firmly, between the thumb and fingers or against the palm, permitting control in maneuvering them and resistance to their displacement by external forces. Of particular importance here are: (a) a separate long flexor muscle to our thumb (m. flexor pollicis longus), which maintains opposition of its tip to other fingers against strong resistance, (b) well developed and differentiated intrinsic muscles (muscles which both originate and insert in the hand), which fine-tune the orientations of the fingers for optimum precision or power grasp of objects and contribute force to grips through movement at the more proximal joints, and (c) broad distal phalanges which support large friction pads. A third requirement is the ability to tolerate large stresses, particularly at the bases of the fingers and palm, generated by forceful grips and by large impulsive forces associated with many kinds of tool use. These stresses are resisted by distinctive features of our hand bones and ligaments. For example, a styloid process on the back of the base of the third metacarpal bone, and a ligament from the pisiform bone in the wrist to the front of the base of the third metacarpal, stabilize this bone against forces that tend to hyperextend the bone [3]. Load is transferred from this metacarpal to a broad capitate, which is at the center of the wrist proximal to the metacarpal, and proportionately robust bones in the thumb and fifth finger accommodate stresses associated with the squeeze form of power grip [4].

Activities like wood-chopping, and many sports involving throwing and hitting with bats and racquets, require extensive mobility at the shoulder and elbow, as well as in the wrist and hand. They also exploit the principle that acceleration of distal lever segments may be enhanced by well-timed braking of the more proximal lever segments that are linked by joints.

EVOLUTION OF HUMAN MORPHOLOGY RELATING TO REACH AND GRASP MOVEMENTS

The morphological features outlined above did not all appear at the same time in human ancestors with the evolution of tool-use. Many of them were present millions of years ago in our remote primate ancestors and are shared with contemporary nonhuman primate species, in which they serve different purposes. Others unique to humans appeared at different stages in our evolution, after we diverged from apes. To trace the sequence of appearance of features in the human forelimb, and to reconstruct the circumstances under which they evolved, anthropologists use two approaches. One is to compare the morphology of contemporary humans with that of nonhuman primate species, looking for features that are shared with these species and for those that are unique to ourselves. We may assume that the shared features were inherited from a common ancestor, unless there is evidence that the they evolved in parallel. If they were inherited from a common ancestor, the behavior of the nonhuman species that shares the feature with us may throw light on the behavior of this ancestor, and thus indicate the behaviors to which the feature was adapted before it was exploited for modern human manipulative behaviors.

The second, related approach is to examine the morphology of fossil primates that appear to be near the ancestral lineage of humans. Applying our knowledge of behavior/morphology relationships in living animals to the study of fossils, we are able to draw some inferences about ancestral behaviors from features in the fossils, together with evidence from the environmental context in which the fossils are found.

Both approaches contribute to the following account of human forelimb evolution, which begins with the analysis of features shared by humans with most living and fossil members of the Primate Order, acquired at an early stage of our evolution, and proceeds to features shared by humans with progressively smaller groups of primates, reflecting our closer relationships and the behaviors to which our immediate ancestors were adapted later in human evolution.

Early stage in the evolution of human forelimb morphology

Our mobile forelimb and five-fingered grasping hand are fundamental elements of a morphological pattern that assured our earliest primate ancestors success in foraging for fruits and insects in the trees. All members of our Order share the retention of a clavicle, which positions the forelimb away from the midline of the body, enhancing reach of the limb. We also share the retention of two bones in the forearm, allowing rotation of the hand as the radius moves around the ulna. (The amount of rotation permitted by the joints between the radius and ulna varies among primate species with the relative demands of locomotion for stability at the elbow and wrist.

Some degree of rotatory facility is essential to the ability of primates to orient the hand for optimal grips of branches that diverge in three dimensions. However, excessive rotational capability compromises stability during quadrupedal locomotion. In our early ancestors, rotation was limited to less than 180 degrees, and the forelimb did not fully extend). A third feature common to primates is a divergent thumb. The grasp of a branch is achieved, not by parallel insertion of sharp claws into the bark, but by convergence of the thumb and fingers around the branch. A characteristic orientation of the primate hand on a branch is one in which the axis of the branch runs between the thumb and index finger [5]. Well developed intrinsic contrahentes muscles are oriented in directions that allow them to draw the thumb and fingers toward the midline of the hand to encompass the branch [6]. Moreover, friction pads on the palmar surface of the distal phalanges secure the grasp of branches, and of objects retrieved by the hand. The pads are supported by nails, in place of the claws of other animals. Ridges on the pads and sweat glands contribute security to the grip [5]. The pads are also supplied with Meissner's corpuscles [7] which, together with the ridges, provide a well developed tactile sensitivity. (For a review of research on the friction and tactile functions of ridge patterns and sense organs in the skin of the hand, see Martin [8]). All the skeletal features described above, shared by humans and other contemporary primates, appear to have been present in primates by the time of the Eocene epoch, which began about 57 million years ago [9]. For example, the hand of Notharctus, a fossil genus related to modern lemurs, had a somewhat divergent thumb and distal phalanges with a shape reflecting the presence of nails [10]. These shared elements of the locomotor apparatus were part of a functional complex that also included grasping feet and features of the face and brain relating to a greater emphasis on vision than existed in other contemporary mammals.

Models explaining the origin of primate reach and grasp morphology. What were the advantages of these features to early primates? Contemporary rodents with claws and a greater emphasis on the nose and tactile vibrissae for exploring their environment were successful in the trees as they are today. The prevailing view is that the mobile forelimb with the grasping hand allowed these primates to concentrate their activities among slender branches of bushes and trees, which provide bountiful supplies of fruits and insects but which are not effectively gripped with claws. Models relating primate features to this small branch environment vary slightly in the proposed primary dietary elements that attracted primates to this arboreal niche. Cartmill [11, 12] has argued that nocturnal predation on insects accounts for the association of grasping hands and feet with an emphasis on the visual system in primates. Sussman's model [13] emphasizes fruit-eating, while Rasmussen [14] considers both fruit and insect food sources to have been significant factors in the evolution of early primate morphology.

Manipulative capabilities of early human ancestors. Many of the living prosimians provide us with clues as to how these general primate features functioned in ancestral species. Prosimii form a suborder of Primates which includes lemurs, lorises and tarsiers. They are more primitive in many ways than the other suborder Anthropoidea, comprised of monkeys, apes and humans, and retain many features of the cranial and locomotor skeleton found in Eocene fossil primates. All the species are arboreal, many are quite highly specialized for leaping, and they do not achieve the large body size found in the majority of Anthropoidea.

Bishop [5] found that in prosimians the positions of the fingers during reach do not vary with the shape of the object to be grasped, and there is no tendency for elements within the hand to function independently. In the prosimian subgroup, Lemuriformes, the extended fingers are always held in parallel as they approach objects, their tips touch the object first, and then the fingers draw the object against the base of the palm. In Lorisiformes the fingers are always spread apart as they approach an object, the distal palm touches the object first, and the fingers then converge around it, pressing it against the distal palm. The grips of branches vary within these subgroups, however, exhibiting some convergence among species in different subgroups. For example, the lemuriform Microcebus murinus and the lorisiform Galago senegalensis both use an adhesive grasp posture, in which the pads of the palm and the fingertips maintain the grip while the middle finger joints are flexed. This may be contrasted with a clinging posture in the lorisiform Loris tardigradus, in which the full surface of the fingers lies against the branch. There is also considerable variability in the orientation of the hand on the branches, relating in part to the size of the substrate. The Lorisiformes appear to be more dexterous than the Lemuriformes, probably (in Bishop's view) because of the greater control required for retrieving insects by the more insectivorous Lorisiformes. Neither group, however, appears to examine surfaces with the tactile pads, indicating that tactile sensitivity is not well developed.

In summary, it appears from Bishop's study of living prosimians that in the early stage of arboreal life, human ancestors had acquired a rather generalized ability to retrieve objects with one hand, but probably did not engage in extensive tactile exploration of the environment or in dexterous manipulative behavior requiring individual control of the fingers.

Intermediate stage in the evolution of human forelimb morphology

Among Anthropoidea, it is the apes (gibbons and orangutans of Asia and chimpanzees and gorillas of Africa) that are most similar to humans in morphology of the forelimb and provide us with a model for the next stage in our evolution. In contrast to monkeys and the more primitive prosimians, apes and humans (which together form the superfamily Hominoidea) have proportionately longer forelimbs and more mobile shoulder, elbow, and wrist joints. The shoulder joints are located at the side of a broad thorax, positioning the limb for optimal reach in all directions. The structure of joints in the shoulder region allows the limb to be drawn fully upward, forward, backward and to the side. The elbow joint is capable of full extension, and the radius is capable of 180 degrees of rotation around the ulna. The wrist articulates only with the radius in humans and the great apes (orangutans, chimpanzees and gorillas), allowing for ample ulnar deviation of the wrist on the forearm, as well as rotation at this joint. With all this length and mobility, the hand is capable of reaching and retrieving objects over a very large range in three dimensional space. (For an excellent review of hominoid forelimb morphology, see Aiello and Dean [15]).

In the hominoid hand, the joints between the metacarpals (in the palm) and the fingers lack features which are related to habitual extension of the fingers during locomotion in prosimians and monkeys. The thumb is not only opposable (as it is in all Anthropoidea, according to Lewis [16]), but it also has an incongruity at its joint with the wrist which allows it to be pulled further away from the palm, allowing these larger-bodied animals to grasp large branches [17].

Grasp of objects obtained by the long reach of the arm may be extremely strong in apes. Their four fingers are curved, with joints specialized for flexion. The two sets of extrinsic muscles that flex the middle and distal elements of the fingers, and the intrinsic muscles that flex the proximal elements at the knuckles, are very large relative to the muscles that extend the fingers [18]. The fingers are longer relative to the thumb than they are in humans, and with their strong flexor apparatus are capable of functioning as a hook to support all the body weight during suspension by the hand in the branches of trees.

We see the first evidence of a few elements of this forelimb pattern in the elbow region of an early ape, <u>Aegyptopithecus</u>, dating from the Oligocene epoch, which began about 37 million years ago [9]. There appears to have been a radiation of primitive ape species in the early and middle Miocene (23 - 15 million years ago) with many species exhibiting several of the features indicative of suspensory behavior. However, the common ancestor of humans and the African apes, which probably had the full suite of suspensory features shared by these groups today, probably dates no further back than 6 - 10 million years ago [9].

Model explaining the origin of ape and human forelimb specializations for reach and grasp. Apes are distinctive among Anthropoidea in their large body size, and in their preference for feeding on fruits located on slender branches that cannot support their full weight. Suspending the body by the hand from a stable portion of a branch overhead, they are able to bridge the tapering sections of branches with the other long arm to reach fruits at the tips and bring them to the mouth for consumption. Grand [19] has shown that this suspensory form of feeding by reaching opens up a large source of food from each support position.

Molecular and fossil evidence of a very close relationship between African apes and humans [20] indicates that our recent common ancestor probably occupied an arboreal niche similar to that of the living African apes, in which suspensory feeding was an important component of behavior. It is very likely that most (if not all) of the morphological features of the human forelimb shared with apes appeared at a stage in our evolution when we were still arboreal, and had become large enough to require suspensory postures for feeding in the tracery of the peripheral branches.

Manipulative capabilities of apes ancestral to hominids. Although the thumb/finger proportions of apes render the precision grip of objects between the distal pads of the thumb and fingers less controlled than in humans, objects are retrieved, maneuvered, and supported quite effectively by other grips, principally the ***key*** precision grip and a cradle grip, in which the palm supports objects which are held against it by the flexed fingers, propped laterally by the short thumb (personal observation). One of the most intriguing applications of the key grip is to the manipulation of sticks for retrieving termites [21]. Noticeably lacking, however, is the ability to maintain a firm pinch of objects with the thumb, even against the side of the index finger. The thumb is relatively short, and lacks a well developed extrinsic flexor muscle for strong flexion of the thumb tip against the pressure of the finger. When fruit is removed from rind, for example, the key pinch is not sufficient to resist the pull by the teeth, and a second hand (occasionally even a foot, with its opposable hallux) is called into service. The weak thumb also reduces the effectiveness of manipulatory behavior requiring maneuvering of objects by the thumb and fingers. For example, when they throw stones, chimpanzees do not control the trajectory of the stone with the thumb, index and third fingers.

Comparison of apes with prosimians reveals a considerable advance among the former in the individual control of the thumb and fingers, facilitating some maneuvering of objects using a variety of grips. This variety contrasts with the stereotyped fingertip/palm grasp of objects by prosimians. The index finger is used alone extensively in the exploration and grasping of objects. Chimpanzees often carry small objects in the flexed index finger during quadrupedal locomotion on the ground, while the knuckles of the third and fourth fingers support their weight. The long, deep flexor muscle to the index finger is relatively independent of the muscle giving rise to the tendons for the third, fourth and fifth fingers.

A well developed tactile sensitivity of the thumb and fingers is evident in the tendency of apes to explore and maneuver objects with the fingers. This is consistent with the extension of epidermal ridges over the entire volar surface of the fingers, and with varied orientations of these ridges in anthropoid primates [8].

Monkeys as models for the intermediate stage in the evolution of human reach to grasp. There are other anthropoid species that share with humans and apes the individual control of thumb and fingers and in some cases also thumb/finger proportions facilitating a precision grip between the thumb and pad of the index fingertip. MacNeilage [22] has noted that these behaviors in nonhuman primates appear to be associated with the procurement of food requiring extraction from sources and/or modification by the hand prior to ingestion. New World capuchin monkeys manipulate food, and even tools used in food procurement, with a variety of grips [23]. Hamadryas baboons, African monkeys whose ancestors diverged from ancestors of modern apes and humans about 20 million years ago, use the index finger extensively for probing, and retrieve small objects effectively by pinching them between their relatively short index finger and long thumb (Marzke et al., in preparation). In a systematic study of manipulative behavior, Torigoe [24] observed use of the fingers independently in apes and in several species of the Old World monkey subgroup, Cercopithecinae. Use of the index finger specifically was observed only in the gelada baboon (Theropithecus gelada) in his sample of Old World monkeys, a species which in the wild forages on grass, picking up individual blades between the thumb and index finger [25]. Very little is known about species differences in the motor and sensory neural supply to muscles of the hand. Neural correlates of independent finger movements have been investigated in nonhuman anthropoid primates but have not yet been fully determined [26].

Recent stage in the evolution of human forelimb mobility

Although modern humans do not have an extremely long forelimb relative to hindlimb length, or the highly specialized long, hook-like fingers of apes, there is fossil evidence that early hominids had more ape-like proportions, with fingers that supported large flexor tendons, even when they were in the initial stages of bipedality. Well preserved skeletal remains of Australopithecus afarensis, the earliest known member of our family Hominidae, include forelimb bones that are longer than ours relative to the hindlimb, hands with longer fingers relative to the thumb, and features in the wrist and fingers reflecting well developed flexor muscles [27]. Forming a mosaic with the ape-like features in the hand are a few distinctively human details of joints involving the second metacarpal bone, which lies at the base of the index finger in the palm [28].

Hindlimb morphology of this fossil species combines pelvic shape and joint configurations for bipedal posture with limb proportions and a distribution of muscle attachments compatible with climbing [27]. Use of the hand for hoisting the body in the trees seems to have been a component of early Australopithecine behavior, prior to full commitment of the hindlimb to bipedality and of the hand to manipulative behavior.

Evidence for a strong flexor apparatus in the fingers persists to at least 1.75 million years ago, in finger bones of an early species of our genus Homo (Homo habilis) from Olduvai [29, 30].

Manipulative behavior of our fossil hominid ancestors. The modern human features found in the hand of Australopithecus afarensis are concentrated in the region of the index finger, and appear to have enhanced its rotation toward the thumb, facilitating the control of objects between the thumb and the side of the index finger in a ***key*** pinch and maneuvering of objects with a ***stone-throwing*** precision grip by the thumb, index finger and third finger [28]. The thumb was shorter relative to the index finger than it is in modern humans, but proportionately longer than in modern apes, indicating that it may have been more effective in manipulating objects with the index and third fingers. A groove on one of the wrist bones (the hamate) may have been occupied by a ligament from the pisiform to the third metacarpal, which in modern humans stabilizes this bone in tool-using activities involving the three-jawed chuck (***stone-throwing***) grip [28]. A chimpanzee-like joint between the thumb and the trapezium (a wrist bone) would have stabilized this grip, but also probably would have limited effective opposition of the thumb to the fourth and fifth fingers in the grip of large objects. Although manufactured stone tools have not been found in association with these early Australopithecines, it is reasonable to infer from the overall morphology of their hands that they not only shared with chimpanzees the ability to manipulate sticks, stones, and bones, but also were able to maneuver these unmodified tools with more control by the thumb, index finger and third finger. This ability should have enhanced their ability to secure and process underground foods, hard-shelled fruits, and small game.

The hand and wrist bones of Homo habilis from Olduvai indicate that by 1.75 million years ago our ancestors were capable of greater control of large objects, by the thumb and all the fingers. The trapezium had a broader, flatter surface than in Australopithecus afarensis, which would have allowed the thumb to oppose all four fingertips. A marked attachment area on the thumb for the tendon of the flexor pollicis longus muscle, which flexes the tip of the thumb, indicates that a firm pinch could be maintained between the thumb and index finger. Bones of the fingertips were very broad, reflecting large friction pads that would have facilitated firm holding and effective maneuvering of large objects by the thumb and fingertips. These features would have been of advantage in the manufacture of tools found at the same level as the hand bones [2]. The tools were made from large stones, which were held in one hand and stabilized against blows applied by a hammerstone in the other hand.

Hominid hand bones from the Swartkrans site in Southern Africa, of approximately the same age as the Olduvai bones, exhibit similar features of the thumb and fingertips [31]. Interestingly, a skeleton from approximately the same time level at Olduvai, attributed to Homo habilis [32], is reported to have short hindlimbs relative to the forelimbs, indicating that even at this stage of human evolution there may not have been a full commitment of the forelimbs to manipulative behavior.

Unfortunately there is no substantial evidence available for hominid hand structure in the time period from 1.75 million years ago until quite recently. Hand bones of several Neandertal individuals, from sites dating to approximately 40,000 - 60,000 years ago, are essentially modern in the length of the fingers relative to the thumb, and in the presence of a styloid process on the third metacarpal and a broad capitate bone at the base of the third metacarpal. Clearly by this time the human hand was structured to tolerate the large internal and external forces associated with the manipulation and manufacture of hand-held stone tools. Proportionately large joints and extensive insertion areas for muscles are testimony to activities requiring the ability to grasp and pinch objects with considerable force. There was a trend after the time of Neandertals toward more gracile hands in modern humans, probably associated with changes in tool design that reduced stress on the hands [33].

LATERALITY OF REACH TO GRASP FUNCTION

There is considerable evidence of a bias in humans toward preferential use of the right hand; approximately 90% of humans exhibit this preference. The bias toward functional laterality is correlated with asymmetry of the human brain; specialization for manual activity tends to occur in the left hemisphere [34].

Evidence for the origin of functional laterality. Evidence has been found in several prosimian species for a significant bias toward use of the left hand in reaching for food, while the right hand maintains stability of the animal on a vertical branch, with the strongest bias manifest in bushbabies, lorisiformes that cling to vertical branches and catch insects with one hand [35]. MacNeilage [22] interprets this left-hand bias for predation and right-hand bias for postural maintenance as evidence that human functional laterality has a long antiquity, with our right handedness originating from a postural specialization in early primates.

Evidence for functional laterality in apes. Investigations into hand preference among apes have been the subject of several recent reviews [36]. At present the lack of an extensive body of systematic comparative data cautions against support of claims that functional laterality exists among apes [37]. However, several recent studies are considered by MacNeilage [22] to be suggestive of a right-hand bias for manipulative tasks and a left hand preference for tasks involving strong visuospatial demands, and Hopkins and Morris [36] find some evidence for hand preferences in carrying, object manipulation, and leading limb in locomotion to be compelling.

Central to investigations of functional laterality recently has been an hypothesis proposed by MacNeilage et al. [34], that right-hand bias for postural support during left-hand reaching in prosimians may have led to a right-hand bias among apes and monkeys for retrieving and processing food requiring force and control by the hand.

Evidence for functional laterality in fossil hominids. Toth [38] has found evidence in prehistoric stone tools of wear and tool-manufacturing practices reflecting use predominantly of the right hand by fossil hominid species. His data indicate that the relative percentage of right-handed stone tool-users increased during a period of prehistory from 1.9 million years ago to

400,000 years ago.

CONCLUSION

A trend in human evolution may be seen in the reach to grasp movement as it relates to both locomotor and manipulative behavior. Our small, early ancestors resembled living prosimians, and were exceptional among placental mammals in the range of reach and in the ability to grasp objects with one hand. The range of reach increased with body size as our more immediate ape-like ancestors exploited fruits in the terminal branches of trees. By the time our early hominid ancestors left the trees, they had acquired a facility for positioning the hand in as many varied locations relative to the trunk as we are capable of today. They also had the ability to position the fingers independently relative to the thumb and palm of the hand, to accommodate the thumb and fingers to objects of varying shapes and sizes, to maneuver these objects with the thumb and fingers, and to adjust these movements according to an exceptionally rich source of sensory information derived from the skin. What remained after our separation from the line leading to African apes was the ability to achieve a firm grip of objects between the thumb and fingers, and to maneuver them with control by the distal pads of the thumb and the palmar surface of each of the four fingers. Our distinctive hand proportions and neuromuscular correlates to grasp emerged with the evolution of tool-use and tool-making, as hominids became increasingly dependent upon these abilities to apply stone, bone and wood to the acquisition of foods that could not be retrieved and processed by the hands alone.

REFERENCES

1 J.R. Napier, J. Bone Jt. Surg., 38B (1956) 902.
2 M.W. Marzke and M.S. Shackley, J. Hum. Evol., 15 (1986) 439.
3 M.W. Marzke and R.F. Marzke, Am. J. Phys. Anthropol., 73 (1987) 415.
4 M.W. Marzke, K.L. Wullstein and S.F. Viegas, Amer. J. Phys. Anthropol., 89 (1992) 283.
5 A. Bishop, in: J. Buettner-Janush (ed.), Evolutionary and Genetic Biology of Primates, Vol. 2, Academic Press, New York, 1964, p. 133.
6 R.W. Haines, Proc. Zool. Soc. Lond., 125 (1955) 761.

7 R.K. Winkelmann, in: J. Buettner-Janush (ed.), Evolutionary and Genetic Biology of Primates, Vol. 1, Academic Press, New York, 1963, p. 230.
8 R.D. Martin, Primate Origins and Evolution, Princeton University Press, Princeton, 1990.
9 J.G. Fleagle, Primate Adaptation and Evolution, Academic Press, San Diego, 1988.
10 W.K. Gregory, Mem. Am. Mus. Nat. Hist., n.s. 3 (1920) 49.
11 M. Cartmill, in: R. Tuttle (ed.), Functional and Evolutionary Biology of Primates, Aldine, Chicago, 1972, p. 97.
12 M. Cartmill, Science, 184 (1974) 436.
13 R.W. Sussman, Amer. J. Primatol., 23 (1991) 209.
14 D.T. Rasmussen, Amer. J. Primatol., 22 (1990) 263.
15 L. Aiello and C. Dean, An Introduction to Human Evolutionary Anatomy, Academic Press, London, 1990.
16 O.J. Lewis, Functional Morphology of the Evolving Hand and Food, Oxford University Press, Oxford, 1989.
17 M.D. Rose, J. Hum. Evol., 22 (1992) 255.
18 R.H. Tuttle, J. Morph., 128 (1969) 309.
19 T.I. Grand, J. Mammal., 53 (1972) 198.
20 P. Andrews, in: C. Patterson (ed.), Molecules and Morphology in Evolution: Conflict or Compromise, Cambridge University Press, New York, 1987.
21 J. Goodall, Animal Behav. Monogr., 1 (1968) 163.
22 P.F. MacNeilage, in: M.A. Goodale (ed.), Vision and Action: The Control of Grasping, Ablex, New Jersey, 1990, p. 1.
23 N. Masataka, Folia Primatol., 55 (1990) 189.
24 T. Torigore, Primates, 26 (1985) 182.
25 R.I.M. Dunbar, J. Hum. Evol., 12 (1983) 647.
26 R.B. Muir, Exp. Brain Res. Suppl., 10 (1985) 155.
27 J.T. Stern (Jr.) and R.L. Susman, Am. J. Phys. Anthropol., 60 (1983) 279.
28 M.W. Marzke, J. Hum. Evol., 12 (1983) 197.
29 J.R. Napier, Nature, 4853 (1962) 409.
30 R. Susman and N. Creel, Am. J. Phys. Anthropol., 51 (1979) 311.
31 R.L. Susman, in: F.E. Grine (ed.), Evolutionary History of the "Robust" Australopithecines, Aldine, 1988, p. 149.
32 D.C. Johanson, F.T. Masav, G.G. Eck, T.D. White, R.C. Walker, W.H. Kimbel, B. Asfaw, P. Manega, P. Ndessokia and G. Suwa, Nature, 327 (1987) 205.

33 E. Trinkaus, in: E. Trinkaus (ed.), The Mousterian Legacy, British Archaeological Reports Int. Ser. 164, p. 165.
34 P.F. MacNeilage, M.G. Studdert-Kennedy and B. Lindblom, Behav. Brain Sci., 10 (1987) 247.
35 J.P. Ward, Behav. Brain Sci., 11 (1983) 744.
36 W.D. Hopkins and R.D. Morris, Int. J. Primatol., 14 (1993) 1.
37 L.F. Marchant and W.C. McGrew, J. Hum. Evol., 21 (1991) 425.
38 N. Toth, J. Hum. Evol., 14 (1985) 607.

Insights into the Reach to Grasp Movement
K.M.B. Bennett and U. Castiello (Editors)

CHAPTER 3

DEVELOPMENT OF THE REACH TO GRASP MOVEMENT

A.M. GORDON

Department of Physiology, 6-255 Millard Hall, University of Minnesota, Minneapolis, 55455, U.S.A.

SUMMARY

Coordinated prehension in humans develops relatively late during ontogeny. For example, the precision grip between the thumb and forefinger does not emerge until around 10 months of age. Once it appears, skilled manipulation of objects requires precise control of finger-tip forces using both feedback and feedforward control. Young children first beginning to use the precision grip use excessive forces with large variability. Furthermore, they exhibit prolonged delays between movement phases and are unable to use information gained during previous experience with the object to scale the force output in advance. During the second year, children are capable of using sensory information more reliably to adjust the force output and anticipatory control emerges. However, it does not approximate adult performance until six to eight years of age and subtle improvements continue until adolescence. Emerging evidence suggests that the development of the grip coordination may coincide with the development and maturation of CNS structures underlying prehension.

INTRODUCTION

Dexterous reaching and grasping movements are not present at birth in humans or other higher primates. They develop in a series of steps during ontogeny. Reaching serves to propel the hand to a desired location in space. Thus grasping and manipulating objects requires appropriate goal-directed reaching. Grasping involves coordinating the digits according to the intrinsic properties of an object and precisely controlling finger-tip forces once contact with an object is established. In the following chapter, I will first review the behavioral steps which underlie the development of reaching and grasping movements. Next, I will relate the ontogeny of these behaviors to the development of the corticospinal tract. While the maturation of the CNS does not necessarily induce changes in prehensile behavior, I will present evidence suggesting that maturation of specific structures is required for certain behaviors to occur and that what is known about these maturational processes largely coincides with the ontogeny of prehension. Finally, I will describe a recent series of studies outlining the development of the coordination of finger-tip forces and the use of somatosensory and visual information for the feedback and feedforward control of grasping movements.

ONTOGENY OF REACH TO GRASP MOVEMENTS

Development of reaching and grasping behaviors

Although the seemingly random and aimless arm movements in newborn infants may in fact be partially directed toward fixated objects moving in their environment [1], the hand is not postured in a manner which is conducive to grasping. The hand is always opened before or during the reach and the fingers do not close when the object is approached (see [2]). While some infants during the first weeks of life may exhibit crude hand shaping to an object [3], these finger movements are highly immature. The movements may represent a preformed proximo-distal motor pattern eventually evolving into prehension [4]. Around four to five months of age both the distance and direction of the reach improve, but the hand orientation and finger closure are still rather limited [5, 6]. By nine months of age, the hand begins to be shaped according to an object's size [6].

Early grasping in infants is largely governed by tactile and proprioceptive reflexes [7, 8]. Stimulation of the palm induces closure of the fingers, while stimulation of the dorsal hand may prevent a grasp reflex or elicit a reflexive opening of the hand. Traction of the arm, imposing stretch on the flexor muscles, also induces hand closure as part of a flexor synergy in which all flexor muscles about the shoulder, elbow and wrist are contracted in addition to finger flexors. The reflexive behaviors gradually become less stereotypical and normal prehension develops in several steps.

It has been suggested that the grasp reflex dissipates before voluntary grasping emerges [9], although others postulate that the reflex remains and purposefully interact with motor commands (e.g., see [8]). Interestingly, in contrast to monkeys deafferented surgically in utero, monkeys that have experienced reflexive grasping during infancy recover manipulative functions when the limb is surgically deprived of sensory information by dorsal rhizotomy [10]. The palmar grasp, in which all fingers are formed around an object, emerges around four months of age in humans [11]. Around this time, there is usually a progressive dissociation of the grasping reflex and traction response [8]. Frequently such reorganization of movements is accompanied by a decline of the motor skills, which are then reformed at a more adequate level [12]. Gradually, relatively independent finger movement (RIFM) emerges and children can grasp between the tips of the thumb and index finger (precision grip) by ten months of age. Slowly, the grip becomes more defined and children are capable of using several grip patterns [13, 14]. Once children are able to perform RIFM and adapt their behavior to the intrinsic properties of objects, they are able to establish relationships between objects and develop two-handed collaboration behaviors (see [15] for review).

Anatomical changes during development

Based on anatomical, electrophysiological and behavioral studies, it appears that control of precision grip and manipulation of objects with RIFM is dependent on the motor cortex and corticospinal pathways ([16-18]; see [19, 20]). The development of cortico-motoneuronal connections occurs relatively late and parallels the development of finger movements in monkeys [21, 22]. Sectioning of the pyramidal tracts at the medullary level in infant monkeys results in a failure to develop precision grip and normal patterns of RIFM [21, 22].

Electromagnetic stimulation of the cortex in an infant monkey does not elicit short-latency responses in hand and forearm muscles, but does in a ten month old animal [23]. Using both cross-sectional and longitudinal studies, Lemon and his colleagues [24, 25] recently characterized the development of the corticospinal tract in infant monkeys. They were unable to elicit short latency responses to electromagnetic stimulation in newborn infant monkeys until RIFM emerged around four to six months of age [24, 25]. Together, these studies suggest that the development of independent finger movements in primates is dependent on completion of cortico-motoneuronal connections.

In humans, electromagnetic stimulation of the cerebral cortex can evoke activity in the descending pathways, even in premature and term newborns, if accompanied by isometric contraction, but the onset of the muscle action potential is delayed compared to older children and adults [26]. Myelination of the corticospinal tract is thought to continue until the age of one or two years [27, 28]. Isometric contraction is required to elicit muscle action potentials until the age of eight years [26, 28]. This suggests that cortical interneurons or pyramidal tract neurons have higher thresholds for activation or that α-motoneurons are less readily excited before this age. Reduction of the thresholds to adult levels by 11 years of age is thought to reflect an uncharacterized process of maturation.

COORDINATION OF THE GRIP-LIFT MOVEMENT

Skilled manipulation of objects involves precise control of finger-tip forces based on mechanisms integrating somatosensory and visual information with sensorimotor memories relating the object's physical properties to past performance. In recent years, such control in adults has been characterized extensively (e.g., [29-35]; for review, see [36]). In a series of studies summarized below, we have recently examined the development of precision grip using a small test object instrumented with force transducers (Fig. 1**A**). The results from these studies suggest that skillful manipulation requires control mechanisms regulating: 1) the proper initiation of the grasp and lift components and smooth transitions between subsequent movement phases; 2) the proper coordination between the isometric grip (squeeze) and load (vertical lift) forces; 3) anticipatory scaling of the force output based on the physical properties of the object; and 4) the adequate adjustment of the force level to the object's physical properties using sensory feedback.

Control of finger-tip forces

The reach to grasp/lift movement consists of several movement phases (Fig. 1**B**). During the first, reaching, phase (not shown), the fingers are preshaped according to the size and shape of the object ([37]; for review, see [38]). In adults, preshaping the fingers results in only a short delay between the finger and thumb contact with the object (T0-T1). During the subsequent *preload phase* (T1-T2), there is a small grip force increase, which establishes a hold on the object, before the onset of positive load force increase (T2). Following this, there is a parallel grip and load force increase during the *loading phase* (T2-T3). When the load force overcomes the gravitational force on the object (T3), the object is lifted from its support and the force increase is terminated. The object is then transported to the desired position during the *transition phase* (T3-T4) and held stationary in the air during the *static phase*. Inputs from various mechanoreceptors in the finger-tips are involved in linking the phases of the lift by signalling the occurrence

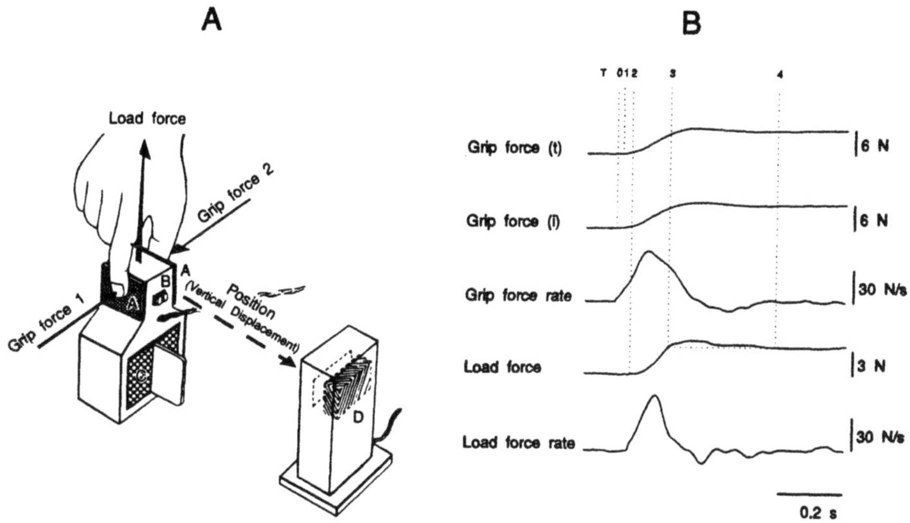

*Figure 1. **A**. Schematic drawing of the grip instrument: A. exchangeable grip surfaces covering stain-gauge force transducers for measurement of grip force and load force, B. infra-red light emitting diodes, C. exchangeable mass and D. photo-electric position sensing unit. **B**. Grip force from the thumb (**t**) and index finger (**i**), grip force rate, load force and load force rate as a function of time (adult). The object is contacted first by one finger (**T0**). Both grip forces begin to increase (**T1**), before the onset of positive load force (**T2**). The grip force and load force increase in parallel during the loading phase until lift-off (**T3**).*

of particular mechanical events (e.g., that the digits have made contact with the object or that the object has started to move; [29]). A similar interaction between feedback and motor commands occurs during locomotion, where somatosensory information during hip extension terminates the stance phase and releases flexor activity (e.g., [39]), as well as during throwing, where sensory information may indicate that the arm has reached a certain angle, triggering the next phase of the movement (e.g., [40]).

Young children show prolongations between the various phases of the grip-lift movement. Even if some hand preshaping is evident by nine months

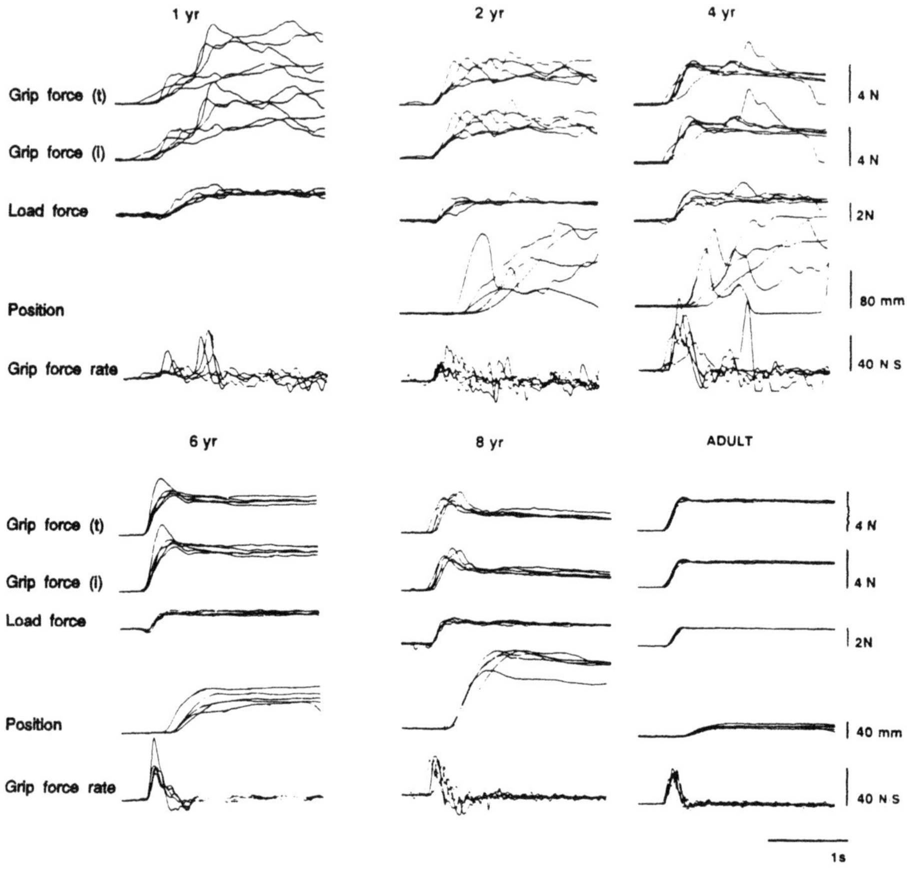

Figure 2. Grip force from the thumb (t) and index finger (i), load force, vertical position and grip force rate as a function of time during several trials for children of various ages and an adult. Note the large variability and excessive grip forces for the young children compared to the adults. Position signals are omitted for the one year old children since they often grasped the object from the experimenter's palm. Modified from [41].

of age during reaching [6], children at this age still exhibit dramatic delays between the contact of the finger and thumb, often requiring several touches before a stable grasp is achieved [41]. Furthermore, there is a substantial delay between the establishment of the grip and the onset of positive

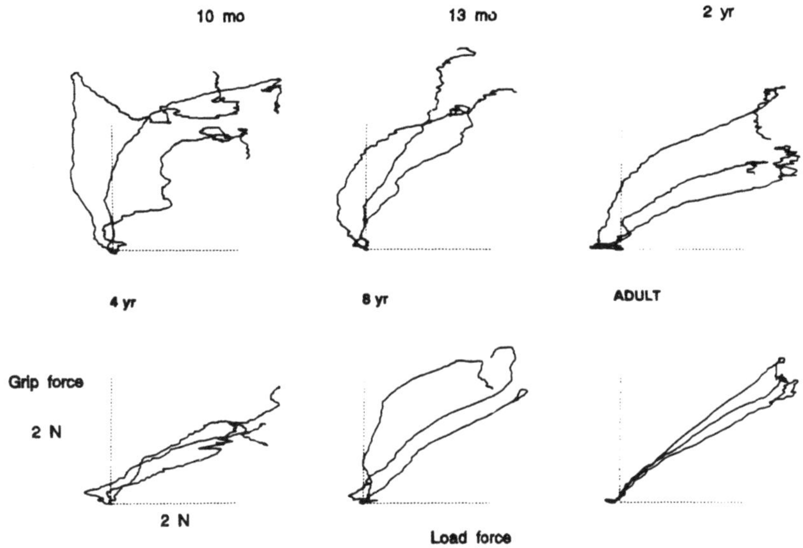

Figure 3. Grip force during the preload and loading phase as a function of load force for children of various ages and an adult. Note the negative load force and non-parallel grip force/load force increases for young children compared to adults. Modified from [41].

load force; that is, a prolonged preload phase. During this delay, young children produce large grip forces (Fig. 2; see also Fig. 3). Thus, instead of the smooth transitions between movement phases characterizing adult coordination, children clearly have a sequential activation of the forces. During development, the forces decrease and there are fewer delays between movement phases. Reflex latencies are known to decrease during development [42], which may contribute to the decrease in the duration of the phase transitions.

While the force generation is highly automatized and invariant between trials for adults, young children exhibit marked variability between trials, with excessive grip forces which fluctuate during the static phase when the object is held in the air (Fig. 2). There is a large intra-subject variability in the force amplitudes and temporal parameters. A similar large variability of motor output in children is also seen in other motor tasks, including postural control and locomotion [43, 44]. The increased variability may allow the

CNS to explore and evaluate different response patterns [45].

Coupling of grip and load force

Adults typically generate the grip force and load force in parallel after the initial contact with the grip surface (Fig. 3; [29]). This reflects a functional synergy coupling grip and load force generators, which serves to simplify the movement. When the precision grip first emerges, children do not generate the forces in parallel (Fig. 3; [41]); instead they initiate the grip force in conjunction with a negative load force, pressing the object against its support. By the onset of positive load force, there is already a further grip force increase, and subsequent increases in the isometric grip and load force during the loading phase are not in parallel [41]. The sequential activation of the forces and excessive grip forces likely allow children to obtain additional information regarding the surface friction and to stabilize the grasp, as well as to provide a strategy less dependent on anticipatory control. During the latter part of the second year, the grip and load forces begin to increase more in parallel, and the coupling between these increases until adolescence.

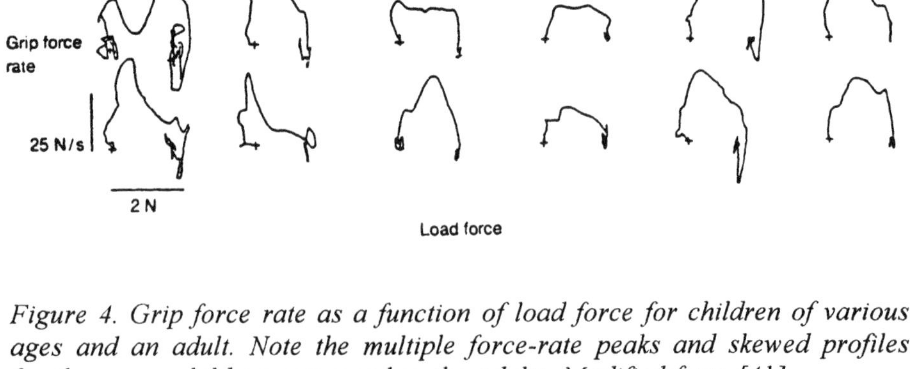

Figure 4. Grip force rate as a function of load force for children of various ages and an adult. Note the multiple force-rate peaks and skewed profiles for the young children compared to the adults. Modified from [41].

Force-rate profiles

The prolonged delays between movement phases may reflect a lack of anticipatory control of the motor output and increased dependence on sensory feedback. Anticipatory control is typically characterized by bell-shaped force-rate profiles (c.f. [46]), which are appropriately decreased prior to lift-off to harmonize with the physical properties of the object (Fig. 4). These resemble the velocity profiles for closure of the thumb and index finger in grasp [47, 48], as well as the rate profiles obtained during programmed isometric contractions [49, 50].

In contrast, young children exhibit force increases which occur in small increments, with multi-peaked force-rate profiles (Fig. 4; [41]). Similar force-rate profiles are observed when adults employ a ***probing strategy*** in which coordinated grip and lifting force commands that yield low force rates are repeatedly issued until terminated upon receipt of somatosensory signals that indicate the start of the vertical lifting movement [51]. The ***probing strategy*** reduces dependence on memory at the expense of more slowly paced lifting, but it prevents large positional overshooting during the lifting movement. This strategy is used by adults when they are not confident about the object's physical properties [33] or when they encounter objects that are heavier than expected [31, 35]. During the later part of the second year the force-rate profiles become increasingly bell shaped with small irregularities. Subsequent development is gradual and approximates adult-like coordination by the time children reach six to eight years of age, with subtle improvements until adolescence [41].

ANTICIPATORY CONTROL OF FINGER-TIP FORCES

Influences of an object's weight

When the weight of an object is varied while its visual appearance remains constant, adults typically use an anticipatory control strategy in which the amplitude (and duration) of the isometric grip and load forces during the loading phase are scaled based upon a representation of the object's weight in previous lifts (Fig. 5**A**; [31]). This results in force rates which are critically damped at lift-off, providing similar vertical accelerations independent of the object's weight (Fig. 5**B**).

In contrast, the force development for young children is not influenced by the object's weight (Fig. 5; [52]). They obtain higher forces mainly by prolonging the duration of isometric force increase. Yet, they are capable of adapting the forces to the object's weight during the static phase,

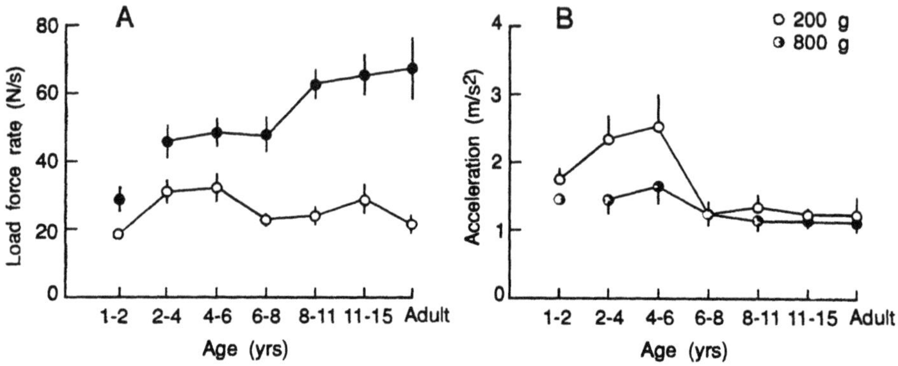

*Figure 5. **A**. Peak load force rate and **B**. vertical acceleration (mean ± SEM) for consecutive lifts with a 200 g and 800 g weight (400 g for the 1-2 year old children) for various age groups. Note that the relative differences between the force rates do not achieve adult levels until the age of 8-11. This results in slower vertical accelerations of the object after lift-off for the heavier object. Modified from [52].*

when the object is held in the air, indicating that somatosensory weight information can be used to adjust the force output. Concomitant to the emergence of mainly single peaked force-rate profiles, as well as the grip force/load force synergy, children begin to scale the amplitude of the forces according to the object's weight during their second year of age (Fig. 5; [52]). The influence of the object's weight in the previous trial increases gradually until six to eight years, when the vertical acceleration of the object following lift-off becomes similar for objects of various weights. Interestingly, this is the same age that isometric contraction is no longer needed to elicit muscle action potentials during electromagnetic stimulation of the motor cortex [26, 28]. There is a reduction of the thresholds to adult levels by 11 years of age. Subtle improvements in the force coordination and anticipatory control during precision grip also continue until adolescence. Anticipatory control of the precision grip may develop later than anticipatory control underlying the hand closure to an object's size during reaching [6]. However, visual information is continuously available during reaching, which may guide the hand, as well as initiate and shape its closure. On the other hand, anticipatory control of the force output during grasping is based

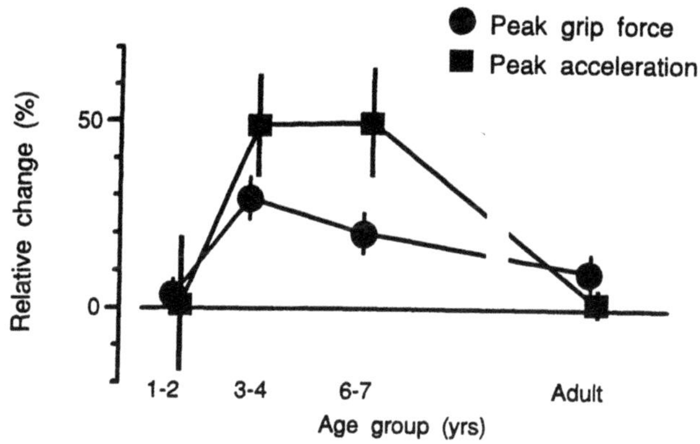

Figure 6. Relative change (mean ± SEM) in peak grip force and vertical acceleration for lifts of a small box and a large box (both 500 g) for various age groups. Note the larger visual influences seen for the 3-4 and 6-7 year old children compared to adults. Modified from [62].

on memories of the object's weights from previous lifts. It is not clear which CNS areas may be involved in storing or representing weight-related information acquired during previous lifts. However, Smith and his colleagues have recently shown that the cerebellum contributes to the object-specific sensory-motor integration in monkeys trained to grasp, lift and hold an object with a precision grip [53, 54]. They also reported that neurons in the hand area of the monkey motor cortex altered their discharge in advance as a function of the object's weight and texture [55]. Indeed, children with damage to cerebral structures and corticospinal pathways, such as in cerebral palsy, retain immature control strategies and never develop anticipatory control of the finger-tip forces during precision grip [56, 57].

Influences of visual size cues

Weight information may also be gained from visual [32, 34] and haptic [33] size cues based on a predicted relationship between the object's size and weight. When both the size and weight of an object attached to an instrumented grip handle vary in tandem (i.e. the weight is kept proportional to the volume), the forces are appropriately scaled by adults toward the expected weight relative to the volume. When only the size of the object attached to the grip handle is changed while the weight is kept the same [32, 33], adult subjects still employ higher forces for the larger object, even though a size-weight illusion is experienced, in which the smaller object is perceived to be heavier [58]. This indicates a dichotomy between perceptual and motor systems. The neural substrates of visual control of object-oriented manual actions in humans are quite distinct from those underlying visual perception [59]. For example, patients with ***visual form agnosia*** may be unable to recognize the size, shape, and orientation of objects, but still are able to accurately guide the arm movement and finger closure toward the same objects [60, 61]. However, there may be some interactions between these systems since the size influences on the force output during the lifting task are relatively small when the weight is unchanged compared to the influences when the size and weight are kept in proportion [34], as well as to the influence of the previous weight [31]. Thus, adults attenuate the influences of size cues when they are no longer meaningful (see Fig. 6).

The force output for children under the age of two and a half to three years is not influenced by the object's size, regardless of whether the weight is kept the same or covaried (Fig. 6; [62]). Furthermore, they are not capable of experiencing a size-weight illusion until after this age (c.f. [63]). This suggests that an additional year of development following the emergence of anticipatory force scaling based on the object's weight is necessary before children can make an associative transformation between the object's size and weight. This likely involves additional demands on cortical processes, requiring further cognitive development. Mounoud and Hauert [64] claimed that children could predict the weight of an object from its size by the age of 14 - 16 months. The yield of the arm was constant when grasping objects covarying in size and weight. This discrepancy may be due to the differences in the tasks or differences in the development of proximal and distal muscle control. In Mounoud's study, a passive receiving task was employed in which mainly the tension of proximal muscles was measured. In contrast, our study involved an active grasping task and the isometric finger-tip forces were measured. In fact Mounoud and his colleagues found that children could partially compensate for the weight of an object using size information by the age of three and a half years when using an active grasping task [64-66].

When the ability to use visual size cues to scale the force output emerges, children exhibit much larger visual size influences than adults if only the size is varied. This suggests that they are incapable of attenuating size information when it is not meaningful (Fig. 6; [62]), perhaps implying that children rely more on vision than adults [67]. However, other studies suggest that children are less likely to integrate sensory information from various modalities than adults [43]. The maintained large but inappropriate size influences may correspond to a later development of mechanisms integrating visual size information and somatosensory weight information.

LATERALIZATION OF GRASPING MECHANISMS

Recently, we examined the force coordination in both the dominant and non-dominant hand of right-handed children of various ages and of right-handed adults [68]. While the force coordination is generally similar between hands, the force output in lifts with the left hand is influenced by the object's weight slightly more than in lifts with the right hand in the younger children (Fig. 7**A** and **B**). In another experiment, we examined the influence of an object's weight when it was previously lifted with the contralateral hand. The influence emerged later during development and was smaller than the influence on lifts performed with the same hand in all age groups. Also, an asymmetric influence was seen during alternations between hands, with an earlier influence during lifts alternating from the right to the left hand than vice versa (Fig. 7**C** and **D**).

In brain-bisectioned subjects, whose corpus callosums and anterior commissures are sectioned, somatosensory information is not transferred to the other hemisphere [69]. Thus during a task such as ours, somatosensory information related to the object's weight and friction is likely transferred from one hemisphere to the other via the forebrain commissures. Impairments in an adult subject with corpus callosum agenesis during performance of the hand alternation task described above support this notion [68]. Since the influence of weight-related information was often reduced when derived from lifts in the contralateral hand even in healthy adults, some information may be lost during this transfer.

The even smaller influence of the object's weight on the contralateral hand in young children may be due to a late maturation of interhemispheric connections [70, 71]. The asymmetry of transfer between hands may be due to an asymmetric development of the forebrain commissural connections (c.f. [72]), which has recently been proposed as a mechanism underlying the development of handedness (c.f. [73]).

TACTILE CONTROL OF FINGER-TIP FORCES

The contribution of cutaneous afferent input to the control of hand and arm movements has long been recognized [74-77]. During the static phase of a lift when the object is held in the air, the amplitude of the grip

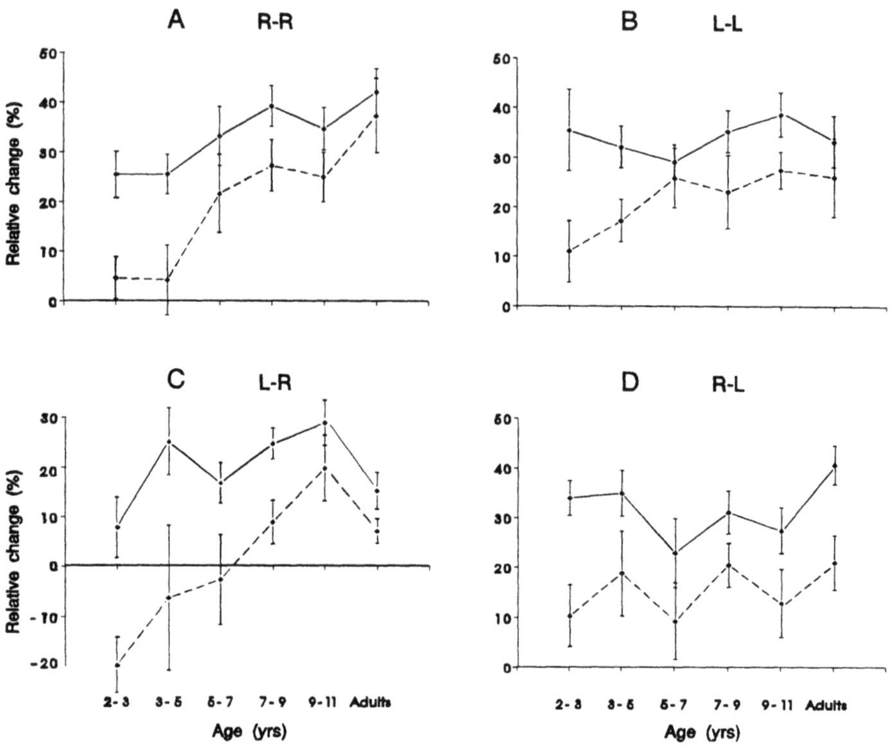

*Figure 7. Relative difference (mean ± SEM) in grip force rate (dotted) and load force rate (solid) between a 300 g and 900 g weight as a function of age during **A**. consecutive lifts with the right hand, **B**. consecutive lifts with the left hand, **C**. lifts with the right hand following lifts with the left hand, and **D**. lifts with the left hand following lifts with the right hand. Note the relatively small influences from the previous weight in the left (non-dominant) hand in young children, as well as in the right hand following a lift with the left hand. Modified from [68].*

force is adapted to the friction between the skin and the object, as well as to the object's weight, producing a grip force just above the minimal force level required to prevent slips [29]. This is dependent on cutaneous afferent input since it is impaired after cutaneous anesthesia [29]. Small localized slips between the skin and the contact surface of the object are sufficient to

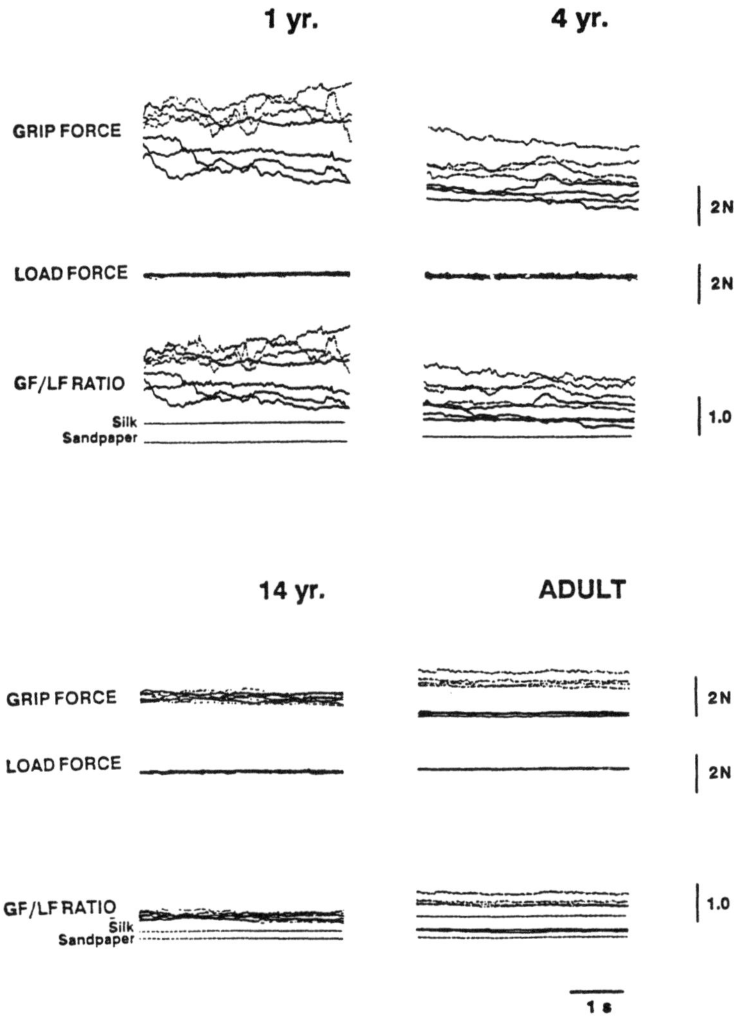

Figure 8. Superimposed recordings of grip force, load force and GF/LF ratio during the static phase as a function of time for three children and one adult. The contact surfaces were either sandpaper (solid lines) or silk (dashed lines). Dotted lines indicate the slip ratio for sandpaper and silk contact surfaces. Modified from [78].

evoke activity in cutaneous afferents, which upgrade the grip force/load force ratio [30].

The tactile influence on grip force regulation in children was recently studied [78]. The grip force/load force ratio was recorded during the static phase when the object was held in the air while children lifted the instrumented test object with either sandpaper or silk contact surfaces. At the end of the trial, the children were asked to gradually release their grip until the object slipped from their fingers, during which point the *slip ratio* was recorded, that is, the grip force at the moment the object slips from the finger-tips. The ***safety margin*** constituted the difference between the grip-force/load force ratio during the static phase and the slip ratio. Young children displayed an excessive grip force with large variability during the static phase, resulting in a large safety margin. The safety margin decreased during development, with the largest reduction already by five years of age. However, slight influences of the contact friction are seen on the level of grip force already by the age of one to two years (Fig. **8A**), and a more mature adaptation is seen by the third year (i.e., higher grip forces for silk contact surfaces). It was suggested that the excessive grip force, increasing the safety margin throughout the static phase, would be a purposeful strategy used to prevent slips when only minimal adaptation occurred [78].

Interestingly, pubescent children exhibited a relatively small adaptation to the object's friction, due to a reduced difference in the slip ratio for silk and sandpaper contact surfaces (Fig. **8C**). This suggests that children's skin is more adhesive during puberty, reducing the differences in friction between various surfaces.

CONCLUSION

Precise grasping and manipulation of objects is dependent on the development and maturation of the corticospinal tract, somatosensory systems that can efficiently monitor sensory information (e.g., from tactile afferents), and perhaps other motor areas since the poor force-coordination in young children partially resembles the deficits observed in patients with various motor impairments (e.g., cerebral palsy, cerebellar lesions, Parkinson's disease). That does not mean that the child is a passive participant waiting for the CNS to mature. Indeed, it is well established that cortical structures are highly plastic during development (e.g., [79]). In fact, cumulating evidence suggests that both sensory and motor maps are dynamically maintained and are capable of extensive reorganization (for recent review, see [80]). Many of the exploratory behaviors and strategies seen in the developing child may suggest that some of the changes seen during development are function driven. However, it is clear from both lesion studies in monkeys and clinical studies in humans that maturation of at least some structures must precede RIFM and precision grip.

Young children also have longer reflex latencies, which may suggest a lack of sensorimotor integration, but the influence of sensory information on subsequent lifts emerges early and gradually increases during development. Although young children can regulate their grip force during the static phase, they appear to have difficulties storing the information or translating the representation into motor commands. A lack of anticipatory control in young children may explain their excessive force generation, prolonged movement phases and large intra-subject variability. When anticipatory control emerges, it appears that children are not capable of appropriately integrating sensory information into the internal representation of the object (see [81]). The ability to deduce one physical property from another probably requires additional cognitive development. It appears that the continued development of anticipatory control, the ability to properly integrate sensory information and the ability to transform size information may reflect important maturations of mechanisms controlling the reach to grasp movement.

ACKNOWLEDGEMENTS

I am grateful to Professor Roland Johansson and the Department of Physiology, Umeå University for providing the equipment and software for the described experiments. I also thank Dr. Didier Flament for helpful comments on an earlier draft of this manuscript.

REFERENCES

1 C. von Hofsten, Hum. Mvt. Studies, 5 (1979) 160.
2 C. von Hofsten, in: M. Jeanenrod (ed.), Attention and Performance, Vol. XIII, Erlbaum, Hillsdale, 1990, p. 739.
3 T.G.R. Bower, Nature, 228 (1970) 679.
4 C. Trevarthen, in: H.T.A. Whiting (ed.), Human Motor Actions: Berstein Reassesed, Elsevier, Amsterdam, 1984, p. 223.
5 C. von Hofsten, Dev. Psychobiol., 18 (1982) 450.
6 C. von Hofsten and L. Rönnqvist, J. Exp. Psychol., Hum. Percep. Perf., 4 (1988) 610.
7 A. Peiper, Cerebral Function in Infancy and Childhood, Consultants Bureau, New York, 1963.
8 T.E. Twitchell, in: K. Connolly (ed.), Mechanisms of Motor Skill Development, Academic Press, New York, 1970.

9 R.S. Illingworth, The Development of the Infant and Young Child, Normal and Abnormal, Churchill Livingstone, Edinburgh, 1975.
10 E. Taub, Ex. Sports Sci. Rev., 4 (1976) 335.
11 H.M. Halverson, Genet. Soc. Gen. Psychol. Monogr., 10 (1931) 110.
12 P. Mounoud and A. Vinter, in: G. Butterworth (ed.), Infancy and Epistemology, Harvester Press, Hassocks, 1981.
13 J.R. Napier, J. Bone Jt. Surg., 38B (1956) 902.
14 K.J. Connolly and J. Elliott, in: C. Blurton-Jones (ed.), Etiological Studies of Child Behavior, University Press, Cambridge, 1989.
15 D. Corbetta and P. Mounoud, in: C. Bard, M. Fleury and L. Hay (eds.), Development of Eye-Hand Coordination across the Life Span, University of South Carolina Press, Columbia, 1990, p. 188.
16 D.G. Lawrence and H.G.J.M. Kuypers, Brain, 91 (1968) 15.
17 H.G.J.M. Kuypers, in: J.M. Brookhart and V.B Mountcastle (eds.), Handbook of Physiology, The Nervous System II, Amer. Physiol. Soc., Bethesda, 1981, p. 597.
18 R.B. Muir and R.N. Lemon, Brain Res., 261 (1983) 312.
19 R.N. Lemon, K.M.B. Bennett and W. Werner, in: G.E. Stelmach and J. Requin (eds.), Tutorials in Motor Neuroscience, Kluwer Academic Publishers, Holland, 1991, p. 477.
20 E.J. Hall, D. Flament, K.M.B. Bennett, and R.N. Lemon, in: G.E. Stelmach and J. Requin (eds.), Tutorials in Motor Behavior II, Elsevier, Amsterdam, 1992, p. 861.
21 H.G.J.M. Kuypers, Science, 138 (1962) 678.
22 D.G. Lawrence and D.A. Hopkins, Brain, 99 (1976) 235.
23 D. Felix and M. Wiesendanger, Exp. Brain Res., 12 (1971) 81.
24 D. Flament, E.J. Hall and R.N. Lemon, J. Physiol., 447 (1992) 755.
25 D. Flament, E.J. Hall, R.N. Lemon and M. Simpson, J. Physiol., 426 (1990) 105.
26 T.H.H.G. Koh and J.A. Eyre, Arch. Dis. Child., 63 (1988) 1347.
27 P.I. Yakovlev and A.R. Lecours, in: A. Minkowski (ed.), Regional Development of the Brain in Early Life, Blackwell, Oxford, 1967, p. 3.
28 K. Müller, V. Hömberg and H.G. Lenard, Electroenceph. Clin. Neurophysiol., 81 (1991) 63.
29 R.S. Johansson and G. Westling, Exp. Brain Res., 56 (1984) 550.
30 R.S. Johansson and G. Westling, Exp. Brain Res., 66 (1987) 141.
31 R.S. Johansson and G. Westling, Exp. Brain Res., 71 (1988) 59.
32 A.M. Gordon, H. Forssberg, R.S. Johansson and G. Westling, Exp. Brain Res., 83 (1991) 477.

33 A.M. Gordon, H. Forssberg, R.S. Johansson and G. Westling, Exp. Brain Res., 83 (1991) 483.
34 A.M. Gordon, H. Forssberg, R.S. Johansson and G. Westling, Exp. Brain Res., 85 (1991) 226.
35 A.M. Gordon, G. Westling, K.J. Cole and R.S. Johansson, J. Neurophysiol., in press.
36 R.S. Johansson and K.J. Cole, Curr. Opin. Neurobiol., 2 (1992) 815.
37 M. Jeannerod, Behav. Brain. Res., 19 (1986) 305.
38 M. Jeannerod, in: G.E. Stelmach and J. Requin (eds.), Tutorials in Motor Behavior II, Elsevier, Amsterdam, 1992, p. 265.
39 S. Grillner and S. Rossignol, Brain Res., 146 (1978) 269.
40 P.J. Cordo, J. Neurophysiol., 63 (1990) 161.
41 H. Forssberg, A.C. Eliasson, H. Kinoshita, R.S. Johansson and G. Westling, Exp. Brain Res., 85 (1991) 451.
42 H. Issler and J.A. Stephens, J. Physiol., 335 (1983) 643.
43 H. Forssberg and L.M. Nashner, Neuroscience, 2 (1982) 545.
44 H. Forssberg, Exp. Brain Res., 57 (1985) 480.
45 B. Touwen, in: B. Touwen (ed.), Neurological Development in Infancy, SIMP and Heinemann Med. Books, London, 1978, p. 99.
46 V.B. Brooks, in: Bloedel et al. (eds.) Cerebellar functions, Springer-Verlag, Berlin, 1984, p. 1.
47 K.J. Cole and J.H. Abbs, J. Neurophysiol., 55 (1986) 1407.
48 W.G. Darling, K.J. Cole and J.H. Abbs, Exp. Brain Res., 73 (1988) 225.
49 C. Ghez, in: H. Asanuma and V. Wilson (eds.), Integration in the Nervous System, Igaku-Shoin, Tokyo, 1979, p. 305.
50 J. Gordon and C. Ghez, Exp. Brain Res., 55 (1984) 167.
51 G. Westling and R. S. Johansson, Exp. Brain Res., 66 (1987) 128.
52 H. Forssberg, H. Kinoshita, A.C. Eliasson, R.S. Johansson, G. Westling, and A.M. Gordon, Exp. Brain Res., 90 (1992) 393.
53 E. Espinoza and A.M. Smith, J. Neurophysiol., 64 (1990) 698.
54 C. Dugas and A.M. Smith, J. Neurophysiol., 67 (1992) 483.
55 N. Picard and A. M. Smith, J. Neurophysiol., 68 (1992) 1882.
56 A.C. Eliasson, A.M. Gordon and H. Forssberg, Dev. Med. Child Neurol., 33 (1991) 661.
57 A.C. Eliasson, A.M. Gordon and H. Forssberg, Dev. Med. Child Neurol., 34 (1992) 216.
58 A. Charpentier, Arch. Physiol., 3 (1891) 122.
59 M.A. Goodale and A.D. Milner, TINs, 15 (1992) 20.
60 M.A. Goodale, A.D. Milner, L.S. Jakobson and D.P. Carey, Nature, 349 (1991) 154.

61 A.D. Milner, D.L. Perrett, R.S. Johnston, P.J. Benson, T.R. Jordan, et al., Brain, 114 (1991) 405.
62 A.M. Gordon, H. Forssberg, R.S. Johansson, A.C. Eliasson and G. Westling, Exp. Brain Res., 90 (1992) 399.
63 H.B. Robinson, Child Dev., 35 (1964) 91.
64 P. Mounoud and C.A. Hauert, in: G.E. Forman (ed.), Action and Thought, Academic Press, New York, 1982, p. 3.
65 C.A. Hauert, Arch. Psychol. Monogr., 7 (1980) 185.
66 C.A. Hauert, P. Mounoud and E. Mayer, Cahiers Psychol. Cog., 1 (1981) 33.
67 D.N. Lee and E. Aronson, Percep. Psychophys., 15 (1974) 529.
68 A.M. Gordon, H. Forssberg and N. Iwasaki, in preparation.
69 M.S. Gazzaniga, J.E. Bogen and R.W. Sperry, Neuropsychologia, 1 (1963) 209.
70 W. Hewitt, J. Anat., 96 (1962) 355.
71 A. Salamy, Science, 200 (1978) 1409.
72 A.S. LaMantia and P. Rakic, J. Neurosci., 10 (1990) 2156.
73 S.F. Witelson and C.H. Goldsmith, Brain Res., 545 (1991) 175.
74 F.W. Mott and C.S. Sherrington, Proc. Royal Soc. Lond., 57 (1895) 481.
75 T.E. Twitchell, J. Neurophysiol., 17 (1954) 239.
76 E. Moberg, Brain, 106 (1983) 1.
77 D. Denny-Brown, The Cerebral Control of Movement, Liverpool University Press, Liverpool, 1966.
78 H. Forssberg, H. Kinoshita, A.C. Eliasson, R.S. Johansson and G. Westling, in preparation.
79 D.H. Hubel and T.N. Wiesel, Proc. Royal Soc. Lond., 198 (1977) 1.
80 P.E. Garraghty and J.H. Kaas, Curr. Opin. Neurobiol., 2 (1992) 522.
81 A.M. Gordon, in: H. Forssberg and H. Hirschfeld (eds.), Children with Movement Disorders, Karger, Basel, 1992, p. 130.

Section Two: Central Mechanisms

Insights into the Reach to Grasp Movement
K.M.B. Bennett and U. Castiello (Editors)

CHAPTER 4

NEURONAL SUBSTRATE FOR VOLITIONAL MOVEMENT

A.B. SCHWARTZ

Barrow Neurological Institute, 350 West Thomas Road, Phoenix, Arizona 85013, U.S.A.

SUMMARY

The act of reaching toward and grasping an object is perhaps one of the best examples of volitional movement. Although the control of this movement is poorly understood, the motor cortex has been considered essential in this regard. While neuroanatomical studies suggest which structures may be involved in producing volitional movement, they cannot describe their functional organization. Only recently have neuronal studies been combined with behavioral paradigms to examine the control of this movement. I will review some historical studies of volitional movement and compare these to recent neurophysiological studies of more natural movements which have shown that characteristics of volitional action may be encoded throughout a large portion of the CNS and not restricted to discrete structures.

INTRODUCTION

In our attempt to understand the process subserving volitional limb movement, we must confront a major impediment to this study. The substrate responsible for generating these movements is covert - concealed within the CNS. Unlike sensory stimuli, which can be identified and detailed externally before the corresponding central response is investigated, the fundamental properties of movement control are far from obvious. This is especially true when considering movements rich in behavioral expression, such as reaching and grasping.

Historically this problem has been approached in a functional-anatomical way through localization. Critical to this approach is the assumption that specific qualities of movement reside in specific structures within the CNS. The motor cortex, for instance, has been considered to be the primary region of the cerebrum responsible for the control of voluntary movement; thus lesions of motor cortex should lead to paralysis. The difficulty with this approach becomes apparent if the assumed quality is not localized. Many cortical and subcortical structures, in fact, participate simultaneously in the control of volitional movement. Indeed, motor cortical lesions result only in temporary paresis, with the ability to make most voluntary movements returning rapidly. The structures remaining after the lesion are similar enough in operation to the motor cortex that they can compensate for its loss.

I will review selectively the history of localization and motor cortical physiology in an attempt to illustrate how we have arrived at our present level of investigation which emphasizes the role of motor cortex in the behavioral aspects of movement. As this chapter concentrates primarily on the neuronal processing taking place in the motor cortex related to reaching, historically important lesion and anatomical studies will not be discussed in detail. Grasping studies, because of the difficulty in measuring the movement are relatively recent and there have been very few physiological experiments examining reaching *and* grasping. The foundation of motor cortical physiology lies in the concept that explicit regulation of volitional movement resides in this region of the cerebrum. It is likely that the principles of motor cortical physiology derived from the reaching and drawing studies will also be relevant to the control of reaching and grasping.

Initially, physiological studies based on the concepts of localization established the motor cortex as a kind of switchboard, mapping cortical activity to specific muscle activation with a discrete one-to-one correspondence. In contrast to the historical localization techniques, recent physiological and psychophysical findings derived from experiments in which animals actually perform complex volitional movement suggest that the cerebral control of these movements is distributed and that movement parameters are encoded coarsely in the activity of single cells. These studies emphasize the idea that information is continuously transformed during the production of volitional tasks. Distributed systems are characterized by parallel processing and multiple representation of information. For instance, neuronal activity encoding a movement parameter may be found in many different parts of the central nervous system (CNS) simultaneously. Within a given anatomical region of the brain, different parameters may be represented at the same instant. Our evolving studies show that simultaneous neuronal activity in a distributed system such as the CNS should be considered as a whole when trying to understand the substrate for volitional acts such as reaching and grasping.

HISTORY OF LOCALIZING MOTOR FUNCTION

Only within the last 130 years has the cerebral cortex been considered to have any motor function. Through most of recorded history the cortex was thought to be the dominion of consciousness, a ***sensorium commune*** without localized function. Motor control was relegated to phylogenetically older parts of the neuraxis. Although the cerebrum was not assigned a large role in motor control, as reviewed by Walker [1], cerebral localization was described at the time of Hippocrates when it was noted that unilateral brain injury resulted in movement loss on the opposite side of the body.

Both stimulation and lesion methods have been used to localize function in the CNS. Lesion studies and clinical pathology have provided insight toward some essential roles for the motor cortex. However, because of space constraints I will concentrate only on a review of the stimulation literature.

Although pinching and pricking were used to stimulate the cortex previously, Aldini [2], in the early 1800s, was the first to electrically stimulate the cerebrum. Using the technique developed by his uncle Galvani who experimented with skeletal muscle, Aldini studied the effect of galvanic stimulation on a wide range of decapitated animals. His enthusiasm extended to human subjects. In one report he applied his electrodes across the exposed cortex and right ear of a freshly beheaded man and observed contraction of the right side of the face.

Fritsch and Hitzig [3] in the 1860s applied more localized galvanic stimulation to the cerebrum in rabbits and were able to elicit movement. Later, using bipolar stimulation in a more detailed study in dogs, they found that activation of the frontal cortex produced contraction in the contralateral limbs. They identified the motor cortex as a localized region most responsive to stimulation. Upon removal of this region, they noted a contralateral paresis of the forelimb. More refined studies at the end of the century carried out in the light of the stimulation studies and confined to motor areas of the cortex showed that monkeys [4] could use their contralateral limbs if prompted. Nevertheless these experiments directly supported the idea that motor centers were localized in the cerebral cortex.

Ferrier extended these experiments using faradic stimulation. Whereas galvanic stimulation activated the cortex only once per application, faradic stimulation activated neurons repetitively. This made it possible to use less current to elicit movement and allowed the cortex to be mapped at a higher resolution. Using frequencies of 30 to 40 Hz, Ferrier observed discrete muscles contracting and was able to map the monkey cortex [5] in detail. These early motor maps spanned both the pre- and postcentral cortices. Horsley and Schaeffer [4] also found the pre- and postcentral cortices to be excitable. However Grunbaum and Sherrington [6] using monopolar stimulation later found that only the precentral cortex elicited movement when stimulated.

These stimulation studies resulted in a map suggesting that the motor cortex was somatotopically organized. Neurons projecting to a portion of the body tended to be located within the same region of the cortex. Adjacent body segments were juxtaposed on the cortical surface. Woolsey [7] later found this pattern to be detailed enough to form a distorted figurine or "simiuculus" on both the pre- and postcentral cortical surfaces. The muscles of the lower face and distal extremities were represented by a disproportionately large cortical area.

About the turn of the century, Jackson was publishing his observations made on epileptic patients with cerebral pathology [8], which were used to support the concept of cerebral localization. As these patients began to seize, he noted that the toe muscles would contract, followed by contractions sequentially up the leg to the trunk and shoulder and finally to the hand. This "march of spasms," he theorized, resulted from the spread of seizure activity across the cortex affecting adjacent regions sequentially. Jackson posited that a relatively direct pathway existed from cortex to the motoneurons and that cortical regions were organized in terms of basic movements that involved widespread activation of multiple muscles (an issue that remained controversial for many years).

At the end of the 19th century the precentral gyrus was considered to be a localized site of motor function. Although there was some controversy whether lesions of discrete areas prevented sensory input from ***releasing*** motor acts or whether the ablations removed motor capability directly, the cerebrum was considered essential for movement. These ideas were prevalent despite the demonstrations that decorticate animals were capable of moderately complex behavior.

Sherrington's school dominated motor physiology at the beginning of the 20th century. Lesions placed at different levels of the neuraxis led Sherrington to view motor behavior, in general, as an elaboration of simpler reflexes. Leyton and Sherrington [9] mapped the ape cortex with faradic stimulation and defined a set of subareas corresponding to five body portions (face, arm, leg, trunk and head). Using graded stimuli, they were able to elicit fractional movements that could be integrated with other primary fragments to produce seemingly purposeful movement. They viewed the cortex as a structure that could assemble and integrate combinations of muscular action fragments into different movements.

At the same time other investigators began to measure the effect of cortical stimulation using electromyographic (EMG) recordings. Cooper and Denny-Brown [10] found that cortical stimulation had a "direct" effect on the EMG and that this was evident when stimulating in the frequency range of 4 to 180 Hz. Chang et al. [11] recorded from several muscles simultaneously and found that cortical loci activated by near-threshold cortical stimulation projected to single muscles but overlapped other projection sites. This overlap was more pronounced with higher intensity stimulation. Bernhard et al. [12] recorded descending volleys in the lateral cortical spinal tract (CST) and in dissected ventral roots elicited from 25 Hz cortical stimulation. Since there was only a seven millisecond latency between the first component of the CST volley and activity in the ventral root, the CST was thought to activate α motoneurons monosynaptically. Preston and Whitlock [13] followed by Landgren et al. [14] used intracellular recording to elaborate this finding. They found that excitatory postsynaptic potentials (EPSP) were elicited in α motor neurons at monosynaptic latencies with single shocks applied to the surface of the motor cortex.

These experiments suggested that patches of cortex projecting to particular motoneuronal pools were not discrete but were large (5 to 20 mm^2) [14] and coextensive with cortical areas projecting to different pools of motor neurons. The results of these surface stimulation studies did not support the idea that discrete patches of motor cortex projected to individual forelimb muscles. However, it could be argued that these findings were compromised by current spread through the cortex since relatively strong (2 to 5 mA) stimuli were used in these studies.

In 1968, Stoney et al. [15] reported the results of a study using ICMS (intracortical microstimulation), a technique which was originally developed by Landau et al. [16]. A microelectrode was inserted into layer V or VI of the motor cortex. Motor neurons could be excited with about 1% of the current used for surface stimulation. Using repetitive stimuli (300 Hz) individual muscle contraction was observed and the volume of cortex from which these contractions could be elicited was small and discrete [17]. Using the idea of the cortical column described for sensory cortex, Asanuma and Rosen [17] described the composition of motor cortex in terms of efferent zones.

The idea of a columnar organization as a basis of motor cortical anatomy was developed by Collonier [18]. Two classes of cells were found in the primary motor cortex, stellate and pyramidal cells. The stellate cells with round cell bodies and non-oriented dendrites were considered interneurons because their axons did not enter the white matter. These were further divided into two types. The double bouquet cells have two major dendritic branches oriented upward and downward, spreading extensively in the vertical plane. These cells are found throughout the thickness of the cortex except in layer I. Their axons entwine the apical dendrites of pyramidal cells and were thought to be excitatory. The other type of stellate cell, termed the basket cell, is found in layers III and V [19]. Its dendrites are star-like and their axons extend horizontally, ending in basket endings (presumed to be inhibitory) on pyramidal cell bodies. Pyramidal cells are found predominantly in layers III and V with the largest cells in the deeper layer. These cells have triangular cell bodies and large apical dendrites that ramify extensively (several mm) in layer I. Their axons may form recurrent collaterals that spread 0.5 to 1 mm horizontally that may also ascend vertically. An important property of pyramidal cells is that their axon collaterals may project horizontally many millimeters, for instance, from area 3b to 4 [20]. A given collateral may generate several terminal patches, oriented radially from layers II to IV, separated by runs of 800μ with no terminals. Thus, an axon originating in 3b may terminate in area 3a as well as in area 4. Since thalamocortical afferents from a given nucleus are thought to terminate within a discrete cortical area, these corticocortical axons may be the most important means of distributing information to different cortical areas. Pyramidal cells also provide the corticofugal output from the motor cortex projecting to most of the subcortical nuclei and to the spinal cord. The distinctive Betz cells of the primary motor cortex are pyramidal cells located in layer V and are among the largest cells in the brain with diameters on the order of 100μ.

More recent studies [21, 22] classified cerebral neurons as either pyramidal or non-pyramidal. The non-pyramidal cells are thought to be intrinsic interneurons possessing spiny and non-spiny dendrites. The non-spiny cells are thought to contain gamma-amino butyric acid (GABA) producing inhibitory postsynaptic potentials (IPSP) in pyramidal cells while the pyramidal cells and spiny interneurons contain the excitatory transmitter glutamate. These anatomical studies suggest that the inhibitory axons are arranged in restricted vertical zones through the cortex and that the basket cell axons, for instance, form an inhibitory surround around a core of excitation produced by the output of the spiny cell axons and recurrent collaterals of the pyramidal cells which would receive input from thalamic afferents. This arrangement would then result in a columnar organization.

Although the horizontal extent of the axon fields of the spiny interneurons were thought to be only 50μ and the ***columns*** in visual cortex were thought to be 300-500μ wide, the efferent zones were thought to be one millimeter wide. Input arriving in the upper layers of cortex was thought to be integrated and eventually output via one or a few pyramidal tract cells in layer 5. According to Asanuma and Rosen [17], this output projecting via the CST to motoneuronal pools would activate muscles individually. However the ICMS results showed that agonist finger muscles usually contracted together as did the antagonist muscles about the wrist. Although these cortical efferent zones overlapped those to other muscles, the boundaries between zones were considered sharp. The sharpness was believed to be the result of the low currents used for ICMS in contrast to the less distinct boundaries resulting from the higher current surface stimulation studies.

Columnar organization within the cerebral cortex has been a fundamental concept since the work of Mountcastle [23] showed that cells in cat primary sensory cortex along a radial tangent tended to respond to the same modality of sensory stimuli applied to the same part of the body. Thus, the idea of cortical efferent zones and their similarity to columns described in the sensory and visual cortex was attractive. The entire cerebral cortex was thought to be constructed of basic modules or columns [24]. The differences in the operation of the different cortical areas was attributed to the different afferent projections that each area received. The processing within each column was restricted to the vertical dimension taking place within a restricted volume of cortex.

At that time, in the early 1970s, the controversy over whether movements or muscles were represented in the motor cortex resurfaced. Both rationales were based on the idea that topographical areas of motor cortex would be active sequentially throughout a movement. For instance, during a reach, cells in the shoulder, elbow, wrist and finger areas would change their activity patterns as the corresponding body parts moved. Based on previous studies showing that the motor cortex was somatotopically organized, it was thought that cells in a particular somatotopic division were all active simultaneously. I will refer to this as the ***chunk*** rationale. If the cortex was organized so that projection areas were large and overlapping, then activation of a given *chunk* would always activate multiple muscles. However if the motor cortex was organized in discrete efferent zones, then muscles would be separately activated from each *chunk*. Though Asanuma and Rosen [17] supported the latter concept, their data showed that the cortical efferent zones overlapped extensively. Those zones projecting to the wrist overlapped in such a way that if a *chunk* of cortex in this area became uniformly active, antagonist muscle groups would be co-facilitated. Although this antagonist co-facilitation was not observed for the finger areas (perhaps because EMG activity was not recorded) more than one muscle was always activated from a particular *chunk* of cortex.

The single muscle - single column hypothesis, as reviewed by Humphrey [25], soon ran into theoretical obstacles. Only the pyramidal cell axons of layer V were shown to project to subcortical sites [26], and more intense ICMS in the upper cortical layers is required to elicit motor responses. This would reduce the efferent column to a disk. Another problem lies in the selection of the cortical columns during a movement. Since multiple muscles are used in any movement and the combinatorial activity of the selected muscles is dynamic during the movement, the afferents selecting the columns to be activated would contain all the information required to activate the proper muscles at the correct intensities in the proper temporal sequence. This would require that all the kinematic and dynamic parameters of the movement be computed before the columns were selected. However, based on the movements elicited by activation of motor cortical afferent systems, this appears unlikely. Recording studies in behaving primates [27, 28] showed that adjacent cells often covaried with antagonistic muscles or muscles about different joints. Very few pairs showed a constant covariation with the activity pattern of the same muscle.

In addition to these shortcomings of the efferent zone concept, ICMS was shown to be a problematic technique. Jankowska et al. [29] showed that ICMS led predominantly to indirect, transynaptic activation of pyramidal tract cells. She and her colleagues compared the latency of an antidromic response recorded in layer V elicited by stimulation of the lateral funiculus to the orthodromically elicited descending volley recorded from the same site on the dorsolateral cord. The orthodromic volley resulted from either surface or intracortical stimulation. The ortho- and antidromic activation had comparable latencies with surface stimulation. However the predominant component in ICMS elicited volleys was of longer latency showing that the activation was transynaptically conducted to the corticospinal tract. This type of indirect cortical activation was also noted by Asanuma and Rosen [30] and most likely resulted from excitation of recurrent collaterals and/or interneurons. ICMS elicits activity over a wide area of cortex and it is likely (especially with repetitive stimuli) that this activates a complex cortical network that produces the inhibitory surround of the efferent zone. This activation pattern is not likely to resemble that which takes place during volitional movement. In fact, a recent report by Lemon et al. [31] found that unitary activity of neurons projecting from the motor cortex to hand muscles almost always facilitated muscle activity while ICMS at the same cortical site was more likely to produce suppression in the post-spike average of EMG.

Although the general somatotopic organization in the motor cortex represented by the inverted figurine on the convexity of the hemisphere is still a good anatomical summary, more recent mapping experiments in non- or lightly anesthetized monkeys dispute the original details of this organization. Murphy et al. [32] trained monkeys to relax their arm muscles and observed invariant movements in response to low intensity ICMS. They found that cortical loci from which movements could be elicited were organized in a nested manner. Those sites corresponding to movements of the most distal part of the arm were surrounded by successively more proximal movements. The areas corresponding to movements about adjacent joints were contiguous. A multiple representation of individual joint movements was found. Humphrey [25, 33] using ICMS and EMG recordings also found a multiple representation of muscles in the motor cortex. Low threshold sites that elicited activity in wrist flexors were coincident with those eliciting activity in wrist extensors. The same site also elicited cocontraction of elbow extensors and flexors. Thus activation of a wrist extensor and flexor as well as an elbow extensor and flexor could be elicited from the same low threshold site. The only observed movement at this site was wrist extension even though the other muscles were active simultaneously. These results were interpreted as support for the movement representation concept. During a natural movement, multiple muscles are simultaneously active and, for instance, when the fingers are moved the more proximal joints act to support the hand so that muscle contraction about multiple joints is also required. Thus activation of a given *chunk* of motor cortex results in simultaneous activity in a combination of muscles to yield a particular movement.

PHYSIOLOGICAL STUDIES

Bernstein, in 1935 [34], developed a set of arguments describing the interrelation of movement control and localization. These arguments were based on his detailed observations of human kinematics. He found that even in stereotyped repeated movements, such as with hammer pounding, that the trajectory of each repetition is different. Using a simplified equation of motion for a single muscle and joint, he showed that the displacement resulting from a single force impulse is dependent on the angle of the joint and its angular velocity. If the CNS is to generate displacement by generating muscular force, then the signal used to excite the muscle is also dependent on joint angle and velocity. The control signal must rely on proprioceptive information. Since the proprioceptive signal changes continually (dynamic environment) a static control signal would lead to different movements each time it is repeated. In other words, to repeat the same movement a different control signal would need to be produced for each repetition. For example, as a hammer strikes a nail, the nail is displaced, altering the starting location for each repetition. If the same movement command was emitted, not only would the hammer miss the nail because the nail had moved, but with a constant stroke displacement the location of the termination of the return stroke would also shift as the nail was driven. Bernstein showed that the entire trajectory changed for each repetition and argued that the CNS must be able to calculate joint and external forces continuously.

Bernstein was interested in the "structural physiology" of movement. This was defined as the temporal pattern of the coordinated activity of multiple muscles across multiple joints during movement. He postulated that many central structures contributed to this control. Since the efferent activity of many structures projected to the spinal cord independently, this was a parallel scheme emphasizing not the activity of single neurons but the organization of their common features. Although he believed in the existence of localization, this was based on functional organization, not topology. "Thus, in the problem of localization what is important for our purpose is not precisely where in the cortex one or another peripheral object or function is reflected, but *what* is represented, and *how*." He further described the fallacy of movement representation in specific sites of the cortex ("push-button control board model"), arguing that since the efferent command for a movement must change at each instance, the structure underlying this control cannot be rigid. If individual muscles are represented in the cortex by a one-to-one mapping, then each repetition of the same movement would entail activation of different parts of the cortex since the muscle activation pattern differs for each repetition. Obviously the boundaries of the movement representation would have to change for each repetition. Muscle localization would deny movement localization and vice versa. "One of the two chess pieces must here be taken, and it is here a very pertinent question which of the two the old-fashioned localizationalist would rather sacrifice." The topographical organization of motor cortical neurons is not a relevant functional factor. Rather, the information contained within the neuronal activity and the way that it is distributed are the key features of motor control. It should be further emphasized that the demonstration of anatomical connectivity is not sufficient for elucidating functional distribution. Whether information from one cell is transmitted to another is a non-stationary, non-linear probability function that is determined by the total state of the system at a given point in time.

Very few studies have examined the neuronal substrate for reach to grasp movements (see Gibson et al. this volume). I will review briefly the literature associated with single-joint, multi-joint and drawing movements. Finally I will discuss the studies that have examined the neuronal activity associated with grasping. The concept of functional localization has been an important issue as these behavioral paradigms have evolved. As we develop more natural paradigms and study the neuronal activity associated with them, our concept of *what* function consists of and whether it is localized in a single anatomical entity has also evolved.

Single-joint paradigms

Since few neurophysiological recording studies employed animals that actually moved, until the 1960s, knowledge of movement-related information contained in central structures was minimal. The first investigator to carry out an experiment in this regard, that was germane to reaching and grasping, was Evarts, who recorded single cell activity in the motor cortices of awake, behaving monkeys. His initial study [35] showed that most pyramidal tract neurons in the arm area of the motor cortex were active during spontaneous arm movement. Monkeys were trained to make alternating wrist extension-flexion movements against a telegraph key in his next study. The animals were cued by a signal light to make the movement. He found that the motor cortical neurons fired in response to the cue if the cue was associated with the movement. Although there was a loose relation between neuronal onset latency and movement onset, the correlation was "far from perfect". The neurons tended to respond prior to wrist EMG onset. These initial studies showed that many neurons from a wide topographical area responded for each movement and although their activity was clearly associated with the movement there was no obligatory linkage between the discharge pattern of individual neurons and the movement onset. A later study [36] addressed the question of whether the neuronal discharge was correlated to the force used to displace the limb or to the displacement itself. In this study, monkeys performed a wrist flexion-extension task against a handle that could be loaded to assist or resist the movement. In theory this was to dissociate force from displacement since the latter was the same in each trial while the force required to produce the displacement was varied by changing the load. Evarts intended to test the theory that the motor cortex acts as a displacement controller, specifying the position of the limb regardless of the force required to get it there. In this theory other CNS structures would transform the displacement signal into the appropriate muscle contractions. The results, however, supported the opposite conclusion. A neuron that fired as the wrist was flexed in the case when the handle was not loaded would increase its discharge rate when the load opposed flexion and decrease when the load assisted flexion. This suggested that the cell was load sensitive and was not coding displacement or, in this case, wrist angle. The discharge pattern of the cells was similar to the EMG activity of the primary muscles used to flex and extend wrist. However, as the author pointed out, even for this relatively isolated wrist movement many muscles about the fingers, wrist elbow and shoulder are active simultaneously. This makes it difficult to ascertain to which specific muscle or combination of muscles the cell is **exclusively** related. In this type of experiment it isn't possible to completely dissociate force from displacement since a change in force is required to accelerate the handle during the movement. Examination of discharge rate and force showed that there was not a direct relation between the two variables. For instance, a cell that would start to fire as flexor force was applied to the handle would be silent as maximal flexor force was generated which coincided with the time that the handle was arrested by a mechanical stop. Although motor cortical activity was related to static force, the cell activity was also related to other parameters, such as the rate of change of force (dF/dt), and displacement.

In the following years, other investigators basically confirmed these findings using similar experimental paradigms. Humphrey et al. [37] modified the paradigm by requiring the monkey to terminate the movement in a target zone instead of against the mechanical stops. Using a small population of motor cortical responses they showed that this activity was related to force, velocity, position and dF/dt. The population response was constructed from the activity of three to eight simultaneously recorded cells and the spike frequency of each was weighted by a regression coefficient to the parameter being considered. This weighted factor was then summed with those from the other cells of the population and scaled before being compared to the time profile of the movement parameters. It was found that the correlation between the population response and a particular parameter increased with the number of cells in the population. The latency between the population response and, for instance, force, was found to be about 100 ms. Interestingly, the correlation between individual cells and a given movement parameter was quite labile and varied between trials. In agreement with Evarts, these investigators found force to be the parameter best encoded in the neuronal activity, but velocity and displacement were also well represented. Although the time course of force was well represented in a population of cells, when the load was changed, the coefficient for each cell had to be scaled. It was concluded that steady force was not well represented in the population. In contrast, a given set of coefficients calculated for velocity and displacement yielded accurate predictions under various load conditions.

Schmidt et al. [38] modified the basic paradigm by replacing the weight and pulley loading system used in the previous studies with a torque motor, giving them the advantage of being able to change loads continuously without disturbing the monkey. The torque was adjusted to resist the movement in a spring-like manner and the movement divided into three portions: an initial hold, a transition and a terminal hold phase. They found that almost all the cells responded in the transition phase where the opposing forces reversed. Although the cells responded in this phase, the magnitude of their discharge was unrelated to the magnitude of the transition. This led the authors to conclude that the "motor cortex is involved in specifying the muscles to be activated for a given movement and not the level of force produced by these muscles."

Using a subset of motor cortical efferents, Fetz and Cheney [39] later reached a similar conclusion. They examined the responses of corticomotoneuronal cells (CMN) that were motor cortical neurons tested with spike-triggered averaging (STA - a technique that measures the influence of a spike on muscle excitability) and assumed to have monosynaptic connections with α motoneurons. Monkeys performed either an isometric wrist task or movements against an elastic load. The researchers found that although the average responses of these cells covaried with tonic force, the responses correlated poorly with the temporal profile of force.

Other investigators using single-joint movements found that motor cortical activity was related to joint displacement and stiffness [40, 41] and the expected direction of the next movement [41]. A recent study [42] compared the activity of cells in the putamen, supplementary motor cortex and primary motor cortex while static loads were applied to the arm in an elbow flexion-extension task. Cells that responded preferentially to load or the direction of displacement were found with equal frequency in all three structures. The lead time between the onset of activity to the start of the movement was shortest in the supplementary motor area, followed by the activity in motor cortex with the latest responses in the putamen. However there was a large overlap in these times so that many of these neurons were simultaneously active.

In general, these studies showed that the motor cortical activity of single cells was related in a complex manner to the production of force. All of these studies examined wrist torque resulting from a multitude of muscular activity throughout the arm. This line of experimentation originated by Evarts can be viewed historically as a transition from the long history of topographical localization toward the identification of relevant movement parameters represented in the motor cortex. The parameters examined with these restricted movements were all mechanistically related to the physical properties of the rotating joint. As Bernstein suggested, these may not be the *what* that is represented in the activity of, in this case, the motor cortex. The concept of a higher order representation of the movement as a ***behavioral output*** was developed as investigators began to examine motor cortical activity in less restricted movements.

Multi-joint paradigms

As unrestrained arm movements began to be employed in cortical studies it became evident that simple parameters related to individual joints or muscles were not well represented by motor cortical activity. Although Evarts [35] described motor cortical activity as monkeys reached spontaneously, Porter and Lewis [43] reported the results of the first recording experiment where a multi-joint reaching task was used. Monkeys were trained to reach out, grasp and pull a spring-loaded handle. Pyramidal tract neurons recorded in the motor cortex were found to be modulated continuously throughout the task [43]. Different neurons appeared to be recruited for specific portions of the task. Unfortunately, individual neurons were not classified as to their somatotopic correspondence.

One of the first studies to detail both motor cortical activity and movement parameters during reaching was carried out by Murphy et al. [32]. They recorded motor cortical activity as monkeys reached forward to a control panel to push one of six buttons. The three-dimensional position and orientation of the arm was measured throughout the task with an optoelectronic device and joint angles about the shoulder and elbow were calculated. The researchers found that there was no simple relation between EMG patterns of the major arm muscles and motor cortical activity. Pairs of single cells that responded to joint rotation in opposite directions were more reciprocal in their activity patterns than supposedly antagonist muscle pairs about the same joint. Although shoulder-related motor cortical units consistently varied their discharge patterns for movements to the different targets, their activity patterns were not related to a particular joint angle profile. It was concluded that "the production of any movement, however complex or discrete it may seem peripherally, engages a complex population of precentral neurons, such that any one neuron may behave similarly for overtly different movements."

At the same time, Georgopoulos et al. [44, 45] were recording motor cortical activity as monkeys performed two-dimensional reaching movements by moving a manipulandum over a planar work surface. The task required that the animal move the manipulandum from a center start target to one of eight equally spaced targets that formed a circle around the start location (center→out task). The movement parameter examined in this work was direction. Movements in different directions require changes in multiple joint angles and muscles. Despite the underlying mechanical complexity of the movement, a simple relation between discharge rate and movement direction was found for most of the precentral cells that were active in the task. The relation between discharge rate and direction was described with a cosine formula that spanned all movement directions. The ***preferred direction*** of a given cell was the movement direction that corresponded to the peak discharge rate. Thus each cell's activity encoded all movement directions. Such coarse coding of a movement parameter is somewhat contrary to the "push-button control board" concept of motor control. Instead of specific cells active only in a narrow range of movement direction, many cells in the motor cortex are simultaneously active, encoding each movement direction as a population. A vector algorithm [45, 46] was developed to describe the emergent directional information represented in this population. Each cell's average discharge rate was calculated for a movement to a specific target. This rate was used to scale a unit vector in the cell's preferred direction. This operation was carried out for movements to each target and for each cell in the population. The resulting contributions from each cell to the population was illustrated as a cluster of vectors pointing in different directions. The vectors representing cells whose preferred direction coincided with the movement direction tended to be longest and the vector resulting from summing the contributions, the ***population vector***, pointed in the direction of the target for each of the movements.

This approach was shown to be valid for the more general case of arm movements through free space [47-49]. Monkeys were trained to reach from the middle of a cube to each of its corners and the resulting single-cell activity could be described with a tuning volume based on the cosine function. Population vectors again closely predicted the movement direction. When the population vectors were calculated at 100 intervals throughout the time course of the movement, it was evident that the population vectors appeared and pointed in the movement direction about 60 ms after the presentation of the movement stimulus which was 120 - 140 ms before the movement began. This showed that the directional information in the motor cortex was predicting the movement direction well in advance of the actual movement.

The population vector algorithm has been used to better understand what type of information is represented in the motor cortex. A clear dissociation between cortical and muscle activity was demonstrated in the 3D task [50]. A set of parallel movements between the buttons on the front of the cube showed that single cells had similar discharge rates for movements in the same direction carried out in different portions of the workspace. EMG activity recorded for the same movements was quite different. The neuronal response seemed to code direction relative to the initial position of the hand regardless of its location.

If direction is coded instantaneously in the activity of motor cortex cells, it should be possible to predict movement direction continuously throughout a movement. This was tested in an experiment where monkeys drew sinusoids [51]. Direction changed continuously as the figure was drawn. The discharge activity of single cells changed through the task in a way that corresponded to each cell's tuning function. The directional activity of these cells coded for a direction of movement that occurred with a latency of about 120 ms. A portion of the discharge activity also was related to movement speed (the speed of the movement also changed continuously). This speed coding was most evident for movement directions near each cell's preferred direction.

A time series of population vectors calculated from this activity showed that the population activity corresponded to the tangential velocity of the drawing movement [52]. The direction of each population vector corresponded to the continuously changing movement direction while the magnitude of each vector was well correlated to the movement speed. The relation between the length of the population vector and movement speed was due to the direction dependence of speed coding in the activity of individual cells. Cells tend to make large contributions to the population vector if their preferred directions are near the movement direction. These contributions will be reduced for low speeds in the preferred direction and enhanced for high speeds. Reduced contributions in the direction of movement tend to make the cluster of constituent vectors more symmetrical and this results in a shorter population vector. The constituent vectors are longer for higher speeds when the movement is in their preferred direction and results in more asymmetrical clusters and longer population vectors. Thus both direction and speed are parameters that emerge from the population. This is a good example of how multiple parameters may be simultaneously encoded in the same cell population.

These results also confirmed those findings suggesting that the information represented in motor cortical activity is related in an *instantaneous* way to the ongoing process of movement. Since the population vectors are well correlated to tangential velocity, the population activity codes for trajectory on a piecewise basis. The origin of each vector is spatially aligned with the tip of the previous vector. The vectors code for direction and speed relative to this origin.

These results show that the trajectory of the arm is well represented in the population activity. This seems to contradict the findings of Evarts showing that force rather than hand position was represented in the activity of single motor cortical cells. Besides the obvious differences between the experimental tasks these findings might be explained if populations are considered. As outlined earlier, even these original experiments were unable to resolve a clear relation between force and discharge rate. Humphrey et al. [37] found that the combined activity of a small group of motor cortical cells was well related to the applied load in this type of behavioral paradigm. It is important to note that a kinematic parameter, velocity, was almost as well related to the neuronal activity.

Investigations that considered direction as a parameter have yielded insights into this issue. Kalaska et al. [53] used a two-dimensional manipulandum that could be loaded in different directions. Monkeys performed the center→out task with the loaded manipulandum. Loads applied in a direction opposing the motor cortical cells' preferred directions tended to increase the activity. The directional tuning curve did not change shape when loads were applied, rather the entire curve shifted along the axis representing discharge rate. Thus the static load seemed to bias the directional tuning curve and both parameters were simultaneously represented in a given cell's discharge rate.

Isometric tasks where force is exerted in the absence of displacement is the only situation where these variables are completely dissociated. Georgopoulos et al. [54] have shown how motor cortical activity is related to force in this condition. Monkeys were trained to exert a force pulse on a handle in response to a set of targets on a computer monitor. Force feedback was provided by a cursor on the monitor. The eight targets around a center start position were arranged in the same way as those of the center→out task. As previously shown for non-isometric movements, the discharge rate of these motor cortical cells was broadly tuned to the direction of force. A constant bias force could be introduced by altering the relation between the cursor and the applied force, so that the subject was required to generate different directions of force, depending on the bias, to move the cursor in the same direction. The shape of the tuning function was unaffected by the bias force. The investigators considered net force to be the difference between the force generated by the subject and the bias force. The population vectors corresponded to the net force. In a different part of the analysis, the change in force between ten millisecond bins was found to be similar to the net force when calculated continuously as a time series of vectors. A time series of population vectors calculated for each bin matched a corresponding sequence of net and force change vectors which coincided in direction with a visually derived intention vector. These were clearly different from the total and bias forces. The authors point out that although the mechanical factors and EMG are quite different in the moving and isometric tasks, the underlying motor cortical activity is very similar and may relate "to an abstract representation of spatial trajectory", a conclusion supported by the drawing study [52]. The activity in the motor cortex seems to be related to behavioral conditions and is somewhat removed from the physical and mechanical constraints of the task.

Two sets of experiments support the role of the motor cortex in abstract processing. One set relates to the invariant rules subserving drawing. Objects are drawn in segments defined by a zero-crossing in acceleration. For instance, figure eights are generally drawn in two segments with each loop of the eight defined as a segment [55, 56]. Monkeys were trained to draw figure eights on the touch screen and single motor cortical units were recorded during the task [57]. The animals produced the same segmentation as humans. Population vectors calculated through the task were added tip-to-tail producing a neural representation of the trajectory. This neural trajectory had the same segmentation as the actual movement showing that this behavioral invariant was encoded in the activity of motor cortical cells. Another invariant, the isogony principle, was also tested. This rule was originally found in handwriting [58] and showed that angular velocity was directly related to the radius of curvature. Subjects slow down in more curved regions of the drawing. This was tested on monkeys trained to draw spirals. These animals produced the spirals obeying the isogony principle and the neural trajectory derived from their motor cortical activity also followed this rule. The invariants derived from drawing movements are incorporated into the activity of the motor cortex, suggesting that the behavioral aspects of the task are an important factor represented in the output of the motor cortex.

An example of the cognitive role that the motor cortex can play in the processing of spatial information was provided in an experiment where a monkey was required to perform a spatial transformation [59]. This experiment employed a variant of the two-dimensional center→out task where the monkey was required to move 90° counterclockwise to the illuminated peripheral target. For example, if the target appeared at the two o'clock position, the animal was required to move the manipulandum to the 11 o'clock location. Population vectors, calculated at two millisecond intervals, initially pointed in the direction of the illuminated target early in the reaction time. They then rotated toward the counterclockwise location until reaching the 11 o'clock position about 35 ms before the animal began to move at this target. This response took place during the reaction time, in the absence of movement, and is a further illustration of how a population of motor cortical cells may subserve the higher-order processing associated with the performance of spatial motor tasks.

Grasping studies

The pyramidal tract and motor cortex have long been established as critical structures for precision grip. This conclusion has been reached using lesion studies [60-65]. The common result of this work is that there is a permanent inability to form a precision grip using the thumb and index finger. Another conclusion from experiments using split-brain monkeys is that there is a dichotomy in the mechanisms used for grasping (distal) from those used for reaching (proximal). Vision-supplied information to the contralateral cerebral cortex is required for grasping, but is not necessary for reaching [66, 67]. Differential control using peripheral and foveal visual fields for the transport and target acquisition phases of pointing movements has also been found [67].

Recording experiments also suggest that motor cortical cells are involved in regulating precision grip. This has been shown in isometric tasks where there is a monotonic relation between force and discharge rate especially for low ranges of force [68, 69]. However, although these cells were thought to be corticomotoneuronal projections based on post-spike facilitation, there was poor correlation between firing rate and EMG in the target muscle during maintenance of static force. Only a few cells were found where there was a high correspondence between discharge rate and grip force. Surprisingly, about the same number of neurons that showed an increasing monotonic relation showed a decreasing relation between grip force and discharge rate. It was hypothesized that the cells with negative correlation may recruit motor units that generate smaller forces. Another study by these investigators [70], shows that the post-spike facilitation of this projection is task dependent. Monkeys performed either a precision grip or a rotation with their index fingers and thumbs. A "dramatic" difference in the post-spike facilitation was found in the two tasks.

The control of precision grip is most certainly aided by vision in normal behavior. Cells responsive to both visual input and motor aspects of manipulation have been recorded in the posterior parietal cortex [71, 72]. These responses were studied quantitatively as monkeys manipulated different objects in light and dark conditions [73]. Neurons were found to be active during manipulation in the dark (hand movement units), only in the light (visual dominant units) or responsive in the dark with an increased response in the light (visual and motor units). The animals were trained to manipulate a variety of objects and each neuron tended to be preferentially active during a particular manipulation. It was suggested that integration of motor and visual information takes place in this cortical area. Visual afferents from extrastriate visual cortical areas (i.e., parietal-occipital and middle temporal) terminate in this region. There is also a reciprocal connection with premotor area 6. This region of area 6, on the posterior bank of the arcuate sulcus, has direct projections to the motor cortex and is another site where neurons responsive to grasp have been found [74].

Although, detailed studies examining neurophysiological activity during reaching *and* grasping have not yet been carried out, the results of the gripping experiments are quite similar to those for reaching. Many structures are active simultaneously. Posterior parietal, arcuate premotor and motor cortex seem to be part of a circuit that utilizes visual information during manipulation. Behaviorally, there is an apparent dichotomy between the use of the proximal joints to transport the hand to the target and the actual grasp of the target. It should be noted that these phases overlap since pre-shaping of the hand occurs during the transport phase. Similarly, the motor cortex, in addition to the cells projecting to distal motor pools, contains many cells related to the displacement of the proximal arm segments and the activity of these cells accurately reflects the arm's trajectory. Interestingly, these cells, although related to proximal joint displacement, may, as a population, encode the trajectory of the hand better than that of the more proximal segments [75, 76].

It is difficult to assign a specific function to the motor cortex, even though ablation of this cortical region leads to a loss of precision grip. In the one study [43] that looked at reaching and grasping, motor cortical neurons were "recruited" continuously throughout the task, that is, it was not possible to conclude that activity in this structure was related to grasping exclusive of reaching. An anatomical study [77] employed ICMS to determine motor cortical somatotopy and small HRP injections to identify cells projecting to the injection site. Wide spread interconnectivity between different parts of the forearm representation was found. Injections made at a site where ICMS elicited thumb movement were found to label cells at sites where elbow, wrist and shoulder movements were elicited. The HRP was transported by horizontal axons in layers III and V. These findings were interpreted as evidence that activity within the motor cortex is distributed and helps to explain how simultaneously active cells communicate. This also argues against the idea that individual cells are controlling aspects of movement (i.e., single muscles) in isolation.

Thus, as with reaching, it is likely that different neurons within the motor cortex are transmitting information related to different aspects of gripping and that the information content within a given corticifugal axon is task dependent [70]. To consider these neurons as controllers of muscle groups is also nebulous since few neurons were simply correlated to total force. It will be necessary to consider the simultaneous action of many neurons within the same structure and the distributed activity of different structures before the nature of the control process can be understood. As we have begun to show with the reaching studies, it is likely that the cerebral processing associated with reaching and grasping is composed of information related to the cognitive strategies used to achieve the behavioral goal in addition to the control of specific mechanisms employed to produce the movement.

CONCLUSION

Historical concepts tend to resurface in cycles. Until the 1800s the cerebral cortex was considered to be removed from the mechanics of somatic function and to house what philosophers called the "vital force". The cerebral cortex was not thought to be electrically excitable until the work of Fritsch and Hitzig and the demonstration of intrinsic current by Canon in the 1870s. Although the conceptual roots of localizing theory can be traced to the ancient Greeks, it was this period in the late 1800s that ushered in the foundations of what we consider to be cerebral localization. As illustrated by the motor cortex, our understanding of cerebral function was driven by successive technological advances. Just as the development of the electrochemical cell and its application to biology by Galvani led to the discovery of the essential nature of the reflex arc by Mueller and the development of the galvanometer made it possible for du Bois-Reymond to describe the action potential, the experiments of the 1870's led initially to a rapid establishment of the motor cortex as a "push-button control board" for motor control. These ideas were to remain deeply established for more than a century. Most of the experiments supporting this concept relied on the notion that one *chunk* of cortex was active to the exclusion of others. Another nuance of this reasoning was that all cells in a particular *chunk* were active simultaneously. The controversy this elicited was whether all the cells in a *chunk* were coding for activation of the same muscle or coding for different muscles all involved in the same movement. It was not until moving subjects were studied that this controversy became a non-issue.

In fact, Bernstein, who developed a cinematic technique to measure human movements in three-dimensional space, argued against this type of localization in the 1930s. The development of chronic recording in monkeys by Evarts made it possible to study actual arm movements as motor cortical activity was recorded. In this transition out of the localization era these first chronic experiments examined only single-joint movement and their results were unclear. The muscle versus movement controversy was still unresolved. As the experimental paradigms advanced to movements in two-dimensional followed by three-dimensional space, it became clear that higher order movement descriptors such as direction and trajectory were represented in the activity of motor cortex. Insights from cognitive experiments have shown that neural activity subserving spatial problem solving takes place in the motor cortex. These experiments are transforming our ideas of the motor cortex away from connectionist dogma toward an understanding of how complex distributed systems control the behavioral expression found in such acts as reaching and grasping. Perhaps our understanding of cortical processing is evolving toward a more philosophical description of the complex and ***vital*** way we behave in our surroundings.

ACKNOWLEDGEMENTS

The author is supported by NIH NS26374

REFERENCES

1 A.E. Walker, J. Neurophysiol., 20 (1957) 435.
2 J. Aldini, Essai Theorique et Experimental Sur le Galvanisme, Fournier, Paris, 1804.
3 G. Fritsch and E. Hitzig, Arch. Anat. Physiol., Lpz., 37 (1870) 300.
4 V. Horsley and E.A. Schafer, Phil. Trans. Roy. Soc. Lond., 179 (1888) 1.
5 D. Ferrier, Proc. Roy. Soc. Lond., 23 (1875) 409.
6 A.S.F. Grunbaum and C.S. Sherrington, Proc. Roy. Soc. Lond., 72 (1903) 152.
7 C.N. Woolsey, in: H.F. Harlow and C.N. Woolsey (eds.), Biological and Biochemical Bases of Behavior, University of Wisconsin Press, Madison, 1958, p. 63.
8 J.H. Jackson, Medical Times and Gazette, 1 (1875) 578.
9 A.S.F. Leyton and C.S. Sherrington, Q. J. Exp. Physiol., 11 (1917) 135.
10 S. Cooper and D.D. Denny-Brown, Proc. Roy. Soc. Lond., 102 (1947) 222.
11 H.-T. Chang, T.C. Ruch and A.A. Ward, J. Neurophysiol., 10 (1947) 39.

12 C.G. Bernhard, E. Bohm and I. Peterson, Acta Physiol. Scand., 29 (1954) 70.
13 J.B. Preston and D.G. Whitlock, J. Neurophysiol., 24 (1961) 91.
14 S. Landgren, C.G. Phillips and R. Porter, J. Physiol., 161 (1962) 112.
15 S.D. Stoney, W.D. Thompson and H. Asanuma, J. Neurophysiol., 31 (1968) 659.
16 W.M. Landau, G.H. Bishop and M.H. Clare, J. Neurophysiol., 28 (1965) 1206.
17 H. Asanuma and I. Rosen, Exp. Brain Res., 14 (1972) 243.
18 M.L. Collonier, in: J.C. Eccles (ed.), Brain and Conscience Experience, Springer-Verlag, Berlin, (1966), p.1.
19 M. Marin-Padilla, Brain Res., 14 (1969) 633.
20 J. DeFelipe, M. Conley and E.G. Jones, J. Neurosci., 6 (1986) 3749.
21 E.G. Jones, J. Comp. Neurol., 160 (1975) 205.
22 E.G. Jones, in: Motor Areas of the Cerebral Cortex, Ciba Found. Symp., No. 132, John Wiley and Sons, Chichester, 1987, p. 21.
23 V.B. Mountcastle, P.W. Davies and A.L. Berman, J. Neurophysiol., 20 (1957) 374.
24 G.M. Edelman and V.B. Mountcastle, The Mindful Brain, MIT, Cambridge, 1978.
25 D.R. Humphrey, Federation Proc., 45 (1986) 2687.
26 E.G. Jones and S.P. Wise, J. Comp. Neurol., 175 (1977) 391.
27 E.V. Evarts, J. Neurophysiol., 29 (1966) 1011.
28 E.V. Evarts, in: M.D. Yahr and D.P. Purpura (eds.), Neurophysiological Bases of Normal and Abnormal Motor Activities, Raven Press, New York, 1967, p. 215.
29 E. Jankowska, Y. Padel and R. Tanaka, J. Physiol., 249 (1975) 617.
30 H. Asanuma and I. Rosen, Exp. Brain Res., 16 (1973) 507.
31 R.N. Lemon, R.B. Muir and G.W.H. Mantel, Exp. Brain Res., 66 (1987) 621.
32 J.T. Murphy, H.C. Kwan, W.A. MacKay and Y.C. Wong, Brain Res., 246 (1982) 141.
33 D.R. Humphrey and W.S. Corrie, J. Neurophysiol., 41 (1978) 216.
34 N.A. Bernstein, in: H.T.A. Whiting (ed.), Human Motor Actions, Bernstein Reassessed, Advances in Psychology, No. 17, Elsevier, Amsterdam, 1984, p. 77.
35 E.V. Evarts, J. Neurophysiol., 28 (1965) 216.
36 E.V. Evarts, J. Neurophysiol., 31 (1968) 14.
37 D.R. Humphrey, Brain Res., 40 (1972) 7.
38 E.M. Schmidt, R.G. Jost and K.K. David, Brain Res., 83 (1975) 213.
39 P.D. Cheney and E.E. Fetz, J. Neurophysiol., 44 (1980) 773.

40 D.R. Humphrey and D.J. Reed, in: J.E. Desmedt (ed.), Motor Control Mechanisms in Health and Disease, Raven Press, New York, 1983, p. 347.
41 W.T. Thach, J. Neurophysiol., 41 (1978) 654.
42 M.D. Crutcher and G.E. Alexander, J. Neurophysiol., 64 (1990) 151.
43 R. Porter and M.McD. Lewis, Brain Res., 98 (1975) 21.
44 A.P. Georgopoulos, J.F. Kalaska, R. Caminiti and J.T. Massey, J. Neurosci., 2 (1982) 1527.
45 A.P. Georgopoulos, R. Caminiti, J.F. Kalaska and J.T. Massey, Exp. Brain Res. Suppl., 7 (1983) 327.
46 A.P. Georgopoulos, J.F. Kalaska, M.D. Crutcher, R. Caminiti and J.T. Massey, in: G.M. Edelman, W.M. Cohen and W.E. Gall (eds.), Dynamic Aspects of Neocortical Function, Neurosci. Res. Found. Inc., 1984, p. 501.
47 A.P. Georgopoulos, A.B. Schwartz and R.E. Kettner, Science, 233 (1986) 1357.
48 A.B. Schwartz, R.E. Kettner and A.P. Georgopoulos, J. Neurosci., 8 (1988) 2913.
49 A.P. Georgopoulos, R.E. Kettner and A.B. Schwartz, J. Neurosci., 8 (1988) 2928.
50 R.E. Kettner, A.B. Schwartz and A.P. Georgopoulos, J. Neurosci., 8 (1988) 2938.
51 A.B. Schwartz, J. Neurophysiol., 68 (1992) 528.
52 A.B. Schwartz, J. Neurophysiol., in press.
53 J.F. Kalaska, D.A.D. Cohen, M.L. Hyde and M. Prud'homme, J. Neurosci., 9 (1989) 2080.
54 A.P. Georgopoulos, J. Ashe, N. Smyrnis and M. Taira, Science, 256 (1992) 1692.
55 F. Lacquaniti, C. Terzuolo and P. Viviani, Acta. Psychol., 54 (1983) 115.
56 J.F. Soechting and C.A. Terzuolo, Neuroscience, 23 (1987) 39.
57 A.B. Schwartz, A. Kakavand and J.L. Adams, Soc. Neurosci. Abstr., 17 (1991) 308.
58 P. Viviani and C. Terzuolo, Neuroscience, 7 (1982) 431.
59 A.P. Georgopoulos, J.T. Lurito, M. Petrides, A.B. Schwartz and J.T. Massey, Science, 243 (1989) 234.
60 S.S. Tower, Brain, 63 (1940) 36.
61 D.G. Lawrence and H.G.J.M. Kuypers, Brain, 91 (1968) 1.
62 M.-C. Hepp-Reymond, E. Trouche and M. Wiesendanger, Exp. Brain Res., 21 (1974) 519.

63 C.N. Woolsey, T. Gorska, A. Wetzel, T.C. Erickson, F.J. Earls and J.M. Allman, Brain Res., 49 (1972) 119.
64 D.G. Lawrence and D.A. Hopkins, Brain Res., 40 (1972) 117.
65 R. Passingham, H. Perry and F. Wilkinson, Brain Res., 145 (1978) 410.
66 J. Brinkman and H.G.J.M. Kuypers, Brain, 96 (1973) 653.
67 J. Paillard, in: D.J. Ingle, M.A. Goodale and R.J.W. Mansfield (eds.), Analysis of Visual Behavior, MIT Press, Cambridge, 1982, p. 367.
68 T.M.J. Wannier, M.A. Maier and M.-C. Hepp-Reymond, J. Neurophysiol., 65 (1991) 572.
69 M.A. Maier, K.M.B. Bennett, M.-C. Hepp-Reymond and R.N. Lemon, J. Neurophysiol., 69 (1993) 772.
70 R.N. Lemon, K.M.B. Bennett and D. Flament, Soc. Neurosci. Abstr., 16 (1990) 42.
71 J. Hyvarinen and A. Poranen, Brain, 97 (1972) 673.
72 V.B. Mountcastle, J.C. Lynch, A. Georgopoulos, H. Sakata and C. Acuna, J. Neurophysiol., 38 (1975) 871.
73 M. Taira, S. Mine, A.P. Georgopoulos, A. Murata and H. Sakata, Exp. Brain Res., 83 (1990) 29.
74 G. Rizzolatti, R. Camarda, L. Fogassi, M.. Gentilucci, G. Luppino and M. Matelli, Exp. Brain Res., 71 (1988) 491.
75 J.L. Adams and A.B. Schwartz, Soc. Neurosci. Abstr., 16 (1990) 423.
76 A. Kakavand and A.B. Schwartz, Soc. Neurosci. Abstr., in press.
77 G.W. Huntley and E.G. Jones, J. Neurophysiol., 66 (1991) 390.

Insights into the Reach to Grasp Movement
K.M.B. Bennett and U. Castiello (Editors)

CHAPTER 5

GRASPING CEREBELLAR FUNCTION

A.R. GIBSON, K.M. HORN, and P.L.E. Van KAN

Barrow Neurological Institute, 350 West Thomas Rd., Phoenix, Arizona, 85013, U.S.A.

SUMMARY

In this chapter we examine the role of intermediate cerebellum in the control of a ***reach to grasp****. Our arguments rest on data from two nuclei: nucleus interpositus, the output nucleus of intermediate cerebellum, and the magnocellular red nucleus (RNm), which receives signals from interpositus and projects to the spinal cord. Interpositus and RNm contain neurons that increase discharge rate during movement of large body parts, such as the forelimb or hind limb. However, when forelimb neurons are tested with movement about specific joints, most cells show little or no change in discharge rate. During a* ***reach to grasp*** *the same cells show large increases in discharge rate. In monkey interpositus, the grasp component is necessary to elicit discharge modulation. Variation in the reach component of a* ***reach to grasp*** *changes neither the pattern nor amplitude of discharge modulation. Temporary inactivation of the cat RNm results in a severe deficit in the ability to grasp a lever, but only a mild deficit in the ability to reach. We conclude that intermediate cerebellum is specialized for the control of specific hand movements. The grasp in the* ***reach to grasp*** *relies on this control circuitry. Jeannerod's [1] hypothesis that the* ***reach to grasp*** *is composed of transport and grasp components controlled by separate channels is strengthened by our findings: intermediate cerebellum may provide a neural substrate for the grasp channel. Presumably additional neural circuitry, possibly involving other divisions of the cerebellum, is specialized for the control of the transport phase of the* ***reach to grasp****.*

INTRODUCTION

Over the past ten years we and our co-workers have concentrated on trying to understand how cerebellar output codes movement. At the outset, there was no reason to believe that ***reaching to grasp*** had any special significance for cerebellar control of movement. After all, the output nuclei of the cerebellum contain representations of the entire body, and individual nuclear neurons are said to discharge selectively to movements of specific joints [2, 3]. In theory, any movement could be produced by activating the correct set of neurons in sequence. However, as our work progressed, the findings could not be reconciled with this view of cerebellar output organization. Grasping has special significance for intermediate cerebellum, and the ***reach to grasp*** may provide a key to understanding the organization of cerebellar output as well as to the function of major divisions of the cerebellum.

Our study of cerebellar output began in the laboratory of Dr. James Houk at Northwestern University Medical School. Dr. Houk's goal was (and still is) to understand the neural processing occurring between the input and output of the cerebellum. Our first studies concentrated on the magnocellular red nucleus (RNm), which receives its major input from interpositus, the output nucleus of intermediate cerebellum. There are many good reasons for beginning with a study of RNm, but the primary reason is that RNm signals reflect the product of cerebellar processing immediately prior to activation of spinal motor circuitry. Therefore, discharge properties of RNm cells should represent the output of intermediate cerebellum in a form closely related to movement. Understanding the output of a circuit is a critical and necessary step for understanding transformations between input and output. Although our understanding is incomplete, several pieces of evidence are now converging to provide a new concept of the output organization of intermediate cerebellum. The purpose of this chapter is to present major pieces of that evidence as well as our perspective of how they fit together.

The output pathways of intermediate cerebellum are schematically illustrated in Fig. 1. Intermediate cerebellum is defined by nucleus interpositus (NI) and cerebellar cortex that projects to interpositus (paravermal cortex). Interpositus in the monkey and cat is divided into two sub nuclei, nucleus interpositus anterior (NIA) and nucleus interpositus posterior (NIP). Although there is evidence that NIP can be further divided along both functional and anatomic lines, in this chapter we will use interpositus or NI to designate all areas that project to RNm (Fig. 1). NI projects to two major motor centers, the RNm and area 4 of the cerebral cortex, or motor cortex. The projection to motor cortex is via ventral lateral (VL) thalamus.

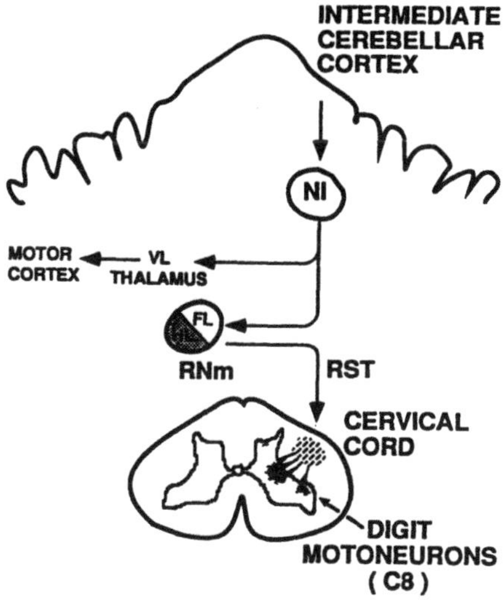

*Figure 1. Output pathways of intermediate cerebellum. **NI**, nucleus interpositus; **RNm**, magnocellular red nucleus; **RST**, rubrospinal tract; **FL**, forelimb; **HL**, hind limb.*

The RNm projects via the rubrospinal tract (RST) to every level of the spinal cord, and, thus, provides NI with its most direct access to segmental motor circuitry. Motor cortex provides less direct access via the corticospinal tract. Since individual NI neurons project to both RNm and VL thalamus, NI gives rise to parallel pathways for motor control. The parallel pathways re-converge at the spinal cord where the termination patterns of the corticospinal and rubrospinal pathways overlap [4, 5]. In both monkeys and cats, the heaviest terminations of the corticospinal and rubrospinal tracts are in Rexed's laminae V-VII. The neurons (interneurons) in these laminae provide input to motoneurons in spinal lamina IX.

Organization of the interposito-spinal pathway

Both anatomic and physiological studies provide ample evidence that NI and RNm contain regions devoted to movement of specific body parts and, therefore, can be considered to contain a form of motor somatotopy. The somatotopic organization of the nuclei has been the major contributor to the concept that neurons in NI are related to movements of specific joints. Yet, when the data are examined critically, it becomes clear that despite convincing evidence for a somatotopy based on large body parts, such as a forelimb or a hind limb, there is little evidence for a finer somatotopy based on joints or muscles. In fact, the data suggest that there is **not** a finer representation of individual joints or muscles.

In both cat [6, 7] and monkey [8, 9] dorsal-medial regions of the RNm project to cervical cord and ventral-lateral regions project to lumbar cord. Only few RNm cells project to both cervical and lumbar cord [10], so there are discrete ***forelimb*** and ***hind limb*** regions within RNm. Since the projection from interpositus to RNm is topographically organized [11, 12], the RNm somatotopy can be extended to NI (although the nuclear subdivisions result in a more complex mapping [12]). Typical tract tracing techniques have sufficient resolution to determine limb regions, but they cannot determine whether or not a finer organization exists within the NI and RNm somatotopies. If a finer organization does exist, one would expect it to be evident in the termination patterns of individual cells.

However, the morphology of individual axon terminations from NI to RNm [13] and the widespread dendritic fields of RNm neurons [14] suggest that individual NI neurons probably contact many RNm neurons within a particular limb region. The high degree of divergence and convergence argues against a fine somatotopy, but it is possible that the projections ***sort out*** in their spinal terminations. Again the available evidence indicates that this is not the case: using electrical activation, Shinoda et al. [15] showed that a single RNm axon can terminate throughout the C4 to C8 extent of the cord; motoneuronal pools innervating the whole forelimb musculature are located in these segments. Intraxonal labeling of rubrospinal fibers confirms collateral branching at widely separated segmental levels and demonstrates that even individual collaterals spread over very wide spinal distances [16].

In an attempt to determine if RNm contains a segmental topography, we injected the retrograde tracer WGA-HRP (wheat germ agglutinin horseradish peroxidase) into the interneuronal region at C6 and plotted retrogradely labeled cells in RNm. If neurons in the forelimb region of RNm terminate in a topographic fashion, one would expect that only a subset of neurons would terminate at a single segmental level. A series of parasagittal sections through the RNm are shown in Fig. 2. Notice that essentially every cell in the dorsal and medial, or forelimb, region of the nucleus is labeled. The result is fully consistent with widespread branching of RNm terminations and suggests little segmental topography in the rubrospinal projection

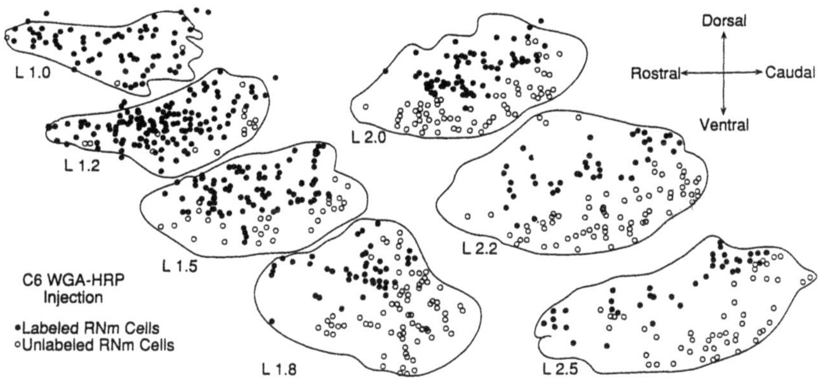

Figure 2. Distribution of retrogradely labeled cells after an injection of WGA-HRP confined to segment C6 of the spinal cord.

to interneurons (a series of segmental injections would be needed to prove this point, but given equally dense rubrospinal terminations throughout the cervical cord [17], it is unlikely that additional cervical injections would reveal a topography).

In marked contrast to the widespread branching of RST terminals in interneuronal regions, the projection of RNm to motoneuronal pools lying in Rexed's lamina IX is restricted and very selective [12, 18, 19]. The lamina IX terminations are largely limited to laterally located pools at C8-T1 (Fig. 1), which innervate digit musculature [20]. These motoneuronal projections, although slight in comparison with interneuronal projections, raise the possibility that movement control by intermediate cerebellum somehow focuses on digit musculature with a less direct influence on musculature of the whole limb.

Single unit recording in the behaving animal offers another way of determining nuclear organization since a map can be constructed from cell properties and recording locations. The recording data provide a close parallel to the anatomic data: a somatotopy based on movements of the forelimb and hind limb is readily recognized, but no somatotopy based on more specific movements within a given limb region is discernible.

In the RNm, cells lying dorsally and medially discharge at high rates when the monkey moves the forelimb opposite to the side of recording, but not when he moves any other body part. Similarly, when recording in the ventral and lateral regions of RNm the cells discharge only when the monkey moves the contralateral hind limb. Within the limb regions it is difficult to tell if cell discharge relates to any specific part of the limb. However, most cells in the forelimb area appear to be related to hand movements, whereas most cells in the hind limb area appear to be related to foot movements. Cells in forelimb RNm discharge at particularly high rates when monkeys reach to grasp raisins or other objects [21].

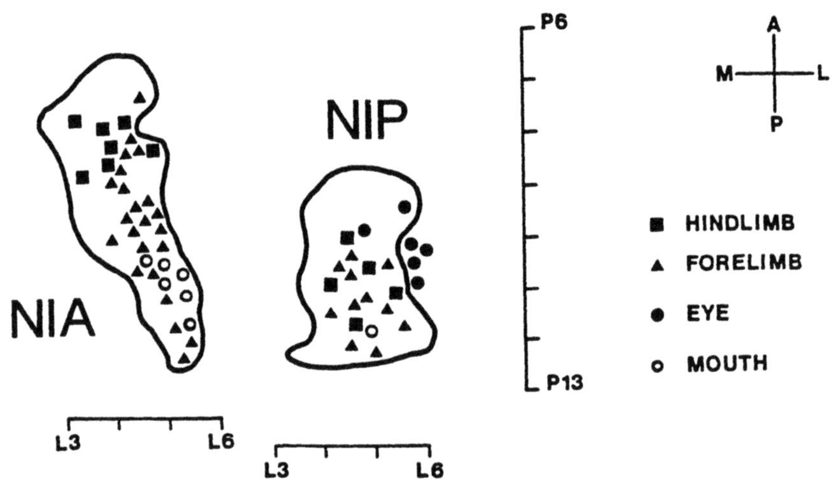

Figure 3. Horizontal reconstruction of recording sites in interpositus. At each site neurons increased their discharge during movement of the indicated body part. ***NIA*** *and* ***NIP*** *have been mapped on independent medial-lateral scales to prevent overlap. Data from [23].*

Since our goal in recording from RNm was to relate discharge to movement parameters, some way was needed to isolate and measure movements of particular parts of the forelimb. Therefore, we trained monkeys to operate a variety of devices that, together, exercised every forelimb joint. Much to our surprise, most forelimb RNm cells failed to fire strongly during operation of any of the devices, despite the fact that they did fire strongly when the monkeys reached out to grasp a raisin. Operation of one device, the twister, did elicit high discharge rates and a tight coupling between discharge rate and movement parameters for some of the cells [22]. Twisting is a complex motion that requires coordinated action of the fingers, wrist, and, possibly upper arm.

Why didn't we find a representation of all forelimb joints in RNm? Since interpositus has parallel output pathways, it seemed possible that the RNm properties represent only a subset of NI movement relations, and we would see the entire limb representation by recording from NI. Therefore, we undertook a recording study of interpositus similar to our study of RNm.

As is the case for RNm, it is easy to determine discharge-movement relations for relatively large body parts in interpositus. A forelimb cell fires when the monkey moves the forelimb on the same side as the recording site (there is a crossing of sides in cerebellar output to RNm) but not when he moves any other limb. Similar specificity exists for movements of other body parts, and only a few cells fire during movement of more than one body part. Fig. 3 illustrates a horizontal reconstruction of cell recording site locations and movement relations in NI.

The physiologically determined somatotopy agrees reasonably well with anatomical determinations of somatotopy [12] and is also in reasonably good agreement with other recording studies of NI [2, 3, 24, 25]: the most rostral portions of NI contain cells related to hind limb movement; cells related to forelimb movement lie more caudally; and cells related to the mouth and/or face lie laterally and caudally. Additionally, there is an area in ventral and lateral NIP related to eye movements that projects to the superior colliculus [26] and not RNm. In general, the somatotopy of NIP is not as well defined as that of NIA.

Once forelimb regions of NI were defined, we then tested the units with devices that were designed to limit movement to one or a few forelimb joints. The devices used for testing are illustrated in Fig. 4 and are the same as we used for testing the properties of mossy fiber input to intermediate cerebellum [21].

The results of the mossy fiber study demonstrated that the testing was effective in determining specific joint relations of mossy fiber discharge, which reflects the activity of muscle receptors. Typically, a mossy fiber has its largest discharge modulation during operation of one device, a distinctly weaker modulation on a device involving movement of the immediately adjacent joints (many forelimb muscles have action across more than

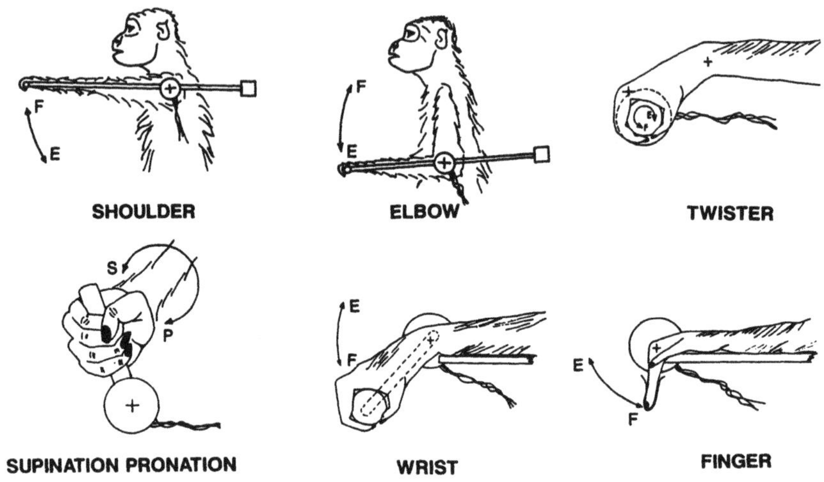

Figure 4. Variety of devices used to test mossy fibers and NI neurons for discharge relations to movements about a specific forelimb joint or set of joints. From [23].

one joint), and little or no discharge modulation during movements of joints separated by one or more joints [27]. Left panels of Fig. 5 illustrate discharge of a single mossy fiber during operation of several devices.

The device specificity that mossy fibers showed was not seen when we tested interpositus cells with multiple devices. In fact, about one-half of NI cells that discharged at high rates when the monkeys reached to grasp failed to modulate discharge during use of any of the devices. Many others that showed some discharge modulation during device use had little specificity for any device. Right panels of Fig. 5 illustrate average discharge records of an NI neuron during operation of different devices. Notice that the cell increases discharge during operation of several devices and shows no distinct preference for one device. Devices as widely separated as the shoulder and fingers elicit discharge of about equal modulation [23]. The result was somewhat of a disappointment, since we were hoping to be able to relate cell discharge rate to specific parameters of movement. No basis existed for selecting the appropriate movement.

Perhaps more disturbing was that even cells that fired during operation of the devices had relatively low discharge rates that did not correspond well to movement. If a cell is related to the task being performed, one would expect large discharge modulation during movement. Otherwise, the discharge may be related to some secondary movement during the task rather than to the device movement itself. Discharge related to a secondary movement could also explain the low selectivity for specific devices. Such suspicions were strengthened when we compared discharge during reaching for a raisin to that seen during device use. Fig. 6 illustrates discharge records of a unit tested on devices and during reaching. The upper histogram of discharge was generated from the spike records recorded on the videotape and then aligned on the beginning of the reaches. Although the alignment is imprecise (since it depended upon our own reaction time), it is

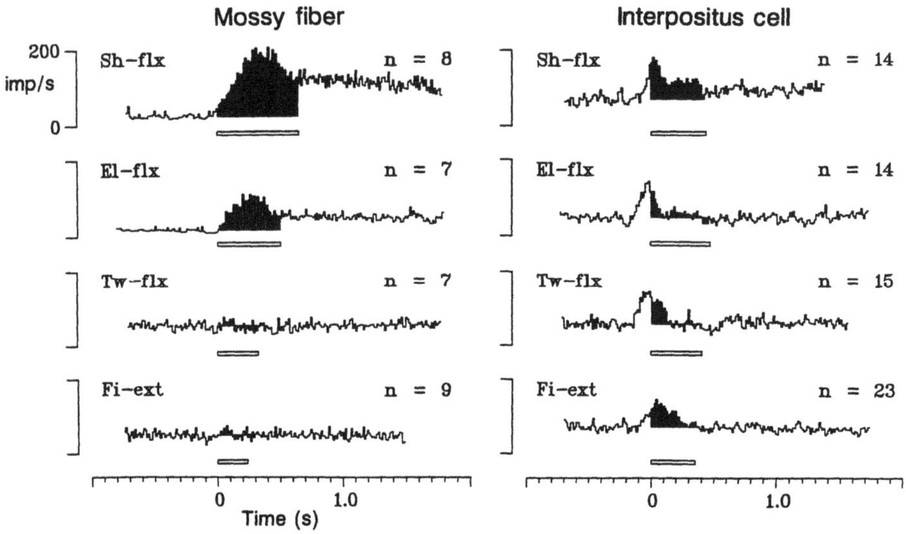

Figure 5. ***Left.*** *Average discharge rate of a mossy fiber while the monkey operated four devices (see Fig. 4).* ***Right.*** *Discharge of an interpositus neuron on same devices. Movement onset, time* ***0****. Shading indicates difference between average discharge rate and discharge during movement (movement is also indicated by bars under records).* ***Sh****, shoulder;* ***El****, elbow;* ***Tw****, twister;* ***Fi****, finger;* ***flx****, flexion;* ***ext****, extension. From [23, 27].*

clear that discharge modulation during reaching and grasping is considerably greater than during operation of any of the devices.

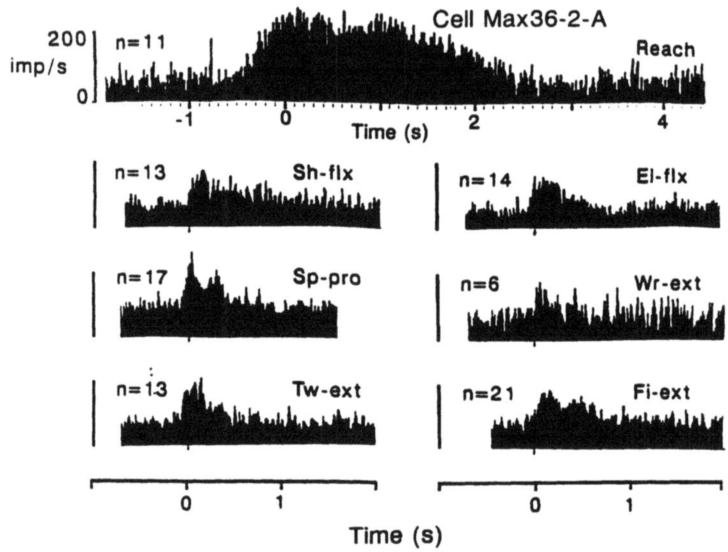

*Figure 6. Discharge of an interpositus neuron during operation of devices illustrated in Fig. 4 and during a reach for a raisin. Time **0** is beginning of movement. From [23].*

Interpositus discharge during reaching and grasping

Why is discharge much stronger during a ***reach to grasp*** than during movements required by use of the devices? Two possibilities immediately come to mind. The first and, seemingly, most likely possibility is that interpositus is important for the control of coordinated whole-limb movements rather than relatively simple movements about one or a few joints. The idea was originally suggested by Flourens in 1824 [28]. Also, recent evidence from inactivation of cerebellar nuclei with muscimol injections indicates that whole-limb reaching movements are much more affected by inactivation than are movements restricted to single joints [29].

The second possibility is that using the hand in a grasping motion is particularly important for interpositus. Some evidence that this might be the case comes from the anatomic and physiological data on RNm, which, as we have seen, suggest that the hand is of special importance.

To test the various possibilities, we trained monkeys to make two types of reaching movement: one consisted of reaching out while the hand gripped a handle of a device; and the other consisted of reaching out to grasp a raisin. If a coordinated limb movement is important to elicit discharge from NI cells, we would expect that both tasks would be accompanied by large modulations in discharge rate. If the grasp is especially important, we would expect only reaching out to grasp the raisin to elicit large discharge modulation.

Fig. 7 illustrates the manipulandum and position of the monkey. There are several important features of the experiment. The device lever was articulated so that there was no restriction of the limb trajectory in 2 planes, and the lever was counterbalanced with weights and springs so that the handle remained stationary when released. A small air-cylinder-actuated drawer was mounted adjacent the upper target position for the device. The drawer could be loaded with a raisin and popped open at any time during the experiment. In addition to recording position of the device and spike discharge, we also recorded handle touch and raisin drawer touch.

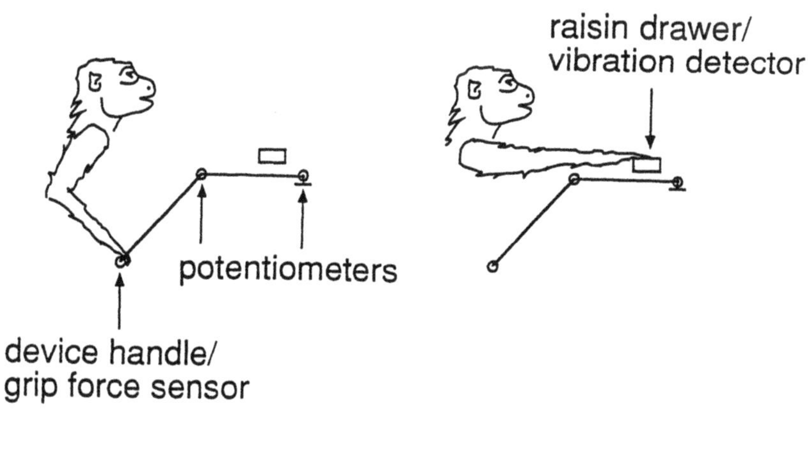

Figure 7. Device and raisin drawer used to test for reaches with and without a grasp. The monkey could move either the handle of the device to a position adjacent to the raisin drawer, or he could release the handle and reach out to grasp the raisin.

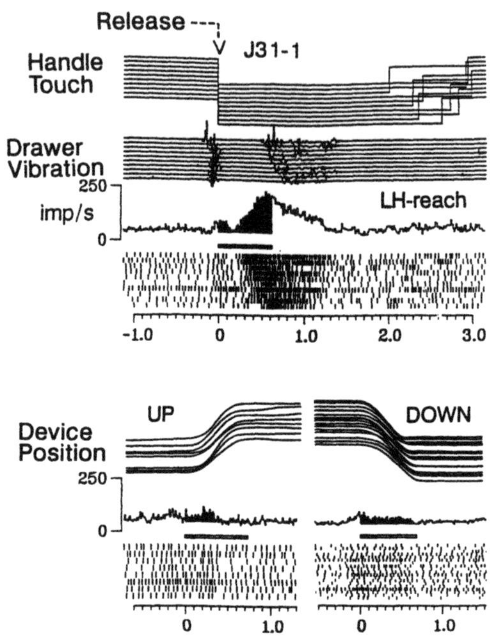

*Figure 8. Discharge of an interpositus neuron during reaching for a raisin and during device operation. Upper rasters are synchronized to the time when the monkey released the handle. Raisin drawer touches are indicated by second disturbances on vibration records. Time from release to average touch, the period when the limb is in motion, has been shaded in the average rate histogram. Lower sets of traces illustrate discharge (or, more accurately, lack of discharge) accompanying up and down movements of device shown in Fig. 8. Time **0** is beginning of device movement. Every 4th discharge has been plotted in spike rasters.*

During testing, the monkey was required to move the device handle from a low position near his waist to a high position with his forelimb extended at approximately right angles to the body axis. Video records showed that the limb described a trajectory similar to that required during reaching out to grasp an object held in front of the monkey. The task required moving the handle from the waist (low position) to the extended position (high) and back in order to obtain water reward. On some trials, we opened the raisin drawer while the monkey held the device handle at his waist. On such trials, the monkey released the handle, reached out, and retrieved the raisin. From the time of release to the time of touching the raisin drawer, the forelimb described a trajectory similar to that described when the monkey moved the device from the low to the high position.

The results were very clear: 93% (N = 85) of forelimb interpositus cells fired more strongly during reaching out to grasp than during moving the device from the low to high (or high to low) positions. In fact, many cells (about half) that had essentially no modulation during the device movement fired at high rates when reaching out to grasp the raisin.

Fig. 8 illustrates data records from one such cell and shows that the cell discharged at high rates during the period of reaching out, but only when the monkey was reaching out to grasp a raisin. Coordinated movement of the limb without grasping failed to elicit high discharge modulation. Therefore, the grasp is a necessary component of the movement in order to involve intermediate cerebellum.

Other characteristics of the discharge during reaching and grasping illustrated in Fig. 8 are also interesting. The peak of discharge occurred in the latter part of the reach out to grasp period. Video review of the movement showed that the limb transport phase occurred early in this period and was not accompanied by a high discharge modulation. During the latter half of reaching out, the monkey formed his hand in order to make the grasp; it was during this time that discharge rate was maximal. Many NI cells showed a discharge pattern that peaked immediately prior to grasping the raisin.

Another characteristic of this cell's discharge is that its firing was closely associated with touching the raisin drawer. The records have been ordered by increasing latency to drawer touch, and the slant of the touch raster is matched by the slant of discharge onset seen in the spike rasters; that is, discharge onset maintains a constant relationship to the time of drawer touch. The discharge rate also continued to be elevated during the period of raisin grasp, which can be seen in the vibration record. Elevated discharge rate ended abruptly when the monkey grasped the raisin and retrieved it to his mouth. Discharge was associated with hand use before and during grasping of the raisin but not with hand use during other behaviors (such as releasing device handle, holding the raisin, or placing it in the mouth). Although high rates of discharge of interpositus neurons required hand use, the discharge associated with hand movements varied between cells. For example, some cells discharged as strongly or more strongly during grasping of the device handle after completing a raisin trial as during the initial grasping of the raisin; other cells only fired during grasping of the raisin.

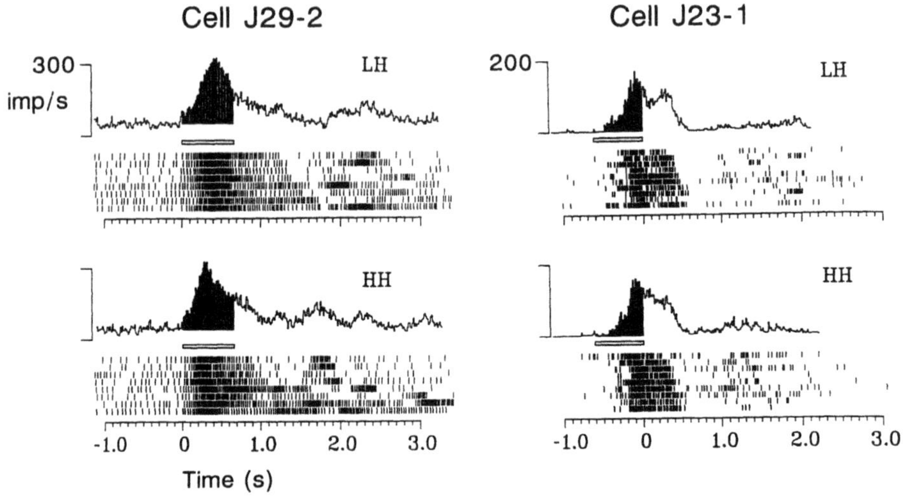

*Figure 9. Discharge of two interpositus neurons during reaches from the waist (low-high reach, **LH**) to the raisin drawer or from an adjacent position to the raisin drawer (high-high reach, **HH**). Discharge occurring between handle release and drawer touch has been shaded. Time **0** corresponds to device handle release.*

The variation in discharge patterns during the sequence of movements were consistent for a cell, thereby suggesting that specific NI neurons discharge during specific types of hand movements.

The reaching test results now raised another question: Do the cells that discharge during a ***reach to grasp*** contribute to the control of the whole arm during the task, or do they contribute only to the grasping component? Since discharge was often low during the early phase but high during the late phase of the ***reach to grasp***, there was already a suggestion that discharge might be preferentially related to grasping. However, some cells did fire during earlier phases of reaching out, and it is possible that the overall output from interpositus controlled both the reaching out and grasping phases of the task.

To test such a possibility, we needed some way to dissociate upper arm control from grasping. One way of doing so is to have the monkeys reach and grasp in different directions. MacKay [30] had already tested NI cells by having monkeys reach out and push buttons in different positions; interpositus discharge did not vary with the direction of reach. Similarly, when we held raisins in different locations in front of the monkeys, we could detect no changes in firing rate that depended upon reach direction. Our monkey experiment offered another way to test: we could open the raisin drawer when the monkey was holding the handle in the position adjacent to the drawer as well as when he was holding it at his waist. By doing this, we could compare discharge records of trials with very different upper limb movements but terminating with similar raisin grasps. Fig. 9 illustrates data from two cells during trials where the monkeys reached out from the waist to grasp a raisin (a low-high reach) or made a short movement sideways to grasp the raisin (a high-high reach).

Although each of the cells in Fig. 9 has its characteristic discharge pattern during reaching and grasping, the patterns repeat for the low-high and high-high reaches. For a given cell, there is only a slight change in discharge between the reaches despite the fact that the transport phases of the movements are entirely different. Therefore, it is unlikely that interpositus can be controlling the trajectory (transport component) of the reach and grasp. It could be controlling the grasp.

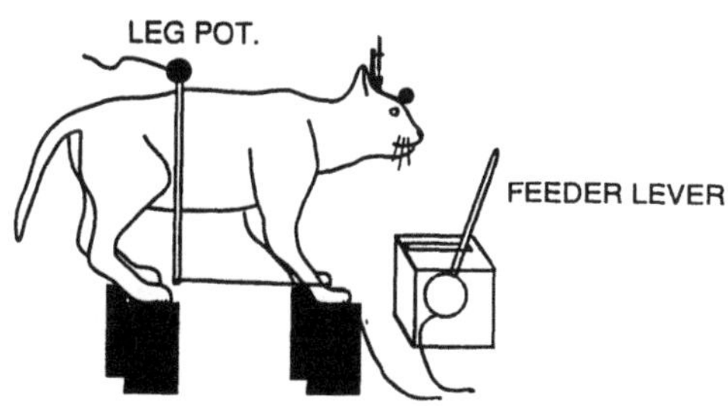

Figure 10. Apparatus for testing ***reach to grasp*** *in the cat. On cue, the cat reaches out, grasps lever, and retrieves it to his mouth. Foot pressure and limb position are recorded.*

Discharge of RNm during reach and grasp

If interpositus is largely concerned with controlling the grasping action of a reach and grasp, the same should also be true for RNm. In our initial studies of RNm [21, 22] we observed that discharge reached its highest rates during reaching for a raisin, but we had no way of recording the free-form forelimb movements (other than video) to correlate with discharge. Therefore, we decided to reexamine RNm discharge during reaching and grasping, but with the use of cats as subjects. We developed a reaching and grasping task for the cat and, surprisingly, found that cats rapidly learn the task and perform in a very consistent fashion.

The cat reaching and grasping paradigm is illustrated in Fig. 10. During the task, the cats stand on a pressure sensing platform, and, after a minimum period of steady stance, a tone sounds, which serves as a cue to reach out and retrieve the lever. The cats are rewarded with pureed chicken with cod liver oil that is extruded from the tip of the lever. Various sensors record limb position, platform pressure, lever position and lever force.

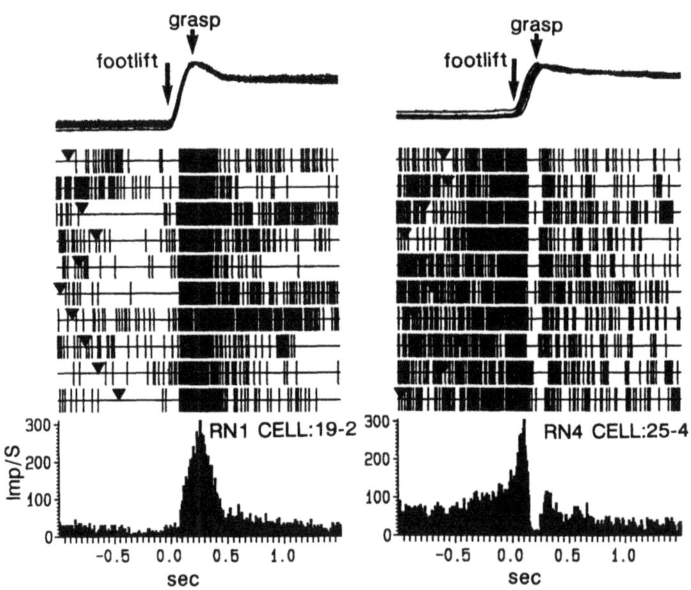

*Figure 11. Discharge of two RNm neurons during reaching out, grasping, and retrieving of the lever. The movements are highly stereotyped, and upper limb position traces have been overplotted. Time **0** is foot lift off. Triangles on spike rasters indicate movement cue tone.*

As we expected, cells in forelimb regions of RNm fire at high rates while reaching out, grasping and retrieving the lever; discharge rasters from two RNm cells are illustrated in Fig. 11. Notice that although individual cells showed characteristic (for these examples, complementary) discharge patterns, the pattern for a given cell was nearly identical from trial to trial. Discharge registered precisely with the movement and showed no dependence upon the movement cue tone.

The large increases in discharge rate and consistent discharge patterns of RNm cells during the task indicate that RNm (and, therefore, intermediate cerebellum) plays a basic role in the performance of the reaching and grasping task.

Inactivation of RNm

Another approach to unraveling the function of neural structures is to inactivate the structures, either permanently or temporarily, and observe the resulting changes in movement. Several inactivation studies have been performed on the RNm. Lawrence and Kuypers [31] sectioned the RST in monkeys and observed a temporary loss in the use of the hand. Similarly, Sybirska and Gorska [32] placed lesions in the RNm of cats and observed a limb ataxia with loss of digit use. The ataxia disappeared after a short recovery period, but the use of the digits in grasping a piece of food showed a persistent deficit. Levesque and Fabre-Thorpe [33] chemically destroyed the RNm in cats, which avoids damage to passing fibers, and found surprisingly small effects on a task that involved batting a spot of light. The batting task has no grasping component, and, therefore, our findings suggest that the circuitry of intermediate cerebellum might not be contributing to this task.

In contrast to batting a spot of light, reaching out and grasping a lever involves RNm, so inactivation of RNm during this task should produce a distinct deficit in the behavior. To test this, we devised an injection electrode with a predetermined spacing between the recording electrode and lumen of the injection cannula. Therefore, when the activity at the recording site is altered, one can assume that the neurons lying between the injection cannula and recording electrode have been affected.

We first tried injecting muscimol, a GABA esterase blocker that has been used to block cerebellar nuclei and RNm [34-36], but we found that the behavioral effects did not correlate well with the changes in activity in the cat RNm. Although behavioral changes occurred as the activity of RNm became silent, the behavioral effects continued to grow in severity for a long period of time. Clearly, behavior was being altered by changes in structures other than the RNm, and we did not feel that one could confidently ascribe the earlier effects to RNm inactivation.

Injections of the anesthetic lidocaine, a sodium channel blocker, produced a much cleaner effect since the time course of altered behavior corresponded with the disappearance and reappearance of activity in the RNm, and there was no severe long-lasting disturbance. However, lidocaine has the undesired property of fiber inactivation as well as cell body inactivation. To overcome this problem we ran a series of experiments using small injections of DGG (g-D-glutamylglycine), an excitatory amino acid antagonist [37, 38].

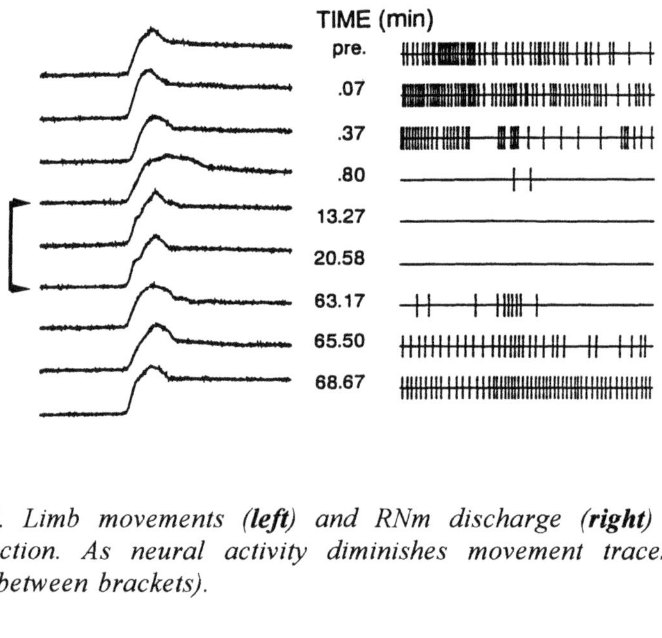

Figure 12. Limb movements ***(left)*** *and RNm discharge* ***(right)*** *following DGG injection. As neural activity diminishes movement traces become distorted (between brackets).*

Fig. 12 illustrates recordings during one experiment using injections of DGG. Notice that within seconds of injection, the activity of the RNm neuron ceases, and, at the same time, the position record of reaching out and retrieving the lever indicates a disturbance in performance. The disturbed performance lasts for the duration of the time that RNm activity is absent and then returns to normal about 1 hour post-injection, which is the time when cell activity returns.

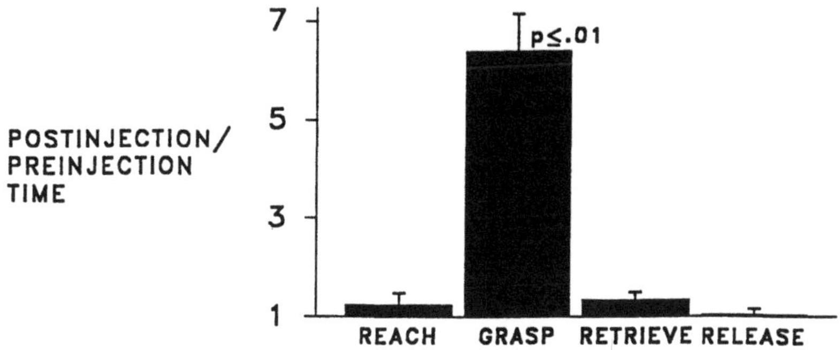

Figure 13. Post-injection / pre-injection movement time ratios for the ***reach to grasp****.*

What is the performance deficit? The answer is not as easy to determine as one would wish, the main reason being that the cats immediately devise new strategies to grasp and retrieve the lever: they have little regard for the scientific issues being addressed. The cats reach out as quickly as before inactivation, but they have trouble clearing the lower part of the apparatus on which the lever is mounted. The bump can be seen as an inflection on the position trace on the way out. The problem seems to be that the wrist and elbow do not bend sufficiently to clear the apparatus.

However, the most severe defect occurs as the cat attempts to grasp the lever. Grasping takes much longer than normal and often requires multiple attempts. The deficit is similar to the loss in the ability to grasp food observed by Sybirska and Gorska [32] following permanent RNm lesions. Once the cat does manage to grasp the lever, lever retrieval to the mouth is about as fast as normal. In Fig. 13, we have plotted the ratio of the time taken post-injection over pre-injection for different phases of the behavior. It can be seen that the time to grasp the lever is much greater than normal, whereas the transport phases of the limb (with the exception of the bump) occur at a relatively normal rate.

What is wrong with the grasp? On video examination the cats seem to over-reach the lever and hook it with their wrist, whereas normally they grip the lever with their toes.

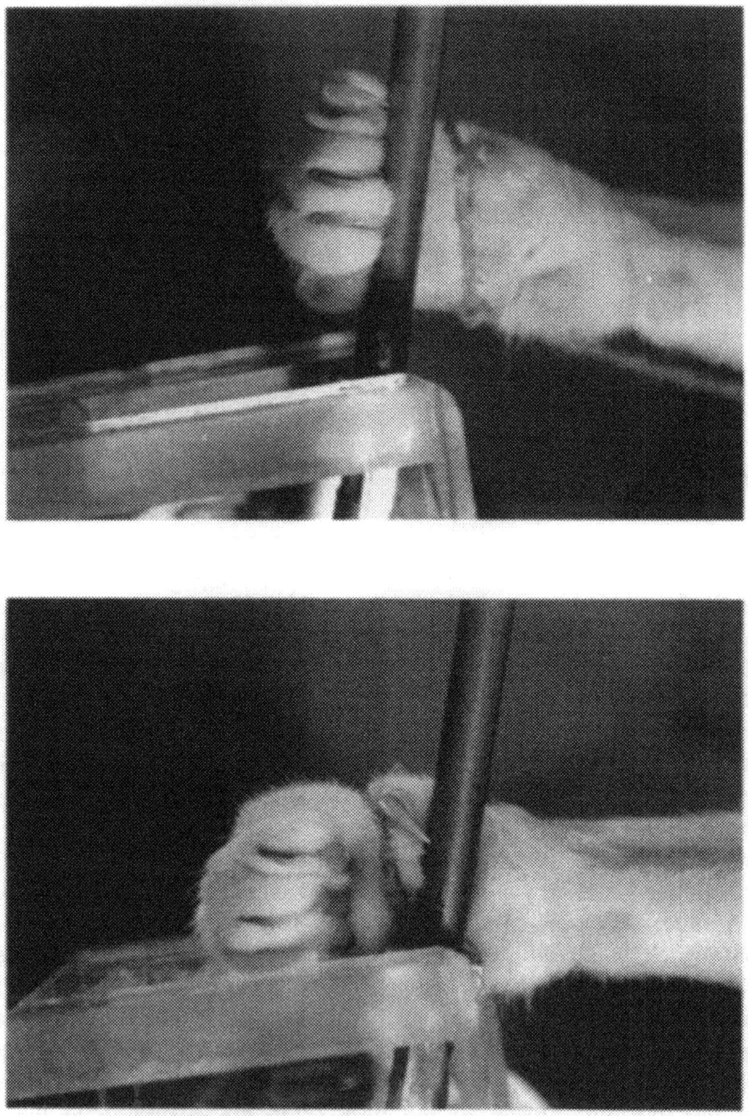

Figure 14. Photographs of cat grasping lever immediately before (top) and shortly after (bottom) a DGG injection into RNm.

Over-reaching is a classically recognized symptom of cerebellar damage, but in this case we feel that the over-reach is a functional adaptation to a deficit, namely, the inability to use the toes to grip. On some trials the cats manage to use their claws rather than wrist to retrieve the bar; claw use is probably another functional adaptation to the deficit. The two photographs of Fig. 14 illustrate the lever grip of a cat pre- and post-inactivation of the RNm. At the time of the photos the cat was holding the lever stationary while licking food off the top. Notice that in the post-inactivation photograph the cat no longer used his toes to grip the lever.

Is gripping the lever uniquely affected by RNm inactivation? We do not believe so: the cats also have difficulty in keeping their forelimbs on the pressure sensing platform. When they place their limb down, they often make multiple placements, as if they do not find a secure footing. After placement, the limb gradually slides forward and off the platform. It appears that the cats cannot get a good bracing action to prevent the sliding. During walking following RNm inactivation, the dorsum of the foot frequently turns under, which never happens in normal cats. These additional observations suggest that many behaviors relying on control of the paw are compromised by RNm inactivation. Along the same lines, it is important to keep in mind that the hind limb representation through intermediate cerebellum is approximately the same size as that of the forelimb. Whereas gripping a lever may only be a forelimb action, other actions, such as providing a secure footing, would be as important for the hind limb as for the forelimb.

CONCLUSION

The evidence presented in this chapter leads to one conclusion: intermediate cerebellum is preferentially concerned with the control of movements that involve hand use. Since interpositus discharge modulation is insensitive to the magnitude, direction, and trajectory of reaching, intermediate cerebellum may be exclusively concerned with hand movements. However, not all hand movements appear to involve intermediate cerebellum. Movements of the hand with the proximal limb stationary fail to produce large discharge modulations in interpositus neurons [23], and some hand movements, such as releasing the device handle to reach out, are not accompanied by large modulations of discharge. Interpositus may be specialized for specific hand movements that occur during grasping, or perhaps for hand movements made in conjunction with movements of more proximal parts of the limb.

However, if intermediate cerebellum is specialized for hand movements, why does the rubrospinal pathway terminate at all levels of the cord? RNm inactivation produces some deficit in reaching out, so it is likely that interpositus output has some contribution to control of musculature of the entire limb. It is possible that proximal limb control is specifically tied to supporting the hand movements being made, which would explain the lack of discharge when the grasp is removed from the reach. Such ideas, however, are clearly speculative and require experimental support.

The hypothesis that intermediate cerebellum is specialized for the control of hand movements used during grasping raises an important question about the functional organization of the cerebellum: are other divisions of the cerebellum specialized for specific types of movement? Some neural circuit must control the reach of the ***reach to grasp*** movement, and it seems likely that reaching would also involve cerebellar circuits. MacKay [30] has reported that neurons in the medial nucleus of the monkey show different discharge patterns depending upon the direction of a reach during a button pushing task, so medial nucleus might control reaching or, at least, direction of reach during a ***reach to grasp***.

On the other hand, the lateral cerebellum receives a large amount of visual input that appears appropriate for limb guidance [39], and it may be involved in the guidance of a reach. Rispal-Padel and Latrieille [40], on the basis of evidence from evoked potential studies in the cat, report that interpositus projects to regions of the motor cortex concerned with distal forelimb movement, whereas the medial and lateral cerebellar nuclei project to regions concerned with proximal movement. Such a scheme might be too simple, since the lateral cerebellum has also been implicated in the control of movements requiring independent finger use [31, 41]. Independent finger movements might constitute a separate class of hand movements requiring specialized control circuitry distinct from the circuitry used to make a grasp.

Clearly, understanding what constitutes a movement for specific regions of the cerebellum is key to understanding its organization. The ***reach to grasp*** is a movement with special importance for cerebellar organization: testing during the ***reach to grasp*** has forced us to completely alter our view of the output organization of intermediate cerebellum. It is likely that understanding what constitutes a movement to other motor pathways will be equally surprising.

ACKNOWLEDGEMENTS

We wish to thank Berber van Duin for anatomical work illustrated in Fig. 2, Anita Rifkin for helpful comments on the manuscript, and Ricardo Meza for technical assistance during many experiments as well as figure preparation and comments on the manuscript.

REFERENCES

1 M. Jeannerod, J. Mot. Behav., 16 (1984) 235.

2 R.J. Harvey, R. Porter and J.A. Rawson, J. Physiol., 297 (1979) 559.

3 W.T. Thach, J.G. Perry and M.H. Schieber, in: S.L. Palay and V. Chan-Palay (eds.) The Cerebellum - New Vistas, Springer-Verlag, Berlin, 1982, p. 440.

4 R. Nyberg-Hansen and A. Brodal, J. Comp. Neurol., 120 (1963) 369.

5 R. Nyberg-Hansen and A. Brodal, J. Anat., 98 (1964) 235.

6 Y. Padel, J. Armand and A.M. Smith, Exp. Brain Res., 14 (1972) 363.

7 O. Pompeiano and A. Brodal, J. Comp. Neurol., 108 (1957) 225.

8 L.W. Kneisley, M.P. Biber and J.H. LaVail, Exp. Neurol., 60 (1978) 116.

9 H.G.J.M. Kuypers, W.R. Fleming and J.W. Farinholt, J. Comp. Neurol., 118 (1962) 107.

10 M.A. Huisman, H.G.J.M. Kuypers and C.A. Verburgh, in: H.G.J.M. Kuypers and G.F. Martin (eds.), Prog. Brain Res., Elsevier, Amsterdam, 57 (1982) 185.

11 J.A. Courville and A. Brodal, J. Comp. Neurol., 126 (1966) 471.

12 F.R. Robinson, J.C. Houk and A.R. Gibson, J. Comp. Neurol., 257 (1987) 553.

13 Y. Shinoda, T. Futami, H. Mitoma and J. Yokota, Behav. Brain Res., 28 (1988) 59.

14 C.J. Wilson, F. Murakami, H. Katsumaru and N. Tsukahara, Neuroscience, 22 (1987) 113.

15 Y. Shinoda, C. Ghez and A. Arnold, Exp. Brain Res., 30 (1977) 203.

16 Y. Shinoda, J. Yokota and T. Futami, Brain Res., 242 (1982) 321.

17 G. Holstege and H.G.J.M. Kuypers, in: H.G.J.M. Kuypers and G.F. Martin (eds.), Prog. Brain Res., Elsevier, Amsterdam, 57 (1982) 145.

18 M.L. McCurdy, D.I. Hansma, J.C. Houk and A.R. Gibson, Soc. Neurosci. Abstr., 10 (1984) 744.

19 G. Holstege, B.F. Blok and D.D. Ralston, Neurosci. Lett., 95 (1988) 97.

20 M.L. McCurdy, D.I. Hansma, J.C. Houk and A.R. Gibson, J. Comp. Neurol., 265 (1987) 367.

21 A.R. Gibson, J.C. Houk and N.J. Kohlerman, J. Physiol., 358 (1985) 527.

22 A.R. Gibson, J.C. Houk and N.J. Kohlerman, J. Physiol., 358 (1985) 551.

23 P.L.E. Van Kan, J.C. Houk and A.R. Gibson, J. Neurophysiol., 69 (1993) 57.
24 G.I. Allen, P.F.C. Gilbert, R. Marini, W. Schultz and T.C.T. Yin, Exp. Brain Res., 27 (1977) 81.
25 R. Wetts, J.F. Kalaska and A.M. Smith, J. Neurophysiol., 54 (1985) 231.
26 P.J. May, R. Hartwich-Young, J. Nelson, D.L. Sparks and J.D. Porter, Neuroscience, 36 (1990) 305.
27 P.L.E. Van Kan, A.R. Gibson and J.C. Houk, J. Neurophysiol., 69 (1993) 74.
28 P. Flourens, in: E. Clarke and C.D. O'Malley (eds.), The Human Brain and Spinal Cord, University of California Press, Los Angeles, 1968, p. 657.
29 W.T. Thach, S.A. Kane, J.W. Mink and H.P. Goodkin, in: R. Llinas and C. Sotelo (eds.), The Cerebellum Revisited, Springer-Verlag, New York, 1992, p. 283.
30 W.A. MacKay, Brain Res., 442 (1988) 240.
31 D.G. Lawrence and H.G.J.M. Kuypers, Brain, 41 (1968) 15.
32 E. Sybirska and T. Gorska, Acta Neurobiol. Exp., 40 (1980) 821.
33 F. Levesque and M. Fabre-Thorpe, Exp. Brain Res., 81 (1990) 191.
34 A. Schmied, M. Amalric, J.F. Dormont and D. Farin, Exp. Brain Res., 81 (1990) 523.
35 J.H. Martin and C. Ghez, Behav. Brain Res., 28 (1988) 217.
36 H.P. Goodkin and W.T. Thach, Soc. Neurosci. Abstr., 17 (1991) 1380.
37 J. Liu and H. McLennan, Exp. Brain Res., 45 (1982) 151.
38 G.L. Collingridge, S.J. Kehl and H. McLennan, J. Physiol., 334 (1983) 19.
39 J.H. Martin, Neurosci. Lett., 127 (1991) 160.
40 L. Rispal-Padel and J. Latreille, Exp. Brain Res., 19 (1974) 36.
41 D.G. Lawrence and H.G.J.M. Kuypers, Brain, 91 (1968) 1.

Insights into the Reach to Grasp Movement
K.M.B. Bennett and U. Castiello (Editors)

CHAPTER 6

BRAIN AREAS RESPONSIBLE FOR THE GENERATION AND CONTROL OF REACHING AND GRASPING. ANATOMY WITH POSITRON EMISSION TOMOGRAPHY

J. DECETY

Institut de la Santé et de la Recherche Médicale Unité 94, 16, Avenue du Doyen Lépine, F-69500 Bron, France.

SUMMARY

This chapter discusses the use of positron emission tomography (PET) in identifying the functional anatomy underlying the generation and control of goal-directed arm movements during the action of reaching and grasping objects. PET allows non-invasive measurements of regional cerebral blood flow (rCBF) changes elicited by stimulation or activation of neurological or behavioural functions and therefore provides a new and powerful investigative tool. Focal increases in rCBF have been demonstrated when subjects perform simple motor tasks and these changes have provided new insights into the functional organization of voluntary movements. Recently the question of the selection of movements has been addressed as well as the identification of the neural pathways involved in visually guided movements. The findings are concordant with clinical reports of patients with brain lesions and physiological evidence that identify a distributed network for performing visually guided movement. Current research focuses on the anatomical basis of ***pragmatic*** *cortical representation and* ***semantic*** *cortical representation for objects.*

INTRODUCTION

Because reaching movements have a clear objective - to bring the hand to the spatial location of an object - they are well suited to study how the central nervous system plans a purposeful act from sensory input to motor output [1]. As we shall see, things are more complex since the decision to move is not necessarily triggered by an external stimulus, but may be internally driven. Indeed, it has been postulated that there are two complementary systems within the prefrontal cortex, a medial system that is driven from an internal model, and a lateral system that operates in a response mode, dependent upon external input [2].

Most of the knowledge in behavioural neurophysiology of reaching and grasping comes from kinematics studies in man [3] and from electrophysiological recordings in alert animals [4-6]. But the many new techniques introduced into clinical neurophysiology as well as in experimental brain research in the past ten years have reawakened interest in the motor areas of the human brain. The physiological role of those areas which participate in skilled voluntary movement has been studied anew, and the disorders produced when these regions are damaged or diseased have been re-evaluated. Magnetic resonance imaging, for example, has allowed the sites of lesions to be much more accurately delineated. Studies of regional blood flow with positron emission tomography (PET) have enable the brain locations associated with different aspects, or processing modules, to be examined in normal humans.

The possibility to quantify ongoing cortical functional activity rests on an intrinsic brain mechanism, which regulates local blood flow in accordance with the functional and metabolic level of the neuronal tissue. Metabolic substrates are not stored in the brain, but are continuously supplied by arterial blood. This tight coupling between brain blood flow, metabolism and neural activity in the normal and chronically diseased brain was experimentally verified by Raichle et al. [7].

Positron emission tomography is a medical imaging technique for measuring the concentrations of positron-emitting radioisotopes within a three-dimensional object by the use of external measurements of the radiation from these isotopes [8]. The technique relies on the injection of radionuclides with excess positive charge that decay by emitting a positron. The positron travels a few millimeters then combines with an electron to produce two 511 KeV gamma annihilation rays emitted at 180 degrees from each other. The origin of the photons, therefore, can be localized directly to the straight line between these coincidence detectors. Tomographic images of the radioactivity distribution are then reconstructed by computer from the many projected counts. The spatial resolution of PET is between 4 and 8 mm full width at half-maximum (FWHM) in the axial image plane, greater than that of electroencephalograms and event-related electric potentials, the other major available methods for probing the dynamics of human brain activity.

PET scanning has been used to study multiple functions within the brain, including receptor distribution and affinity, drug metabolism, tissue pH, blood volume, cerebral oxygen metabolism ($CMRO_2$), cerebral blood flow (CBF) and cerebral glucose metabolism (CMRgl). PET thus provides a unique approach to study the functional anatomy of the highest functions of the brain of unanaesthetized conscious humans. Most of the work reviewed in the present paper reports measurements of rCBF. It is admitted that, in the normal brain, rCBF is adjusted to local metabolic demands. That is, a change in regional cerebral oxygen consumption is followed by a proportional change in rCBF [9]. Because the brain uses glucose almost exclusively as an energy source, changes in the rCMRgl are also linearly correlated with changes in regional cerebral oxygen consumption. Measurements of $rCMRO_2$, CMRgl and of CBF thus provide quantitative measures of local neuronal and glial metabolism. In the normal cerebral neocortex an increase in local metabolism presumably indicates a net local increase in average excitation [10].

Although the reach to grasp movement has not yet been studied with PET, this chapter will use the results from past studies in the field of movement physiology to speculate about the functional anatomy of reaching and grasping.

FUNCTIONAL ANATOMY

From primate studies, one has learnt that the neocortex possesses some degree of regional specialization. According to an old but still valid scheme proposed by Allen and Tsukahara [11], ideas for movements are translated into specific programs by premotor and association areas of the cortex (frontal and parietal), basal ganglia, and the lateral hemispheres of the cerebellum. Motor areas have been divided into an increasing number of subfields. Each subfield is characterized by properties of neuronal activity in response to specific aspects of the stimulus or to the behavioural output [12, 13]. Although it is out of the scope of the current paper to present all the anatomical work (for review see [14]), the main cortical motor areas and their connections will be introduced.

Most of cortical motor areas are located in the frontal lobe. Next to the primary motor cortex, two major fields have been and still are the aim of extensive studies: the premotor cortex on the lateral hemispheric surface and the supplementary motor area in the mesial cortex. Both the premotor cortex and the supplementary motor area send projections to the primary motor cortex and thus act on distal muscles principally through their projections to brain stem neurons. In addition to their interconnections, the premotor cortex and the SMA have reciprocal connections with the same cortical areas including area 5 of the parietal lobe, the prefrontal cortex and the cingulate motor areas [15]. These two motor regions can only be distinguished, one from another, by their subcortical connections. The premotor cortex receives its main thalamic input from nucleus X of the thalamus [16, 17], which in turn is connected to the cerebellum. The SMA receives thalamic projections originating from several nuclei including VLo [18, 19] which is the target of the basal ganglia. Thus, although this scheme is still debated, premotor cortex and the SMA seem to be part of two separate motor circuits, one involving the cerebellum (for premotor cortex), the other the basal ganglia (for SMA). Both areas project to the spinal cord, either directly or indirectly via the medial reticular formation [20, 21] which is involved in axial and proximal muscle control. Anterior to the arcuate sulcus, the prefrontal cortex also plays an important role in integrating temporally and spatially discontinuous information to performance of goal-directed actions [22]. In addition, many neurons in frontal cortical areas are active in relation to motor preparation [23, 24]. There is still a debate regarding the suggestion that the SMA and premotor cortices are involved in internally versus externally triggered action. A recent comparison between premotor and prefrontal cortices shows that the activity of premotor neurons is more likely to reflect motor preparation whereas prefrontal activity more reflects attentional processes [25, 26]. Lesion studies in animals [27], recording of cerebral motor potentials [28], clinical observations in man [29] as well as measurements of rCBF also support these differences in the way these areas participate in the control of movement.

Some cortical motor areas are located in the parietal association cortex. Areas 5 and 7 have direct and indirect links with frontal areas. Parietal area 5 receives its principal input from the primary somatosensory cortex and relates this input to limb position. Area 5 is also informed by the vestibular system about the orientation of the head in space, by the premotor areas about motor plans, and by inputs from limbic cingulate cortices about motivational state. Area 5 in turn projects both posteriorly to area 7 and anteriorly to the premotor cortices. Area 7 receives input from visual, somatosensory (through area 5), auditory (through area 22), and limbic structures and projects to the prefrontal cortex and to the frontal eye field as well as to the lateral cerebellum [30]. Area 7 is involved in the processing of visual information that relates the location of objects in space [31].

At least two classes of neurons can be distinguished in the posterior parietal areas: sensorimotor neurons that are selectively activated during one aspect of visuomotor behaviour, such as reaching or manipulation [32] and sensory neurons that respond to purely visual stimuli, such as that given by moving targets [33]. Lesion studies revealed the motor properties of these areas: the monkeys or humans are unable to perform the movement in an orderly sequence, do not properly adjust the hand for a given object, or simply fail in the appropriate context [3, 34].

RECORDING OF CEREBRAL POTENTIALS

Most of the work seeking the relationships between motor preparation and its neurophysiological basis has been done by recording cerebral potentials preceding unilateral voluntary movements. Moreover, most research concerns simple ballistic movements. The topography of the cerebral potentials is well known and three different cerebral potentials preceding unilateral movements have been demonstrated [35, 36]. The early activity is bilateral from the beginning, even though the following movement is unilateral. A slow rising negative potential, the Bereitschaftspotential (BP) starts with an average premovement onset time of 800 msec in precentral and parietal regions of both sides and in the midline. The second potential, the pre-motion positivity (PMP), is also bilateral and has its maximum over the parietal and the precentral regions. The PMP starts, on average, with a premovement onset time of 90-80 msec. Only the last potential, the motor potential (MP), is unilateral preceding unilateral movements, and is restricted to the motor cortex contralateral to the movement. Kristeva et al. [37] reported a cerebral potential study during bilateral finger movements. They showed that the BP was not symmetrical or larger over the dominant hemisphere; rather it was larger over the non-dominant hemisphere. This difference was pronounced in precentral leads but very small in parietal leads. The PMP was found well developed at the vertex and even larger with bilateral than with unilateral movements. The MP was larger with unilateral than bilateral movements. These results indicate a crossed-hand-brain dominance, that is, amplitudes were larger over the left precentral region when the right index finger was moved, than they were over the right hemisphere when the left index finger was moved. Such findings may also be interpreted in terms of effort. The minor movement (left hand with minor hemisphere) needs more effort (or requires more energy resources) in order to equal the better-trained dominant system in performance.

Brunia [38], in a review of potentials recorded during goal-directed movements makes a distinction between simple ballistic movements and complex movements. The latter are continuously controlled by different kinds of feedback. Proprioceptive, kinaesthetic, and exteroceptive information indeed contribute to an immediate control of ongoing movements. In this context, the negativity over the sensory projection areas may be considered as a reflection of an attentional process directed to the information that is of immediate relevance for the future movement. Brunia suggested that attention to these different kinds of information is reflected in negative waves recorded over the relevant cortical areas.

The principle advantage of cerebral recording techniques is the temporal resolution (about 10 ms) but the spatial resolution is rather poor since the electrical activity is recorded over the cortex according to specific areas of the skull and thus gives crude information about the deeper parts of the brain. Extensive studies are carried out to clarify the actual anatomical localization of cortical generator. It is only recently that the technology has become available for the measurement of magnetic fields outside the skull. This technology, named magneto-encephalography, has promise to offer a significant advance over conventional EEG and ERP, making it possible to detect the minute magnetic field changes in the human brain that occur in relation to electrical cortical activity. Deecke et al. [39] showed that the BP of the EEG has an analogous phenomenon in the MEG, the Bereitschaftsmagnetofield, and that it has, for finger movement, a reversal around the hand representation of the rolandic area. Moreover, the MEG recording has confirmed the distinction between two principal generators, SMA and MI, before a voluntary sequential movement, the SMA generator being upstream of MI in the temporal chain.

CONTRIBUTION OF NON-TOMOGRAPHIC TECHNIQUES

First, it must be remembered that before PET was available, in vivo cerebral imaging became possible in the early sixties when Ingvar and Lassen [40] introduced the radioactive inert gas method. This technique was based on intra-arterial administration of tracer amounts of $^{133-}$Xenon for a short period, and extracranial recording of clearance curves. Already in 1971, Olesen [41] showed that voluntary unilateral simple hand movements were accompanied by an rCBF augmentation in the contralateral rolandic region. This finding was soon confirmed by Ingvar [42, 43]. Furthermore, it was found that internally rehearsed hand movements led to a bilateral rCBF increase in the SMA, whereas during execution of the same motor sequence there were equivalent rCBF increases in both SMAs and increases were also observed in the contralateral sensory motor hand area, the convex area of the premotor cortex, and bilaterally in the inferior frontal region [43-45]. In subjects required to mentally perform a writing task with their dominant hand, not only the SMA and the premotor cortex were found to be significantly activated but, in addition, the lateral cerebellum was highly activated [45]. It is of interest to note that these mentioned studies exclusively involved the distal portion of the dominant hand. In addition, none of these studies have addressed the question of bilateral hand movements which is rather difficult since one arm is immobilized for the tracer injection.

However, there are some limitations inherent to the nontomographic imaging techniques. Data are almost exclusively recorded from the surface of the cerebral cortex and thus cannot account for the deeper parts of the brain. When the administration of the radio-tracer is done with the inhalation method, there are artefacts from the ^{133}-Xenon in the airways. Due to the diameter of the external detectors, the spatial resolution is poor (around 2 cm) and it is mainly the cortical regions close to the detectors that are measured. Thus important parts of the cerebral hemispheres are not seen (such as, parts of the frontal or occipital lobes). Over the last decade there has been an explosive development of tomographic imaging techniques such as single photon emission tomography (SPECT) and positron emission tomography (PET) which enable three-dimensional studies of the rCBF, with a much better spatial resolution (only a few mm in full width at half maximum: FWHM). Oxygen-15 is the radio-isotope mostly used in activation studies to trace cerebral blood flow with PET because of its short half life (2.07 mn) which allows repetitive measurements in rapid succession in the same subject. As a tracer for CBF, oxygen-15 labelled water (H_215_O) is used.

The PET technique has the following advantages compared to the earlier techniques referred to above:

A) It is a tomographic method and consequently all regions of the brain are simultaneously accessible to measurements.

B) Current PET cameras have a spatial resolution of 4 - 5 mm, which is greatly superior to the nontomographic methods

C) With adequate corrections of the positron camera for dead time, random coincidences and attenuation (correction for events lost due to scatter and absorption of the radiation) accurate quantification can be achieved

D) Not only rCBF but also rCMR can be measured.

Functional studies with PET require some type of activation paradigm to produce discrete changes in local blood flow from a resting or control state. A typical experiment involves three (up to six) successive conditions in the same subject. Subtraction of the measurements allows some assessment of the activity distribution implicated by the task. In the following review, PET studies are based on this principle.

PREPARATION FOR ACTION

Motor behaviour can be divided into a preparation or planning phase, and an execution phase. Indeed, as soon as a subject becomes involved in a behavioural sequence ended by a motor act, an adaptive functional reorganization of the systems responsible for successive stages of action is set up [46]. There is evidence provided by experimental psychology and neurophysiology which suggests that the pre-setting processes underlying specific preparatory effects seem themselves to result from a planned and coordinated action.

Investigating the functional anatomy of motor preparation with PET requires the elaboration of activation paradigms during a very short period of time, such as 800 ms before the movement onset. Due to the PET methodological limitations in temporal resolution, 40 to 90 seconds being necessary to acquire a scan, it is very difficult to find suitable paradigms which allow a specific focus on this premovement period. To date, only one study of rCBF measurements with PET, in normal subjects, during the preparation phase for a reaching task has been reported [47]. In this study, subjects were instructed to memorize seven targets (circles of different diameters), briefly visually presented in the reaching distance (60 cm) and after a "point" command, were required to touch the centers of these targets in order of increasing size with the right index finger. The PET measurements were taken in the 90 s interval during which the subjects kept the targets in mind and prepared for the reach. Increases of rCBF were found in the left SMA, the angular gyrus, the ventrolateral thalamus and the cerebellar vermis. No activations were seen in the primary motor cortex. Thus, with the exception of the ventrolateral part of the thalamus and the left SMA, no motor sector of the brain showed consistent increase in rCBF. The thalamus mediates motor functions by transmitting information from the cerebellum and basal ganglia to the motor regions of the frontal lobe. The lack of motor cortex activation in the delay period might seem surprising, since the premotor cortex shows enhanced neuronal activity when a monkey waits for a go signal to reach for a target [48]. One possible interpretation may be the task paradigm, which permitted the subjects to wait for the 'go' signal. Organization of the pointing task may have only begun following this signal. Since the task placed no emphasis on reaction time, due to the PET procedure, but rather on accuracy, fast responses and rapid reaching movements were not required and consequently there was no tuning of the motor structures. Other activations were seen in several remote visual association cortical areas in the parietal lobe, and in the supramarginal gyrus. These fields belong to visual association areas representing especially extrapersonal space and are interpreted as reflecting neural activity associated with the working memory necessary for keeping the targets in mind in the delay period and with the representation of hand position relative to the target positions.

THE GENERATION OF WILLED ACTS

James [49] defined an action to be willed when we consciously pay attention to its selection. Many researchers have suggested that internally driven and stimulus driven responses are associated with different brain systems. Goldman-Rakic [30] proposed that, for monkeys to succeed in a delayed performance task which is based on stored information, an area of the dorsolateral prefrontal cortex (equivalent to Brodmann's area 46) is required to be intact. Whereas most neurones in the primary motor cortex show similar activity before and during sequential movement, regardless of whether movement is visually or internally determined, a large proportion of both premotor and SMA neurones exhibit different activity, depending on how the sequence is instructed [50]. This author (and colleagues) has shown, on the basis of single-cell recordings, that during both the premovement and the movement periods, premotor neurones are more active when the sequential motor task is visually guided while SMA neurones are more active when the sequence is remembered and self determined. The same question was addressed by Deiber et al. [51] in a PET study concerning the neural substrate of movement selection in man. In their experiment rCBF was monitored in subjects required to move a joystick with the right hand in four possible directions (left (L), right (R), forwards (F), backwards (B)) upon hearing a tone. The rate of the movement was paced by tones presented by a computer every two seconds. There were four tasks differing on the basis of movement selection and a control condition in which no selection was required. For the four tasks, the subjects were required to select one of the four movements every two seconds. The conditions differed according to how the subjects selected the next movement. In all cases the instructions were learned prior to performance. In all tasks the tone acted as a trigger to tell the subject when to move the joystick and in some conditions the sounds also instructed the subjects about which movement to make. In task 1, the subjects performed the following learned sequence of eight movements: F, L, L, R, B, R, B, F. In task 2, the subjects moved the joystick in any desired direction. The only requirement was that there should not be long sequences of movement in the same direction. In task 3, the subjects moved the joystick in the direction specified by the identity of a tone every two seconds. The meaning of the tone was taught just prior to the scanning session by moving the subject's hand in the correct direction. Finally, in task 4, the subjects moved the joystick in the direction opposite to that specified by the tone during the previous condition.

As such, the four selection tasks could be classified into two groups: a group of two tasks driven by internal cues and another group of two tasks externally driven by the sounds. A comparison was made between the four selection tasks pooled together and the control task: for the former, a significant rCBF increase was observed bilaterally in the frontal lobe (premotor cortex, SMA and prefrontal areas 46 and 9) and in the superior parietal association cortex. When individually contrasted to the control condition, each of the selection tasks showed a significant rCBF increase in the left superior parietal cortex. Finally, the comparison between internally and externally driven movement showed greater activation in the SMA for the tasks based on internal cues, but no evidence of greater activation in the premotor cortex for the externally cued tasks.

The question of whether differences in brain activity between systems associated with external and internal responses could be measured has also been addressed by Frith et al. [52]. In their study, they contrasted novel and routine tasks that used the same stimulus and response mechanisms. In order to show that this mental process is independent of stimulus and response modality, they selected two different systems: auditory input and verbal response in study one; somato-sensory input and motor response (i.e., lift a finger) in study two. In study one the stimuli were frequent words which had unambiguous opposites. These words were spoken by the experimenter in random order and the subject was required to respond with each word's opposite. In study two, they used a routine task in which either the first or second finger of the right hand was touched in a random sequence with a metal spatula and each response was specified by a stimulus, that is, lift the finger that has been touched or lift the other finger. In the novel tasks, the response was not fully specified and had to be selected by willed action, that is, lift either one of the fingers at will. Increases during performance of tasks involving willed action were identified in the dorsolateral prefrontal cortex (area 46) and in the anterior cingulate cortex (area 32). The major difference between the two studies was that willed word generation was associated with increases in the left dorsolateral prefrontal cortex only, whereas willed movement generation was associated with bilateral increases. In addition, the area activated in the verbal task extended backwards into Broca's area (area 44) whereas the task associated with sensory-motor skills extended downwards into area 10. These findings thus added to the growing body of evidence that selection of movement requires the dorsolateral prefrontal cortex in the region of Brodmann's area 46.

In light of these results is of interest to ask whether the reach to grasp movement is externally or internally driven. This movement requires several information processing stages such as visual identification of the goal, transportation of the limb to the location of the object and preshaping of the hand. Moreover, the grasp movement is influenced by the physical attributes of the object such as its size, shape and orientation (see Weir in this volume). One can thus consider that the reach to grasp movement relies both on external cues (the goal) and internal cues (the intention to grasp and the motivation for it). Therefore, such a movement requires the co-ordination and the control of more than one or two anatomical units and is also much more complex than the motor tasks so far investigated in PET activation paradigms. Furthermore, one must keep in mind that the isolation of the reach component from that of the grasp component is somewhat artificial. Reaching and grasping are indeed functionally interrelated, such that reaching is a precondition for grasping. These general considerations have implications for studying their underlying neural substrates.

EXECUTION OF MOVEMENTS

The neural activity in association with different types of unilateral arm movement was assessed by Colebatch et al. [53] using PET in normal volunteers. The four movements studied were selected to differ in both the joint about which they occurred (proximal-distal) and the degree of *fractionation* of finger movement required to perform them. The four movements consisted of abduction of the index finger, making a fist, sequential thumb to digit opposition, and shoulder flexion. All four movements were associated with significant increases in rCBF in the contralateral sensorimotor and premotor areas and in the SMA. Both these areas project to each other as well as to area 4. Both regions are also sites of origin of corticospinal projections to the cord. Unlike Roland et al. [43] which showed that the SMA was activated when subjects performed a complex sequence of finger movements, Colebatch et al. [53] found no direct relation between task complexity and SMA activation. Indeed, the increase in this area with the finger sequence was the same as for repetitively making a fist. They thus support Fox et al. [54] who postulated that the SMA is active in association with all motor tasks by establishing readiness to move. The average increase in blood flow in the contralateral sensorimotor cortex was significantly greater for the shoulder movement than for the three other movements which is contrary to classical descriptions of sensorimotor cortex organization. The increases with finger opposition and fist making were not different, and both were greater than with index finger movement. These data, according to the authors, indicate that neither fractionation nor distal movement per se cause selective activation of the sensorimotor cortex. Shoulder movement alone was associated with significant increases in rCBF in the ipsilateral sensorimotor cortex. The average location of the centre of excitation in the sensorimotor cortex and SMA differed for the movements and was interpreted as evidence of within-limb somatotopy. Finally, they demonstrated bilateral activation within the premotor cortex in association with the simple finger-touching task. The difference from their findings and the study of Roland [43] might be due to the use of an auditory cue as premotor cortex has been suggested to have a role in sensorially guided movement. It might also be due to the use of different rates of movements performed in the different conditions. Indeed, it has been demonstrated that repetition rate influences blood flow response [55].

Recently, in a further PET study, Matelli et al. [56] compared rCBF changes elicited by proximal (shoulder), distal (fingers and thumb) and whole arm movements in human subjects. Distal movements were associated with significant rCBF increase in the contralateral sensorimotor cortex. In contrast, proximal movements (shoulder and whole arm) were accompanied by a significant rCBF increase both in the sensorimotor cortex and in the mesial motor areas contralateral to the moving arm. In addition, an activation of the ipsilateral mesial motor areas was observed during whole arm movement. Their results thus confirm the old work of Roland. Pure elementary distal movements do not activate the mesial cortical areas. Simple proximal movements activate the SMA, where there is indeed a large proximal arm movement representation but a very small distal one. Additional activations were seen in the parietal lobe during both proximal and distal movements.

The somatotopic representation of the primary motor cortex has been examined by Grafton et al. [57]. Relative CBF was estimated with PET in subjects who performed a motor tracking task with the arm, first finger, tongue, and great toe. PET images were individually matched with magnetic resonance imaging scans for accurate visualization. All subjects demonstrated focal increases that followed the classic sequential somatotopic representation of the motor cortex as defined by Penfield and Boldrey [58]. These findings must obviously be considered with caution because of the spatial resolution of the PET technique. Indeed, intracortical microstimulation studies performed in monkeys by Sessle and Wiesendanger [59] showed that somatotopical representation is more sophisticated than the classical view of Penfield. The organization in monkeys is such that the fingers are represented nearest to the central sulcus, with horseshoe-shaped bands of cortex representing progressively more proximal muscles around the central core of the finger representation. Muscles acting on the elbow are represented partly in the wrist-band, partly in the shoulder-band. Their results also agreed with Wong et al. [60] in the fine organizational pattern. Notably, the presence of multiple microzones representing a particular simple movement is confirmed. Each of these microzones could be considered as a *local concentration* of output cells which form the texture of corticomotoneuronal colonies.

At the opposite end of this network motor plans must be generated and executed. Blood flow studies have confirmed earlier findings concerning the role of the supplementary motor area (SMA), as an executor for sequencing motor plans. Indeed, it was already known, from electrophysiological recording, that neurones in the SMA fire prior to motor neurones in the primary motor cortex [37]. This view was further supported by investigations of cerebral blood flow in human subjects using non tomographic techniques. [43, 45, 61, 62]. These authors reported significant increases in blood flow in a region corresponding to the SMA not only during the actual performance of complex movements but also when subjects mentally simulated a motor act.

CURRENT TRENDS

The great majority of anatomical studies has been directed toward localizing some component of the motor system (see Schwartz this volume). On the other hand, psychological and biomechanical studies have described kinematics rules in a rather precise way. It is now time to combine anatomical studies with behavioural paradigms in humans. This is possible by using neuroimaging techniques such as PET and MRI in activation paradigms which should, in the future, take into account the methodological limitations inherent to the physics as well as controlled behavioural stimulations.

How visual representations of an object are transformed into useful motor commands for grasping is a central issue for neuroscience and psychology. Ungerleider and Mishkin [63] proposed that projections arising from primary visual areas (V1) follow two major cortical pathways. One pathway, necessary for object recognition, is directed into the inferotemporal cortex (***What*** pathway) and relies on visual cues; the other route ends in the posterior parietal cortex and mediates visuospatial perception and visuomotor behavior that is required in the sensorimotor transformations for guided actions (***Where*** pathway). The proposed functions of these two streams were inferred largely from behavioural evidence derived from lesion studies.

These authors noted that monkeys with lesions of the infero-temporal cortex were impaired in visual pattern discrimination and recognition. The opposite pattern was observed in monkeys with posterior parietal lesions.

In order to study the functional anatomy of visually guided movement, Grafton et al. [64] measured relative CBF responses in subjects tracking a moving target with the index finger. Focal responses were located in the primary motor cortex, dorsal parietal cortex, precuneate cortex, SMA and ipsilateral anterior cerebellum. When the spatial complexity of the task was augmented by adding a secondary target, there were additional increases of CBF in bilateral parietal cortex and precuneate cortex. Interestingly, when performing the tracking task with different body parts (toe, tongue) subjects produced somatotopically distributed responses in only the motor cortex. Their results provide direct evidence that the SMA contributes in part to the sequencing of movements and the medial and dorsal parietal cortex participates in the integration of spatial attributes during selection of movements.

Although knowledge of how different cortical regions participate in processing voluntary movements is still sketchy, there have been several important conceptual advances. The motor system has been found to have a modular arrangement similar to the one that exists in sensory systems. Specific cooperative distributed processing exists between widely separate areas of the cortex and some subcortical structures. PET plays a valuable role in identifying such a network in humans. However, so far, none of the PET studies reported have directly explored the neural network underlying reaching and grasping in man. Most of the knowledge comes from animal studies. Indeed, there is strong experimental evidence suggesting that in monkeys two independent parietal-inferior premotor circuits mediate visuomotor transformations underlying reaching and grasping (e.g., [65]). This evidence as well as the different computational requirements necessary for transport and grasping suggest that a similar segregation should exist also in man [2, 66].

SOME DIRECTIONS FOR FUTURE RESEARCH

In order to organize the action of reaching and grasping objects, the nervous system has to solve a series of fundamental operational problems which include: location of objects in space, extraction of objects primitives, creation of a central representation of the objects as a goal for the action and elaboration of the corresponding motor commands. Here are some suggestions for future research:

Since there are strong experimental arguments suggesting that the respective motor systems involved in reaching and grasping can be considered, to some extent, as separate modules, PET experiments should focus on either one (e.g., simple hand transport to the same location, hand preshaping without grasping). In order to determine whether corrective movements are controlled for in the same regions where the normal movements are regulated, perturbation of the reach and preshape could be made after the initiation of movements (by perturbing target location and object size for instance; see Haggard this volume).

Brain asymmetry in the control of reaching is an important topic. While it is clear that, for reaching movements, humans prefer to use their right hand, there are no available PET data on the performance of the right hand, as compared to the left. Goodale [67] has shown that while the right hemisphere may be involved in visuospatial localization of an intended target, its contribution to the organization of movement does not depend on a close coupling between controller and effector, such as that which characterizes the left hemisphere's relationship with the right hand.

Another aspect of cortical motor organization worthy of PET investigation in humans is related to the purpose of the movement. Indeed the work of Gentilucci and Rizzolatti [68, 69] has demonstrated that movements that, from a kinematic point of view, are very similar can activate a neurone in one context, but not in another. Their findings indicate that from a topographic point of view the cortical organization of movement is not based on the similarity among motor patterns, but rather on the similarity of the purpose that can be achieved by the joint action of neurones located in the same cortical area.

ACKNOWLEDGEMENTS

Dr. Driss Boussaoud is thanked for valuable comments on an earlier draft of this manuscript.

REFERENCES

1 J.F. Kalaska and J. Crammond, Cerebral cortical mechanisms of reaching movements. Science, 235 (1992) 1517.
2 G. Goldberg, Behav. Brain Sci., 8 (1985) 567.
3 M. Jeannerod, The Neural and Behavioural Organization of Goal-Directed Movements, Oxford University Press, Oxford, 1988.
4 G. Rizzolatti, R. Camarda, L. Fogassi, M. Gentilucci, G. Luppino and M. Matelli, Exp. Brain Res., 71 (1988) 491.
5 K. Kurata and S.P. Wise, Exp. Brain Res., 72 (1988) 237.

6 H. Sakata, M. Taira, A. Murata and S. Mine, Adv. Neurol. Sci., 35 (1991) 990.
7 M.E. Raichle, R.L. Grubb and M.H. Gado, Arch. Neurol., 33 (1976) 523.
8 E.J. Hoffman and M.E. Phelps, in: M.E. Phelps, J. Mazziotta and H. Schelbert (eds.), Positron Emission Tomography and Autoradiography: Principles and Applications for the Brain and Heart, Raven Press, New York, 1986, p. 237.
9 P.E. Roland, L. Friberg, N.A. Lassen and T.S. Olsen, J. Cereb. Blood Flow Metab., 5 (1985) S205.
10 P.E. Roland and L. Friberg, J. Cereb. Blood Flow Metab., 3 (1983) S244.
11 G.I. Allen and N. Tsukahara, Physiol. Rev., 54 (1974) 957.
12 M. Matelli, G. Luppino and G. Rizzolatti, Behav. Brain Res., 18 (1985) 125.
13 G. Rizzolatti, R. Camarda, M. Fogassi, M. Gentilucci, G. Luppino and M. Matelli. Exp. Brain Res., 71 (1988) 491.
14 C. Ghez, in: E.R. Kandel, J.H. Schwartz and T.M. Jessell, Principles of Neural Science, Elsevier, New York, 1991, p. 609.
15 P.B. Johnson, in: R. Caminiti, P.B. Johnson and Y. Burnod (eds.), Control of Arm Movement in Space. Exp. Brain Res. Series 22, Springer-Verlag, Berlin, 1992, p. 199.
16 G.R. Schell and P.L. Strick, J. Neurosci., 4 (1984) 539.
17 J. Kievet and H.G.J.M. Kuypers, Exp. Brain Res., 29 (1977) 299.
18 C. Darian-Smith, I. Darian-Smith and S.S. Cheema, J. Comp. Neurol., 299 (1990) 17.
19 R. Wiesendanger and M. Wiesendanger, Exp. Brain Res., 59 (1985) 91.
20 K. Keizer and H.G.J.M. Kuypers, Exp. Brain Res., 74 (1989) 311.
21 R.P. Dum and P.L. Strick, J. Neurosci., 11 (1991) 667.
22 P. Barone and J.P. Joseph, Brain Behav. Evol., 33 (1989) 132.
23 J. Tanji, K. Taniguchi and T. Saga, J. Neurophysiol., 43 (1980) 60.
24 S.P. Wise, Ann. Rev. Neurosci., 8 (1985) 1.
25 D. Boussaoud and S.P. Wise, Soc. Neurosci. Abstr., 18 (1992) 216.
26 D. Boussaoud and S.P. Wise, Exp. Brain Res., in press.
27 R.E. Passingham, Behav. Brain Res., 18 (1985) 175.
28 L. Deecke and H.H. Kornhuber, Brain Res., 159 (1978) 473.
29 H.J. Freund and H. Hummelsheim, Brain, 108 (1985) 83.
30 P.S. Goldman-Rakic, in: F. Plum and V. Mountcastle (eds.), Handbook of Physiology: The Nervous System V, Am. Physiol. Soc., Bethesda, 1987, p. 1225.

31 R.A. Andersen, in: V.B. Mountcastle, F. Plum and S.R. Geiger (eds.), Handbook of Physiology: The Nervous System I, Am. Physiol. Soc., Bethesda, 1987, p. 483.
32 V.B. Mountcastle, in: G.M. Edelman and V.B. Mountcastle (eds.), The Mindful Brain, MIT Press, Cambridge, 1978, p. 7.
33 D.L. Robinson, M.E. Golberg and G.B. Stanton, J. Neurophysiol., 41 (1978) 910.
34 M. Wiesendanger, in: J.M. Brookhart and V. Mountcastle (eds.), Handbook of Physiology: The Nervous System II, American Physiology Society, Bethesda, 1981, p. 1121.
35 H.H. Kornhuber and L. Deecke, Pflügers Archiv für die Gesammte Physiologie, 248 (1965) 1.
36 L. Deecke, B. Grözinger and H. H. Kornhuber, Biol. Cybern., 23 (1976) 99.
37 R. Kristeva, E. Keller and H.H. Kornhuber, Electr. Encephal. Clin. Neurophysiol., 47 (1979) 229.
38 C.H.M. Brunia. in: H. Heur and A.F. Sanders (eds.), Perspectives on Preparation and Action, Erlbaum, Hillsdale, 1988, p. 231.
39 L. Deecke, H. Weinberg and P. Brickett, Exp. Brain Res., 48 (1982) 144.
40 D.H. Ingvar and N.A. Lassen, Acta Physiol. Scand., 54 (1962) 325.
41 J. Olesen, Brain, 94 (1971) 635.
42 D.H. Ingvar, in: D.H. Ingvar and N.A. Lassen (eds.), Brain Work, Munksgaard, Copenhagen, 1975, p. 31.
43 P.E. Roland, B. Larsen, N.A. Lassen and E. Skinhoj, J. Neurophysiol., 43 (1980) 118.
44 D.H. Ingvar and L. Philipson, Ann. Neurol., 2 (1977) 230.
45 J. Decety, B. Philippon and D.H. Ingvar, Eur. Arch. Psychiat. Neurol. Sci., 238 (1988) 33.
46 J. Requin, in: G.E. Stelmach and J. Requin (eds.), Tutorials in Motor Behavior, Advances in Psychology I, North-Holland, Amsterdam, 1980, p. 373.
47 J. Decety, K. Kawashima, B. Gulyas and P.E. Roland, NeuroReport, 3 (1992) 761.
48 K. Kurata and S.P. Wise, Exp. Brain Res., 72 (1988) 237.
49 W. James. Principles of Psychology, Holt, New York, 1890.
50 H. Mushiake, M. Inase and J. Tanji, J. Neurophysiol. 66 (1991) 705.
51 M.P. Deiber, R.E. Passingham, J.G. Colebatch, K.J. Friston, P.D. Nixon and R.S.J. Frackowiak, Exp. Brain. Res., 84 (1991) 393.
52 C.D. Frith, K. Friston, P.F. Liddle and R.S.J. Frackowiack. Proc. Royal Soc. Lond., 244 (1991) 241.

53 J.G. Colebatch, M.-P. Deiber, R.E. Passingham, K.J. Friston and R.S.J. Frackowiak, J. Neurophysiol., 65 (1991) 1392.
54 P.T. Fox, J.M. Fox, M.E. Raichle and R.M. Burde, J. Neurophysiol., 54 (1985) 348.
55 P.T. Fox, and M.E. Raichle, Proc. Natl. Acad. Sci. USA, 83 (1985) 1140.
56 M. Matelli, G. Rizzolatti, V. Bettinardi, M.C. Girardi, D. Perani, G. Rizzo and F. Fazio, NeuroReport, submitted.
57 S.T. Grafton, R.P. Woods, J.C. Mazziotta, and M.E. Phelps, J. Neurophysiol., 66 (1991) 735.
58 W. Penfield and E. Boldrey, Brain, 15 (1938) 389.
59 B.J. Sessle and M. Wiesendanger, J. Physiol, 323 (1982) 245.
60 Y.C. Wong, H.C. Kwan, W.A. MacKay and J.T. Murphy, J. Neurophysiol., 41 (1978) 1107.
61 P.E. Roland, L. Eriksson, S. Stone Elander and L. Widen, J. Neurosci., 7 (1987) 2373.
62 J. Decety, H. Sjöholm, E. Ryding, G. Stenberg and D.H. Ingvar, Brain Res., 535 (1990) 313.
63 L.C. Ungerleider and M. Mishkin, in: D.J. Ingle, M.A. Goodale and R.J.W. Mansfield (eds.), Analysis of Visual Behavior, MIT Press, Cambridge, 1982, p. 549.
64 S.T. Grafton, J.C. Mazziotta, R.P. Woods and M.E. Phelps, Brain, 115 (1992) 565.
65 G. Rizzolatti, M. Gentilucci, R. Camarda, V. Gallese, G. Luppino, M. Matelli and L. Fogassi, Exp. Brain Res., 82 (1990) 337.
66 M.A. Goodale and D. Milner, TINs., 15 (1992) 20.
67 M.A. Goodale, in: M.A. Goodale (ed.), Vision and Action, Ablex, New Jersey, 1990, p. 14.
68 M. Gentilucci, L. Fogassi, G. Luppino, M. Matelli, R. Camarda and G. Rizzolatti, Exp. Brain Res., 71 (1988) 475.
69 M. Gentilucci and G. Rizzolatti, in: M.A. Goodale, Vision and Action, Ablex, New Jersey, 1990, p. 147.

Section Three: Behavioural Studies

Insights into the Reach to Grasp Movement
K.M.B. Bennett and U. Castiello (Editors)

CHAPTER 7

OBJECT PROPERTY AND TASK EFFECTS ON PREHENSION

P.L. WEIR

Department of Kinesiology, University of Windsor, Windsor, Ontario, Canada, N9B 3P4.

SUMMARY

The control of the reach to grasp movement (prehension) requires the processing of information related to both object properties and the requirements of the task. Object properties are presented that rely on visual information, haptic information or both for their identification. The task related factors focus on goal related aspects of the movement (time, error, intent). All of these variables have been shown to influence the control and coordination of the grasp and transport components. The control issue is at the heart of these research studies. Different approaches to the control of the reach to grasp movement are discussed, and an integrated approach that allows the relations between the components to change is highlighted.

INTRODUCTION

Prehension: the act of seizing or grasping

When we think of the reach to grasp movement, we conjure an image of a hand reaching out to grasp hold of an object that has specific physical attributes, and a specific relationship in the environment with respect to the body. In order for this to occur, the arm must be transported (transport component) toward the object and the hand and fingers postured (grasp component) to make contact with the object. After contact, forces must be applied and modulated to maintain a stable grasp. The reach, or transport component has been studied in the context of pointing movements for many years [1, 2, 3]. In pointing, the hand is transported as part of the arm. The reach to grasp movement is complicated because the arm and hand must act as separate segments. The hand is postured to accommodate the size, shape and use of the object, and it's relative timing is coordinated with the transport component to ensure that finger closure occurs in synchrony with the approach to the object. The anticipatory preshaping and enclosing is hypothesized to be related to purely visuomotor mechanisms, independent of tactile and kinesthetic inputs which are available once the object is contacted and grasped.

Prehension has been studied from evolutionary [4], computational [5, 6], clinical [7, 8], developmental [9], dynamical [10], and cognitive [11] perspectives. This chapter focuses on the experimental evidence from the cognitive perspective of how object properties (size, weight, texture) and task requirements (movement time, error tolerance, intent) influence the execution of the reach to grasp movement. A great deal of the prehension data has been analyzed using kinematics of the movement patterns. Analyzing the spatio-temporal patterns allow researchers to make inferences as to the underlying control processes. Several different cognitive explanations have been offered for the control of human prehension, such as motor program [12, 13], visuomotor template [11], and multimovement [14, 15, 16]. Kinematic measures allow us to determine whether any of these control theories hold true for these movements. These ideas will be revisited at the end of this chapter.

OBJECT PROPERTIES

Size - visual

Jeannerod [11, 17] classified object properties according to their visual characteristics. Intrinsic object properties refer to the physical attributes of the object, that is, size, shape, texture, weight. Extrinsic object properties are relational attributes between the object and the subject, that is, distance, location, orientation. Linking the object properties to the two components of prehension, Jeannerod [11] suggested that intrinsic object properties should affect only the posturing of the hand (i.e., grasp component), while extrinsic object properties should influence only the spatial path of the arm and hand in locating the object (i.e., transport component).

Experimentally, the object property that has received the most attention is object size. We adapt our hand postures quite readily to successfully grasp drinking glasses of different diameters, pens, bottles, etc. on a daily basis.

According to Jeannerod's visuomotor channel theory [11, 17], object size should influence only the grasp component. This assertion has some conceptual and empirical problems, and will be addressed in the subsequent review. Vision plays a dominant role in controlling grasping movements to objects of different sizes. Jeannerod and Decety [18] had subjects form a grasp aperture for objects of different sizes, without vision of their hand. Subjects proved to be remarkably accurate. Their ability to calibrate based solely on vision of the disk suggests that they have a knowledge base linked to the efferent signals for the grasp aperture. From a kinematic view point several studies have been completed to quantify changes in the grasp and transport components as a function of object size.

Jeannerod's early work [11, 17] was directed at the duality of visuomotor mechanisms (channels) and how their integration reflected the organization of the sensory systems associated with the detection of object properties. He was one of the first to provide a quantitative description of the reach to grasp movement. Jeannerod compared reaching movements directed at objects of different sizes and shapes (cube, sphere, rod, cylinder). As predicted, size had no influence on the transport component. However, the grasp component did change as a function of object size. The maximum opening of the hand (aperture) increased as the size of the object increased. This is an expected finding, because if the hand did not open to accommodate the size of the object it would not be successfully contacted, or contact and grasp would be delayed after initial contact until the hand had reshaped to the appropriate size. Most interestingly, regardless of the size of the object, the time of maximum aperture and the time of the onset of the low velocity portion (maximum deceleration) were highly correlated ($r = 0.76$ to 0.89; [17]). These events occurred between 74 to 81% of the movement duration, leading Jeannerod to hypothesize a central program that imposed temporal constraints on the components of the motor output.

Wing and Fraser [19] collected data from a single subject proficient in using a manually operated artificial hand. The subject using both her natural right hand and artificial left hand made reaching movements towards dowels that were 12 mm and 22 mm in diameter. Supporting Jeannerod [11], maximum aperture increased with increasing dowel size. While movements made with the artificial hand were longer in duration, both hands showed similar usage of the thumb and index finger. The index finger was used to reduce the size of the grasp aperture, while the thumb maintained a stable position, guiding the arm to the location of the object. Wing and Fraser [19] used these data to suggest that the grasp and transport components were spatially linked in that the hand helped to guide the arm to the correct location. However, without having measured the transport component this spatial linkage cannot be confirmed.

Wallace and Weeks ([20] - Exp. 2) also provide support for Jeannerod's findings. Subjects performed reaching movements to dowels 3 mm and 25 mm in diameter, and reported that aperture size increased as dowel diameter increased. While these two studies provide direct support for Jeannerod's visuomotor channel theory [11, 17], a number of studies have also shown that the transport component is influenced by object size.

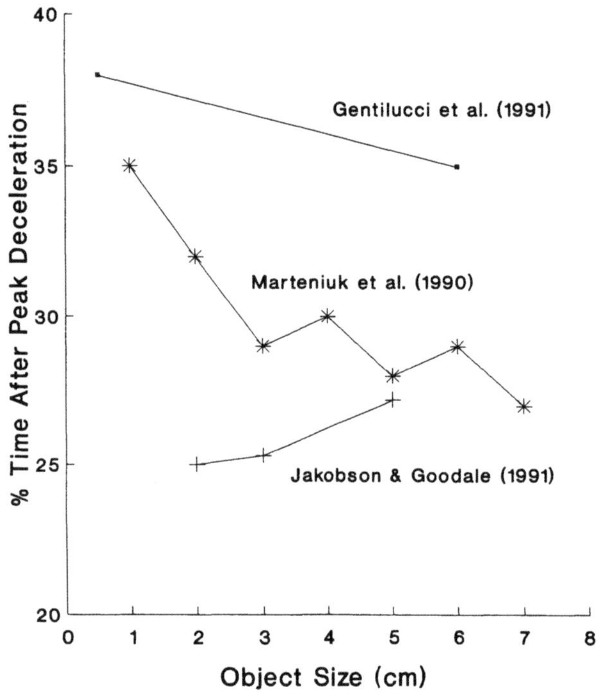

Figure 1. The percentage of movement time after peak deceleration as a function of object size. Note that for Marteniuk et al. [22] and Gentilucci et al. [24] as object size decreases, the percentage of movement time increases. The opposite trend is true for Jakobson and Goodale [23].

Marteniuk et al. ([15] - Exp. 1), measured only the transport component and compared pointing and grasping movements to 2 cm and 4 cm targets and disks respectively. They showed that regardless of the task, peak velocity increased as target/disk size increased. They also showed that the percentage of movement time at which peak velocity occurred was invariant across the two disk sizes. Thus, the velocity profiles belonged to the same family of scalar curves. This finding is the first to suggest that object size influences more than the grasp component. Because the grasp component was not quantified, a full comparison to other studies is not possible.

Extending this work, Marteniuk et al. [21] examined the effects of grasping ten disks that ranged in size from one to ten centimeters. Similar to other studies, maximum aperture increased as object size increased. The transport component was also affected. The percentage of movement time after peak aperture and peak deceleration decreased as object size increased (see Fig. 1). This supports the pointing literature, that the greater the precision required, the longer the time it takes to ***home-in*** on the object. However, this finding seems to be most pronounced among the extreme disk sizes (1 cm = 35%; 10 cm = 28%). There is very little difference in the two to nine centimeter range (27 - 32%).

Jakobson and Goodale [22] have also provided evidence that the transport component is influenced by object size. Subjects reached to objects of three different sizes (2, 3, 5 cm blocks) over three distances (20, 30, 40 cm) under conditions of vision and no-vision. Regardless of the visual condition, or the amplitude of the reach, object size influenced not only the peak aperture, but the overall duration, peak velocity, its relative timing, and the percentage of time after peak deceleration. In all cases, values increased with increasing object size. These relative timing effects are opposite in direction to those reported by Marteniuk et al. ([21]; see Fig. 1). It is puzzling that Jakobson and Goodale [23] do not discuss this opposing finding.

There are a couple of methodological differences between the two studies. First, the shape of the objects, and the shape of the contacting surfaces differ. Marteniuk et al. [21] used a round disk whose diameter was varied while the height was held constant, whereas in Jakobson and Goodale [22], the width and length of the block both changed (2 x 5 cm; 3 x 7.5 cm; and 5 x 12.5 cm). This corresponding change in length may have confounded or overridden the change in width. If the objects were centered over the target position, subjects may have been biased by the changing length. For example, for the smallest block, subjects may have grasped it on the end closest to themselves and still successfully picked it up. However, for the largest block, subjects would have to pick it up in the middle to ensure that it did not slip out of their grasp. Thus, subjects may have moved a greater distance to pick up the largest block, thereby increasing the time spent in deceleration. These ideas are supported by the fact that the maximum wrist height increased as object size increased. With respect to the contacting surface, the round disks used by Marteniuk et al. [21] had curved grasping surfaces, whereas the wooden blocks used in Jakobson and Goodale [22] had flat gripping surfaces. Recent work by Weir and MacKenzie [23] suggests that reaching to grasp a curved cylindrical surface results in a longer percentage of movement time following maximum velocity than reaching to grasp a flat surface. This finding is the result of a lower amount of surface area being contacted on the curved surface than the flat surface. Thus, the findings of both Marteniuk et al. [21] and Jakobson and Goodale [22] may be significant given the different object shapes. Second, the object sizes used by Jakobson and Goodale [22] fell in the middle range of objects explored by Marteniuk et al. [21], where there were few differences in the percentage of movement time spent after peak deceleration. While there are statistical differences in the temporal measures (25 - 28%), they are not large and correspond quite closely to those reported by Marteniuk et al. [21] for the same range of sizes. Thus, the differences in relative timing between the two studies may simply be the result of differences in object shape, size, or both.

In the studies reported, regardless of the size of the object, subjects were required to grasp the object using a precision grasp, stabilizing the object between the thumb and index finger. Gentilucci et al. [24] suggested that the velocity profiles in Marteniuk et al. ([15] - Exp. 1) scaled because while object size was varied, the type of grasp that was used remained constant. The main purpose of their studies was to determine if changes in the grasp component, as a function of object size, produced changes in the transport component. In Experiment 1, two test objects were used that necessitated the use of two different grasps. The first was a large cylinder (5 cm in height, 6 cm in diameter) that was grasped using a whole hand prehension (WHP), and the second object was a small sphere (0.5 cm in diameter) grasped using a precision grip (PG). Similar to previous studies, maximum aperture increased as object size increased and maximum aperture was achieved earlier in the PG (small sphere) condition (58.5%) than in the WHP (large cylinder) condition (74%). Similar to Marteniuk et al. [15], peak velocity increased as object size increased. Of particular interest was whether the trajectories were scalable in time. The percentage of movement time spent after peak velocity was greater for the PG (60.8%) than for the WHP condition (58.85%). While not significant, the percentage of movement time following peak deceleration tended to be greater for the PG (38%) than for the WHP condition (35%; see Fig. 1). Thus, similar to the pointing literature, as the precision requirements increase, the time spent homing in on the target also increases [3]. These data also support the findings of Marteniuk et al. [21]. Thus, regardless of the grasp type, smaller objects result in a lengthened deceleration portion. Similar to Jakobson and Goodale [22] the objects varied in both height and width, however, the shape of the grasping surface was curved. While this first experiment showed changes in both the grasp and transport components as a result of a change in object size, it is unclear as to whether the transport modifications resulted from the programming differences between the two grasp types, or from the different precision requirements of the two tasks. To determine which of these possibilities was a more plausible explanation, a second experiment required subjects simply to point at the objects rather than to grasp them. In this experiment there was no grasp formation. In essence, because the task was to point, there was no grasp component. If the difference in programming between the two grasp types was influencing the transport component in Experiment One, then no differences should be apparent in the movement trajectories here. However, if it was the precision requirements that were necessitating the modifications in the transport component, then changes should result from the subjects pointing at the objects. Similar to the first experiment, the percentage of movement time spent after peak velocity was significantly greater when pointing to the small object (59.5%) as compared to the large object (56.65%), supporting the findings of MacKenzie et al. [3]. Thus, similar patterns of results were present for the transport component regardless of whether it was embedded in a pointing or a grasping task, implying that it is the precision requirement that results in modifications to the transport component.

Concluding remarks. A wide range of papers have been presented dealing with object size. Some offer direct support for Jeannerod's visuomotor channel theory [11, 17, 19, 20], whereas others have suggested that in addition to the grasp component being influenced, so is the transport component [15, 21, 22, 24]. While Jeannerod [11] originally hypothesized that the link between the arm and hand was a temporal one such that the time of maximum aperture correlated strongly with the time of maximum deceleration (low velocity portion), there is little correlational evidence to support this idea. Marteniuk et al. [21] and Gentilucci et al. [24] were the only ones to provide data on within subject correlations between the transport and grasp components. In Marteniuk et al. [21], only 55% were above $r = 0.60$, whereas in Gentilucci et al. [24] there appeared to be a higher degree of correlation for the whole hand grasp ($r = 0.69$ to 0.92) than for the precision grasp ($r = -0.06$ to 0.75). Overall, the evidence is weak to suggest that there is a centre that temporally synchronizes the two components. While Jeannerod's original ideas are not fully supported by the correlational data, this may not be the best measure to use to assess the degree of coordination between the two components. It is a necessity of the reach to grasp movement that the two components be in some way temporally linked. The hand must open while the arm is being transported to the object. If the hand does not open *en route*, the object will not be successfully grasped. Thus, looking for a temporal invariance between the two components may not be the best approach. The exact temporal relationships may be task specific, implying a functional linkage between the two components [21]. The range of object sizes used over the years has been quite broad (0.3 cm to 10 cm), and it could be that the grasp and transport components couple together differently given the specific range of sizes presented in the experiment.

Three of the eight studies presented have shown that some aspect of the transport component is affected by object size. As stated earlier, this is not in contradiction to the pointing literature [2, 3], nor does it entirely disprove the independence of channels controlling the two components of prehension. It is quite easy to accept the fact that the transport component should be affected by changes in object size, given that an object occupies space, and for the arm and hand to be accurately transported to the object, its dimensions must be considered. What is not easy to test experimentally is whether the ***programs*** for each channel are accessible by the other channel. This would more completely address the issue of independence of visuomotor channels. Gentilucci et al. [24] offer some proof for this idea. They suggest that because the hand and the arm both initiate movement at approximately the same time, that if the preshaping and enclosing of the hand was going to influence the transport component in some way, it should happen early in the transport phase. The fact that the time to peak deceleration was relatively constant across the two experiments (grasp versus point) suggests that the use of the hand did not affect the transport component.

Weight - tactile

Lederman and Klatzky [25], in their work on haptic exploration, have also classified object properties as being substance related, for example, hardness and texture; or structure related, for example, size and shape. While both vision and haptics are important to the control of prehension, Klatzky et al. [26] have shown that substance related properties are more easily encoded using haptics while structure related properties are more easily encoded using vision. Employing these ideas to the prehension task, some object properties may affect the grasp and transport component prior to contact (visual) and others subsequent to contact when tactile information becomes available. A potential confound in the studies reported in the previous section was that object weight varied with the changing object sizes. Thus, the results may be reflecting the effects of object weight as well as object size. Because weight is not always detectable by vision, we might hypothesize that prehension movements to objects of different weights may be under the control of two different systems: one related to the visual information so that the object can be located and the grasp planned; and one related to tactile and kinesthetic information needed to stably grasp and manipulate the object.

Weir et al. [27] isolated the effects of object weight by holding the size of the object constant and varying the weight. Subjects performed reaching and grasping movements to metal dowels whose visible characteristics were constant from trial to trial, but whose weight varied (20, 55, 150, 410 grams). Movements were performed under two conditions of weight presentation, random trials (weight unknown) and blocked trials (weight known). The beginning of movement was defined as the time at which the hand was lifted off a mechanical contact plate, and the end of movement as the time when the dowel was lifted off a second contact plate.

Movement time effects revealed a weight by condition interaction. In the random condition, reaching to grasp the 410 gram dowel resulted in longer movement times than the other three dowels. For each weight, random trials were performed significantly slower than blocked trials. This movement time effect was the result of a lengthening of the time after peak deceleration. Only the latter part of the movement was influenced.

In terms of the grasp and transport components, object weight did not affect the maximum kinematic values. This was the anticipated finding based on Jeannerod's visuomotor channel theory, since the visual information was constant from trial to trial. However, the relative timing of the aperture revealed additive effects of object weight and condition of presentation. The 410 gram dowel resulted in a significantly longer percentage of time after maximum aperture, as did the random condition trials in comparison with the blocked trials. Most interestingly, the aperture profile for the 410 gram dowel in the random condition had a ***tail*** on the end of the profile, where the aperture became a constant unchanging size (see Fig. 2**a**). We inferred that this represented the time subjects were in contact with the dowel prior to lifting.

Similar to aperture, the percentage of movement time following maximum velocity and deceleration were greater for the 410 gram dowel. These data suggest that both object weight, and condition of weight presentation influenced the amount of time spent in contact with the dowel prior to lifting it. In a second experiment, a more elaborate contact breaking system was used that defined the time of hand lift, the time at which both the thumb and index finger contacted the dowel, and the time at which the dowel was lifted. This allowed a single trial to be divided into two phases. A **free motion phase** prior to contact, and a **finger-object interaction phase** from dowel contact to lift.

Similar to Experiment One, when the data were analyzed over the entire trial (hand lift to dowel lift) the temporal measures were greater for the 410 gram dowel. The analysis prior to contact revealed that regardless of the dowel weight or the condition of weight presentation, the time taken to contact the dowels was similar. However, the time spent in contact with the dowel prior to lift increased significantly as a function of dowel weight, and was greater for the random trials (see Fig. 2**b**). Similar to the movement time findings prior to contact, there were no differences in the kinematic profiles for either the grasp or transport components as a function of object weight. Thus, the movements were effected in the same way up until the time of dowel contact. This confirms the data from Experiment One that the increased time spent after the kinematic peaks was a function of the time spent in contact prior to lifting the dowel. These data emphasize the dominance of visual information prior to dowel contact and the influence of tactile and kinesthetic information after contact is made.

Texture - visual and tactile

Daily we are faced with having to pick up objects of different textures (e.g., plastic versus glass drinking cup). We perform these movements very adeptly and rarely do we perform them incorrectly. As an extension of the work on object properties that require haptic identification, Weir et al. [28] examined the effects of object texture. The only work done examining its effects on the precision grip is by Johansson and Westling [29] who showed that not only the duration of force application was affected, but also the rate and peak values of the gripping force. From a kinematic viewpoint this variable was of interest to determine if any changes occurred in the kinematics prior to contacting the dowel because, unlike weight which can not always be accurately assessed by the visual system, texture can extracted. Subjects were instructed to reach and grasp one of three dowels covered in either plain metal (Normal), vaseline (Slippery), or rough sandpaper (Rough). Again, the length of the trial was from the time of hand lift to the time of dowel lift. The findings were remarkably similar to the weight experiment. Peak kinematic measures were similar for all three textures, however, the temporal measures were greater for the slippery dowel as compared to the normal and rough. This ordering was always present and not unexpected given the coefficients of friction associated with grasping the

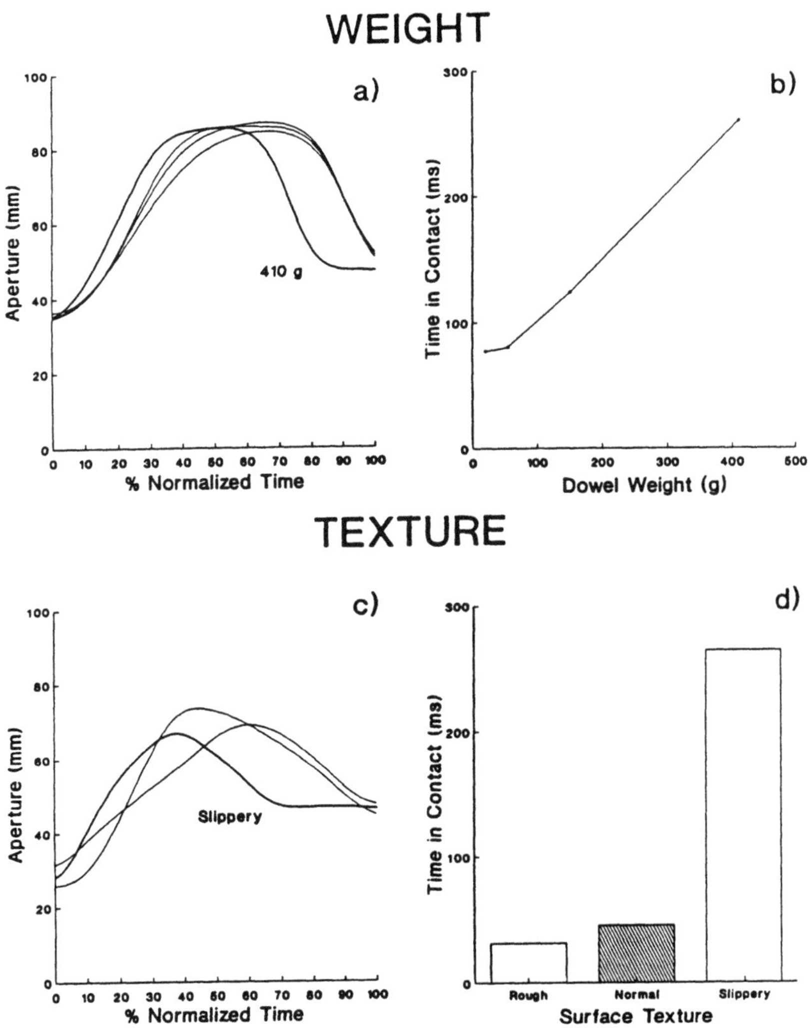

*Figure 2. Aperture profiles from the time of hand lift (0%) to dowel lift (100%) as a function of (**a**) object weight and (**c**) texture. Note the **tails** on the profiles for the 410 gram weight and the slippery dowel. These represented a greater time spent in contact prior to dowel lift as shown in (**b**) and (**d**) for weight and texture respectively.*

three dowels. The coefficient of friction was highest for the rough dowel followed by the normal and slippery dowels. Thus, the lower the coefficient of friction, the longer the subject spent decelerating towards the object. As seen in the aperture and velocity profiles for the slippery condition exhibit ***tails*** similar to those seen in the weight experiment (see Fig. 2**c**). Again, the data was re-analyzed both prior to contact and subsequent to contact. As in Weir et al. [27] neither the time prior to contact nor any of the kinematic measures were influenced significantly by surface texture. However, the time spent in contact with the slippery dowel was significantly greater than the time spent in contact with either the normal or rough dowels, confirming that the ***tail*** did in fact bias the kinematic profiles over the entire trial (see Fig. 2**d**). Thus, although this property could be accurately identified by the visual system prior to contact, it exerts its primary effects on that portion of movement subsequent to contact.

Marteniuk et al. ([15] - Exp. 2) also examined the effect of an object property that can be extracted visually, but that affects tactile interactions with the object. Subjects reached to grasp either a lightbulb or a tennis ball. Thus, they differed in their surface texture, and their fragility. Movements were significantly slower to grasp the lightbulb (469 ms) as compared to the tennis ball (430 ms), which was the result of a lengthening of the time after peak deceleration. Normalized kinematic profiles were not different when reaching to the two objects. Since the time in contact prior to lifting was not quantified, it is not possible to determine if the time in contact varied as a function of the fragility of the object. This study supports the findings of Weir et al. [27, 28] that object properties that require haptic information do not influence the movement trajectories prior to contact.

Concluding remarks. These studies suggest that when the visual characteristics of the object (diameter, location, orientation) remain constant, the weight, texture, and fragility information are not crucial to controlling the movement until contact is made. This is confirmed by the lack of statistical significance in the kinematic profile measures prior to contact. Thus, there is no evidence of anticipation or preplanning of the upcoming contact. Although grip and load forces were not measured, the longer time in contact prior to lifting the 410 gram dowel or the slippery dowel parallels the work of Johansson and Westling [29] and Westling and Johansson [30]. This increased time presumably reflects the larger forces that must be generated to lift the dowel. These studies also confirm the work on haptic exploration by Klatzky et al. [26] who have shown that while both substance and structure related object properties can be encoded haptically with and without vision, substance related properties are more easily encoded using haptics.

Two phases of the prehension task have been introduced. In the first phase prior to contact the hand is postured and oriented to contact the object based primarily on vision. In the interaction phase, kinesthetic and haptic inputs contribute to the generation of functionally effective forces needed to achieve a stable grasp. More recent work by Weir [31] and Weir and MacKenzie [23] has combined the kinematic and force paradigms, and divided the reach to grasp movement into four unique phases. During the **free motion** phase from hand lift to dowel contact, the hand is transported and postured to contact the object based primarily on visual inputs. The **dowel acquisition** phase follows from dowel contact to dowel lift and during this phase a stable grasp is effected based on tactile and proprioceptive inputs of the dowel characteristics (weight, shape). During the **dowel transport** phase from dowel lift to when it is replaced, vision is again active in guiding the dowel to the target location. However, during this phase, the cutaneous receptors are providing information regarding the maintenance of the stable grasp. Lastly, during the **dowel release** phase from dowel replace to let-go, touch is providing the primary information regarding contact with the supporting surface, and the release of the grasp. Vision provides cues as to whether the dowel is upright. Object weight was shown to influence only the force generating phases, while grasping surface area and task intent influenced both kinematic and force generating phases (see [32] for a review).

TASK RELATED

Movement time

Task or goal related factors are also an integral part of the control of prehension. Not only are the physical properties of the object important, but the underlying goal of the task. In prehension the goal is to successfully reach and grasp an object. However, our ability to do this changes on the basis of the demands that are made on the motor system. The object properties become part of the overall task, yet certain parameters of the task can be isolated. Wing et al. [33] were interested in continuing their work on the spatial dependency between the grasp and transport components [19]. Drawing from the pointing literature where the variability around the target increases with increases in the speed of movement, it was of interest to determine whether increases in speed of transport of a prehension task would be compensated for by changes in the grasp aperture. If changes in the transport component were accompanied by systematic changes in the grasp component, this would provide evidence of coordination between the spatial aspects of the two components. Wing et al. [33] had subjects reach to grasp a wooden dowel 130 mm tall and 28 mm in diameter. The subjects performed under three instruction conditions: 1) **normal** reaching where the subject moved at their own comfortable pace; 2) **fast** reaching where the subjects were instructed to move as fast as possible; and 3) **blind** reaching where the subjects reached with their eyes closed. It was hypothesized that the transport trajectories would be more variable in the fast and blind conditions. The measure used to reflect the accuracy of the transport component was the variability in the perpendicular (y-axis) distance of the thumb from the straight line between the start position and the position of the dowel.

The overall movement time in the normal and blind conditions were quite similar, while in the fast reaching condition the movement time was shorter by half. Most interesting was the variability associated with the transport component, and the subsequent changes in grasp aperture. Average variability in the spatial position of the wrist increased significantly as the conditions became more unpredictable: normal (4.0 mm), fast (5.0 mm), and blind (9.3 mm). With respect to the aperture, maximum aperture increased significantly in size as the variability in the transport component increased, with the blind condition having the largest aperture. Thus, the larger aperture was hypothesized to be compensating for the increased spatial variability of the transport component to ensure that the dowel is contacted and grasped. Opening the hand wider gives increased tolerance for position errors prior to contact. Thus, changes in the grasp aperture are associated with changes in the spatial accuracy of the transport. This increase in maximum aperture should be considered predictive and not based on concurrent feedback.

Wallace and Weeks ([20] - Exp. 3) suggested that movement time rather than movement speed may be producing the changes in grip formation. In the Wing et al. [33] study, movement speed and movement time were confounded because the distance of the reach was held constant. Wallace and Weeks ([20] - Exp. 3) corrected for this by varying both distance and movement time. Subjects performed reaching and grasping movements in either 200 or 400 milliseconds (ms) over distances of 15 and 30 cm. Subjects were trained to perform within temporal bandwidths of 15%. Using a movement speed basis, two predictions were made. First, regardless of movement time, movements with greater speeds should have larger apertures, and second, there should be no kinematic differences between movements of equal speed.

Not surprisingly, peak wrist velocity increased as the speed required by the condition increased, supporting the fact that subjects were in fact performing within the temporal bandwidth. Maximum grasp aperture was related to movement time as evidenced by the movement time main effect. Maximum aperture was significantly larger in the 200 ms condition as compared to the 400 ms condition. Within trial correlations were done to determine the relationship between peak wrist velocity and aperture. The Fisher's z values were higher in the 200 ms than in the 400 ms condition again suggesting that the relationship between the reach and grasp components was due more to temporal constraints than to distance or movement speed.

Wallace et al. [34] used a wider range of movement times (300 - 800 ms) to further test this relationship between the arm and hand. A 10% bandwidth was provided around the three goal movement times of 300, 600, and 800 ms. Again, peak wrist velocity increased as movement time decreased. Similar to their previous work, maximum aperture increased as movement time decreased. Correlations between peak wrist velocity and aperture were quite linear prior to peak velocity ($r = 0.913$ to 0.92). However, following peak velocity the relationships became less linear as movement time increased and movement speed decreased (300 ms, $r = 0.812$; 600 ms, $r = 0.533$; and 800 ms, $r = 0.473$). Thus, with longer movement times the two components become more independent.

Error tolerance

Wallace and Weeks ([20] - Exps. 1 and 2) also examined grasping accuracy as a means of looking at the relationship between the transport and grasp components. If similar to pointing, limiting the allowable grasping error decreases the movement time, changes in the movement trajectories should follow [35]. A joystick (1.27 cm diameter) like dowel with a flexible base was used. A potentiometer measured the dowel movement. The three error-tolerances were 7.62 cm, 5.08 cm, and 2.54 cm. Graphical feedback regarding the dowel movement was provided following each trial. Movement times in the medium and large tolerance conditions were significantly shorter than in the small tolerance condition. The resultant dowel movement increased as tolerance increased. With respect to the transport component, peak wrist velocity was significantly greater in the large as compared to the small tolerance condition. The percentage of movement time following maximum velocity was greater for the large tolerance/small dowel condition than all others. The trend for the grasp component was for maximum aperture to increase as the tolerance increased. As predicted, changes in the transport component support the pointing literature. The findings also support Jeannerod's [11, 17] visuomotor channel theory that extrinsic properties affect only the transport component.

Experiment Two was designed to further elucidate the trend in the grasp aperture. The small (2.54 cm) and large (7.62 cm) tolerance conditions were factorially combined with two dowel sizes (3 mm and 25 mm). The movement time and resultant dowel movement findings supported those of Experiment One. The wrist velocity findings also supported previous findings as peak velocity was greater for the larger tolerance condition. Maximum aperture increased with both dowel size and tolerance. While the correlations between the wrist velocity and aperture were high prior to peak velocity ($r = 0.94$ to 0.96) and showed no difference between the large and small tolerance conditions, after peak velocity the correlations were smaller in the small tolerance condition (small, $r = 0.83$ to 0.86; large, $r = 0.94$ to 0.96) where the movement times were longer. These data suggest that the two components become more closely linked as movement time decreases and movement speed increases. Thus, the correlational data from this study and from Wallace et al. [34] provide compelling evidence for the temporal coupling of the transport and grasp components in short duration movements.

Intent/goal

Changing the movement time or allowable error are only two ways of altering the task. Another option is to alter the overall goal movement or task. Marteniuk et al. [15] examined the effects of changing the task intent.

In Experiment One, pointing versus grasping movements were contrasted. Subjects moved over 20 and 40 cm distances to targets two and four centimeters in diameter. Subjects were required to either point to the target or to grasp a disk. Concentrating on the task differences, grasping movements were significantly slower than pointing movements across the different amplitude, target-width conditions. In terms of the transport component, peak velocity was similar for the two tasks. However, the temporal components of transport varied for the two tasks. The time spent in both the acceleration and deceleration portions was greater in the grasping task than the pointing task. The data were also scaled in the temporal domain to determine if the trajectories belonged to a family of scalar curves. The acceleration portion was significantly longer for pointing (67.25%) than grasping (47%). Conversely, the time spent in deceleration was greater for grasping (53%) than pointing (32.75%). The increased demand of having to grasp the disk necessitated a more controlled approach than in pointing where the target was used to decelerate the hand.

Experiment Three also examined task goal effects on grasping. Subsequent to grasping a four centimeter diameter disk, subjects either: a) fitted the disk into a 4.1 cm diameter well placed ten centimeters to the left, or b) threw the disk into a 20 x 40 x 15 cm box placed 15 cm to the left. Movements were measured to disk contact. The movement time was longer for the grasping movement prior to fitting the disk into the well as compared to throwing the disk. This was a result of a lengthening of the deceleration portion. As a percentage of total movement time, the acceleration portion was longer prior to throwing the disk (54.3%) as compared to fitting the disk (48.2%). Again, this results in a longer deceleration portion for the fit task (51.8%) as compared to the throw task (45.7%). Thus, the precision required for the fit task resulted in the subjects spending a longer portion of the movement in deceleration.

Both of these experiments provide evidence that the intent of the task affects the movement planning and control processes. Presumably, had the grasp component been quantified some changes in its control would also have been present [20, 34].

Concluding remarks. In the papers presented the grasp aperture appears to increase in size to compensate for changes in the transport component (spatial inaccuracies) as either the movement time decreases, or the amount of allowable error increases. These papers represent a concise set of findings that are well developed and supported. The findings for goal and task intent support the notion that as the precision requirements of the task increase, the time spent in deceleration prior to contact also increases. This occurs presumably to ensure that the object is stably contacted.

The correlational data presented by Wallace and colleagues [20, 34] is another attempt to look for invariances between the two components. However, instead of looking at temporal data, peak wrist velocity and aperture values were correlated both prior to and after peak velocity. These data suggest that prior to peak velocity, there is a strong linear relationship between the two variables regardless of the duration of the movement. However, after peak velocity, as the movement time increases, the relationship becomes non-linear, showing an independence between the two components. Thus, for short duration movements, the two components are functionally linked.

OVERALL SUMMARY

This review has been concerned with variables that influence the unfolding of the transport and grasp components controlling the reach to grasp movement. Object properties received a considerable amount of attention due to the equivocal nature of the findings. It is difficult to tease out commonalities in the different factorial studies. The papers on task related factors were much more focused, and followed one another logically. Within all the papers reviewed, several explanations have been offered regarding the nature of the control and coordination between the grasp and transport components. Jeannerod's [11] visuomotor channel theory while providing separate input-output channels for each component is presumably " .. represented by a unique programme governing the integrated aspect of the action, or in other words the coordination of the components ([36], p.57)". For Jeannerod, the coordination occurs in the time domain such that the maximum opening of the hand corresponds to the onset of the low velocity portion (peak deceleration) of the transport component. Thus, the first view for the control of prehension relies on a **temporal** coupling between the components.

Wing and colleagues [19, 33] provided the second view on the coordination between the two components. They suggest the link goes beyond timing. They have shown **spatial** coupling between the arm and hand through two different experimental paradigms. Wing and Fraser [19] showed that the index finger is primarily responsible for the closure of the grasp with the position of the thumb remained relatively constant to guide the hand to the object. Using the manipulation of movement speed, Wing et al. [33] showed that as the accuracy demands of the transport component decreased the hand opened wider to compensate for the increased variability. Thus, while there is still a temporal link to ensure the hand opens prior to contact, a spatial controller is also involved.

From these two views it is apparent that timing is an important variable in the unfolding of the reach to grasp movement. It may not be as strict as the view espoused by Jeannerod, but some timing must be present in order for the actions of the arm and hand to be coordinated. The view of Wing and colleagues [19, 33] suggests there is more involved than just timing. These ideas are encompassed in the third view presented by Marteniuk and colleagues [15, 21]. Marteniuk et al. [21] showed variant relative timing in the grasp and transport components as a function of object size. Thus, the two components are not temporally linked in an invariant manner, and they are not controlled by a general abstract ***program***. Marteniuk et al. [21] suggests that the two components may be **functionally** linked such that the exact temporal relationship is specific to the demands of the task.

This functional view does not negate those presented by Jeannerod [11] and Wing and colleagues [19, 33]. It simply speaks to the notion of an integrated approach which provides a broader base. It allows for different couplings between the components based on not only the task, but also on individual differences. These ideas are supported by the equivocal findings for object size, and the correlational data. Some of these theoretical notions are captured in the multimovement theory presented by Abbs et al. [16] for the control of speech. They proposed a theory which involves the integration of information from a number of sensory systems in achieving the goal of the movement. The movement is controlled on the basis of the current circumstances and can be updated through feedforward processes. A feedforward approach is often characteristic of coordinated systems, as it allows intersensory regulation throughout the entire period of movement execution.

What all these approaches have in common is a ***program*** centered explanation that relies on high level representations [15, 18] for its formation and execution. These ideas are also recognized in Arbib's [6] coordinated control program (CCP) where actions are deemed to be determined by knowledge that is greater than can be obtained from the environment through the senses, in this case vision. Perceptual schemas are thought to obtain information regarding the location, orientation and size of the object, and this information is then passed on to two motor schemas that control the transport and grasp component respectively. The CCP is responsible for regulating the time varying actions of the two components. Like Jeannerod [11], the CCP hypothesizes two independent channels controlling the movement. The problem with this idea is that the two channels cannot be strictly independent of one another. For the reach to grasp to be completed, the two components must be integrated with one another so that the arm and hand reach a position that allows the object to be contacted and grasped. The correlational data that has been used by Jeannerod [11], Marteniuk et al. [21] and Wallace and colleagues [20, 34] may not be the appropriate way to look at the relationship between the two components.

The ***program*** controlling the reach to grasp movement is presumably sensorimotor based as evidenced by the literature that has been reviewed in this chapter. Jeannerod and Decety [18] provide evidence of the use of a sensorimotor store such that the visual cues are linked to the efferent signals controlling aperture size. The work by Weir and colleagues [27, 28] also supports these ideas in that the visual cues are linked to the efferent pathways for the execution of the transport and grasp components, while the tactile signals are linked to the efferent pathways for force generation. Thus, in a multimovement system the two components are functionally linked to preserve the overall goal of the task which is to successfully grasp the object. As Marteniuk et al. [21] and Jakobson and Goodale [22] suggest, at some central level the commands to the two components become integrated. As shown in the literature it may be inappropriate to think that this integration occurs at the same point in time in each experimental setting.

Further evidence for a sensory based control system comes from the observation that in 60% of the papers reviewed, changes occurred in the decelerative portion of the movement, with relatively few changes being seen in the accelerative portion. This would seem to suggest that looking for temporal invariances or changes before this point in the movement may not be the most appropriate way to tap into the underlying mechanisms controlling the prehension movement. Perhaps it is more informative to more fully examine this latter part of the movement for insights into what about the movement is changing. Wing et al. [33] measured the variability in the wrist trajectory at the time of maximum aperture and immediately preceding contact. While there was no significant difference between these two times, the trend was for the variability to be smaller immediately preceding contact. Similar findings have been shown by Marteniuk et al. [21] who examined the standard deviation in the wrist and aperture trajectories prior to contact. They reported that regardless of the size of the disk to be grasped, the variability in the wrist trajectory decreased between 300 ms prior to contact and contact with the disk. The same was true for the aperture profiles for disk sizes one to five centimeters, whereas for the six to ten centimeter disks, the variability increased. As Marteniuk et al. [21] offer, this may be due to an increased number of available contact points on the larger objects. This kind of analysis allows a closer examination of the underlying mechanisms. Other measures that may be beneficial are: 1) the spatial variability, 2) the number of zero crossings, and 3) the distance remaining to the target. Future work should be directed at looking beyond invariant relative timing. The identification of phases of prehension will also allow an examination of sensory based influences upon the control of the reach to grasp movement. Variables that influence one phase, may not influence a subsequent phase, allowing a more complete description of the reach to grasp movement [23].

ACKNOWLEDGEMENTS

I would like to thank the two anonymous reviewers and Jack Leavitt for comments on an earlier draft of this manuscript.

REFERENCES

1 R.S. Woodworth, Psychol. Rev. Monogr. Suppl., 3 (1899) 1.
2 P.M. Fitts, J. Exp. Psychol., 47 (1954) 381.

3 C.L. MacKenzie, R.G. Marteniuk, C. Dugas, D. Liske and R. Eickmeier, Quart. J. Exp. Psychol., 39A (1987) 629.
4 J.R. Napier, J. Bone Jt. Surg., 38B (1956) 902.
5 M.A. Arbib, in: V.B. Brooks (ed.) Handbook of Physiology, Section I, The Nervous System, Vol. II, Motor Control, Am. Physiol. Soc., Bethesda, 1981, p. 1449.
6 M.A. Arbib, Hum. Neurobiol., 4 (1985) 63.
7 M. Jeannerod, Behav. Brain Res., 19 (1986) 99.
8 K.P. Bergmann, Am. J. Occ. Ther., 44 (1990) 736.
9 C. von Hofsten and L. Ronnqvist, J. Exp. Psychol., Hum. Percep. Perf., 14 (1988) 610.
10 R.J. Bootsma and P.C.W. van Wieringen, Hum. Mvt. Sci., 11 (1992) 205.
11 M. Jeannerod, in: J. Long and A. Baddeley (eds.), Attention and Performance IX, Erlbaum, Hillsdale, 1981, p. 153.
12 R.A. Schmidt, Psychol. Rev., 82 (1975) 225.
13 S.W. Keele, in: V.B. Brooks (ed.), Handbook of Physiology, Section 1, The Nervous System, Vol. II, Motor Control, Am. Physiol. Soc., Bethesda, 1981, p. 1391.
14 K.J. Cole and J.H. Abbs, J. Neurophysiol., 55 (1986) 1407.
15 R.G. Marteniuk, C.L. MacKenzie, M. Jeannerod, S. Athenes and C. Dugas, Can. J. Psychol., 41 (1987) 365.
16 J.H. Abbs, V.L. Gracco and K.J. Cole, J. Mot. Behav., 16 (1984) 195.
17 M. Jeannerod, J. Mot. Behav., 16 (1984) 235.
18 M. Jeannerod and J. Decety, in: M. Goodale (ed.), Vision and Action, The Control of Grasping, Ablex, New Jersey, 1990, p. 33.
19 A.M. Wing and C. Fraser, Quart. J. Exp. Psychol., 35A (1983) 297.
20 S.A. Wallace and D.L. Weeks, J. Mot. Behav., 20 (1988) 81.
21 R.G. Marteniuk, J.L. Leavitt, C.L. MacKenzie and S. Athenes, Hum. Mvt. Sci., 9 (1990) 149.
22 L.S. Jakobson and M.A. Goodale, Exp. Brain Res., 86 (1991) 199.
23 P.L. Weir and C.L. MacKenzie, submitted.
24 M. Gentilucci, U. Castiello, M.L. Corradini, M. Scarpa, C. Umiltá and G. Rizzolatti, Neuropsychologia, 29 (1991) 361.
25 S.J. Lederman and R.L. Klatzky, Cog. Psychol., 19 (1987) 342.
26 R.L. Klatzky, S.J. Lederman and C. Reed, J. Exp. Psychol, Gen., 116 (1987) 356.
27 P.L. Weir, C.L. MacKenzie, R.G. Marteniuk, S.L. Cargoe and M.B. Frazer, J. Mot. Behav., 23 (1991) 192.
28 P.L. Weir, C.L. MacKenzie, R.G. Marteniuk and S.L. Cargoe, J. Mot. Behav., 23 (1991) 205.

29 R.S. Johansson and G. Westling, Exp. Brain Res., 56 (1984) 550.
30 G. Westling and R.S. Johansson, Exp. Brain Res., 53 (1984) 277.
31 P.L. Weir, Unpublished Doctoral Thesis, University of Waterloo, Waterloo, Canada, 1991.
32 C.L. MacKenzie, in: G.E. Stelmach and J. Requin (eds.), Tutorials in Motor Behavior II, North-Holland, Amsterdam, 1992, p. 371.
33 A.M. Wing, A. Turton and C. Fraser, J. Mot. Behav., 18 (1986) 245.
34 S.A. Wallace, D.L. Weeks and J.A.S. Kelso, Hum. Mvt. Sci., 9 (1990) 69.
35 G.D. Langolf, D.B. Chaffin and J.A. Foulke, J. Mot. Behav., 8 (1976) 113.
36 M. Jeannerod, The Neural and Behavioural Organization of Goal-Directed Movements, Clarendon Press, Oxford, 1988.

Insights into the Reach to Grasp Movement
K.M.B. Bennett and U. Castiello (Editors)

CHAPTER 8

PERTURBATION STUDIES OF COORDINATED PREHENSION

P. HAGGARD

Laboratory of Physiology, Parks Road, Oxford, OX1 3PT, England.

SUMMARY

Recent studies in which either the reaching or the grasping component of movement is perturbed have been influential and informative. This chapter begins with a detailed methodological discussion of what a perturbation experiment involves, and of what it tells us about coordination of movement. The experiments are then considered in three groups, according to whether the perturbation acts at the level of general cognitive processing, at the level of the perceptual environment of the motor task, or at the level of the physical movement of the limb. The range of perturbations reviewed includes secondary movement tasks, vocal responses, apparent changes in the location of the target of a movement, or in the size of the target object, and mechanical loads imposed on the arm.

METHODOLOGY OF PERTURBATION EXPERIMENTS

In a perturbation experiment, an unexpected event, or perturbation, occurs during the performance of the task being studied. Almost any features of the perturbation can be varied to make it unexpected, such as its timing, magnitude, and probability of occurrence. The experimenter is typically interested in the effects of the perturbation on the subject's performance of the task. This involves detailed analysis of the portion of the trial following the perturbation, rather than a single measure describing the entire trial. The behaviour following the perturbation is compared to unperturbed behaviour.

The perturbation may be a physical interference with the moving limb, a perceptual change in the environment of the movement, or a requirement for the subject to perform a secondary task at the same time as the movement. The logic of the experiment is the same in each case, and is schematised in the top panel of Fig. 1. **A** is an ongoing continuous behaviour, and **p** is an unexpected perturbation occurring during the course of **A**. **A'** is the observed behaviour following the perturbation. If **A'** is no different from **A** then the perturbation does not affect the information-processing involved in behaviour **A**. That is, the control of the ongoing behaviour does not involve any channel of information related to the information in the perturbation. However, this null result is rather inconclusive, since more subtle measures might disclose an effect of the perturbation. Conversely, if **A'** differs from **A** then the perturbation does affect the information-processing involved in behaviour **A**. The experimenter may then need to separate out the immediate physical effects of the perturbation on the behaviour from the response to the perturbation. Often, only the latter is of psychological interest. Perturbation experiments have been particularly influential in motor control research, where they have typically been used to show how a number of effectors form a coordinative structure or synergy. Perturbing one effector within the synergy typically elicits compensatory adjustments in the other effectors, so that the movement of the synergy as a whole remains stable. Perhaps the best-known example comes from the speech articulation experiments of Kelso and colleagues and Abbs and colleagues [1, 2]. Both groups found that a mechanical tug on the lower lip or jaw while uttering syllables with bilabial closure (e.g., "*pap*") elicited greater than normal movement of the upper lip, so that the closure was achieved despite the perturbation. That is, the movements of the lower and upper lips were coordinated, so that one compensated for the other.

Some psychologists have been reluctant to accept patterns of perturbed behaviour as evidence of how normal behaviour is controlled. Certainly, when a subject knows they may be perturbed, they will be more attentive, they may co-contract opposing muscles, they may learn to circumvent the perturbation, and so on. Athènes and Wing [3] found that even the possibility of perturbation altered the coordination of hand aperture and hand transport in reaching and grasping. They found that the unperturbed trials in blocks where subjects knew they *might* be perturbed showed greater hand aperture than trials that subjects knew would not be perturbed. Their subjects prepared for the uncertainty of possible perturbations by adopting wider hand apertures. These expectation effects probably increase with the proportion of perturbed trials. However, the motor system always works in the presence of small, unpredictable changes in the environment and in the musculoskeletal system itself, and much of its fundamental neural organisation, such as the stretch reflex, is dedicated to providing efficient motor control despite such perturbations. Further, some studies

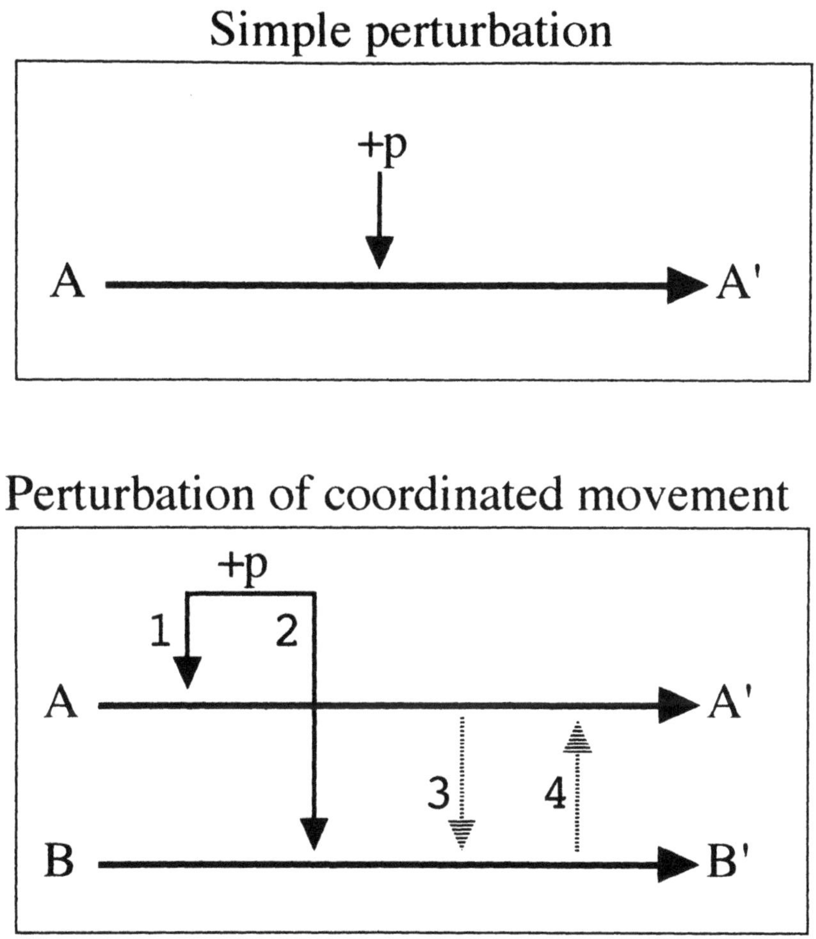

Figure 1. Interpretation of perturbation experiments. See text for explanation.

[1] have demonstrated that subjects compensate successfully on the first perturbed trial that they experience, suggesting that the response to perturbation does not just involve learning a suitable strategy.

INTERPRETATION OF PERTURBATION EXPERIMENTS

The coordination of two related behaviours can also be investigated using perturbations. The lower panel of Fig. 1 shows how perturbations can be used to investigate the coordination between two component tasks, **A** and **B**. The perturbation **p** can now affect the behaviour in four distinct ways. The route numbered **1** represents an informational channel common to both component **A** of the ongoing behaviour, and to the perturbation **p**. The simple perturbation experiment (top panel) shows that this channel exists, so the perturbation will lead to a changed behaviour **A'**. **p** may also directly affect the second component of the behaviour via channel **2**, producing **B'**. However, a change in the second component could also arise from coordination of **B** with **A**. This indirect effect of the perturbation is shown by the flecked arrow of channel **3**. That is, the perturbation **p** could affect component **A**, which could in turn affect component **B**, producing **B'**. Conversely, channel **4** represents the coordination of **A** with **B**. The actual behaviours observed depend on the relative strengths of these four routes, but, for purposes of exposition, I will consider each route as simply being present or absent.

A simple perturbation experiment, as previously described, can be used to investigate routes **1** and **2**. Assuming, for simplicity, that route **1** is present, and route **2** is absent, then the coordination of the two tasks via routes **3** and **4** can be assessed as follows:

1) If the perturbation has no effect on **B**, then components **A** and **B** are independent. That is, channel **3** does not exist. This interpretation is complicated by the difficulty of drawing inferences from a null result.
2) If, by contrast, **B** is altered by the perturbation, then channel **3** must exist. That is, component **B** uses information about component **A**: **B** is coordinated with **A**.

This combination of simple and complex perturbation paradigms is suitable for investigating the coordination between the hand transport and hand aperture components of reaching and grasping movements. If the two components are completely independent, then the response of each component to a given perturbation should be the same whether the other component is operating concurrently or not. On the other hand, if the two components are coordinated, in the sense of using a common informational channel, then a perturbation which affects one component should propagate onwards via that channel, producing subsequent effects on the other component as well.

MEASUREMENT OF RESPONSES TO PERTURBATION

This section considers how to assess responses to perturbations in reaching and grasping movements.

Modern recording techniques allow researchers to represent a multi-effector movement, such as a prehensile movement, as a signal of Cartesian coordinates of each effector as a function of time, rather than in terms of a few summary measures. However, the richness of this new data also requires a method for identifying and quantifying a change in the movement signal which indicates a response to a perturbation. In studies of coordination, any such method must be generally applicable to all the effectors in a multi-effector movement, so that responses to the perturbation can be identified with similar criteria in different effectors, and can be compared across effectors.

Typically, a perturbation will lead to an active adjustment of the movement pattern in an effector being studied. This adjustment may propagate forward to other effectors as discussed above, even if they are not directly affected by the perturbation. If the magnitude of the perturbation is quantifiable (i.e., if the perturbation is a physical rather than a psychological event) then the magnitude of the response in each effector may be measured, and correlated with the magnitude of the perturbation. If two effectors are very tightly coupled (i.e., channel **3** is very important), then the magnitude of the response in the coordinated effector will be large relative to the response in the perturbed effector. The gain of the relation between the two responses indicates the strength of the coordination.

The latency of the adjustment in each effector is also a useful measure for investigating the response to the perturbation by different effectors. Latency measures, however, involve the rather subtle problem of identifying a discrete event in the continuous movement signal at which an adjustment to that signal begins. Wing [4] has highlighted some of the difficulties involved in making such estimates. Further, the latency value obtained depends critically on which movement signal is used to detect the adjustment. This last point is particularly important when calculating the latency from digitally smoothed and differentiated movement signals. For example, the top left panel of Fig. 2 shows two hypothetical movement signals, as plots of the displacement (in arbitrary units) of an effector against time. The solid trace shows an unperturbed movement, while the dashed trace shows a perturbed movement. The perturbation causes an adjustment to the movement, which involves a temporary increase in movement velocity. The movement velocity is shown as changing instantaneously: although obviously unrealistic, this corresponds to the hypothetical case of a single moment corresponding to the onset of the adjustment.

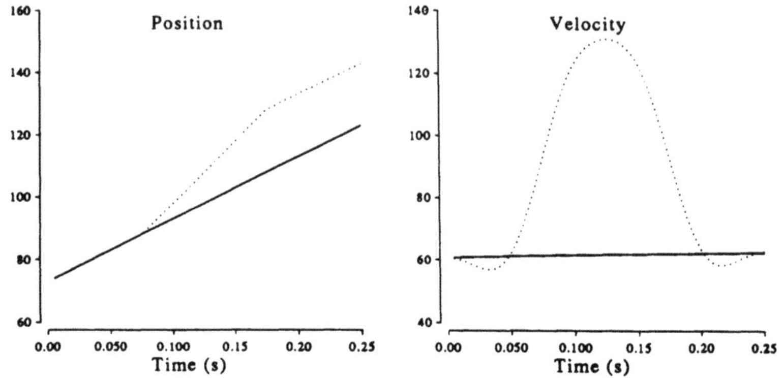

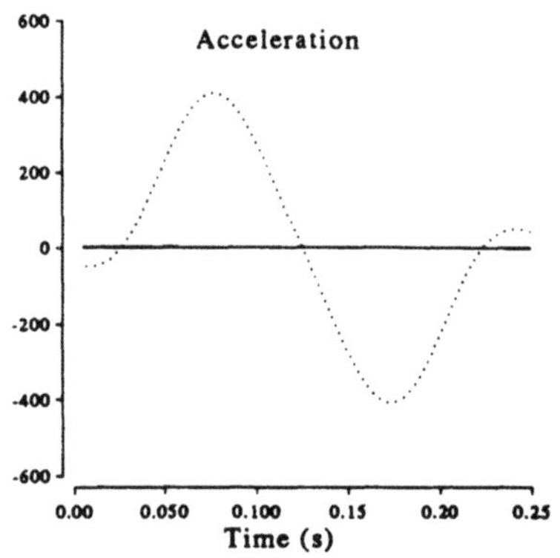

Figure 2. The effects of smoothing a differentiation on estimates of adjustment latency are shown for hypothetical unperturbed data (solid lines) and perturbed data (dashed line). The unsmoothed velocity and acceleration of the perturbed movement (dotted lines) were obtained by one-point differencing. Note that the smoothing accompanying each differentiation alters the apparent latency of the change in the perturbed waveform.

Visual inspection of this hypothetical data might estimate the adjustment latency using the time when the difference between the perturbed and unperturbed signal has exceeded a given threshold value. This might give a latency of, say, 95 ms. The top right panel of Fig. 2 shows the velocity traces obtained for the same signals. The smoothing method of Teulings and Maarse [5], which removes power gradually over a band from 10 Hz to 26.7 Hz, was applied at the same time as the differentiation. The additional dotted line shows the unsmoothed velocity of the perturbed movement obtained by one point differencing. The smoothing has spread out the increase in velocity. Visual inspection of the smoothed velocity waveforms might now suggest an adjustment latency of 55 ms. It is not possible to use the same threshold for detecting adjustments as with the position traces, because the units of measurement differ. If the differentiation and smoothing process is repeated again to obtain the acceleration traces (lower left panel of Fig. 2), the latency of the response might now be 35 ms. The latencies calculated from the displacement and from the acceleration data differ by 60 ms. This represents about one third of a visuomotor reaction time, yet the apparent advance is purely a result of the smoothing process. In fact, the Teulings and Maarse smoothing method is designed to minimise these artefactual phase-shifts.

In conclusion, it is hard to identify the latency at which a motor adjustment begins. Latency estimates depend both on the nature of the adjustment, and the method for detecting it. Although there is no single ***correct*** method of measuring adjustment latency, researchers should be aware that variations in latencies reported for similar responses may reflect differences in methods of analysis rather than in underlying neural processes.

PERTURBATION BY SECONDARY TASKS

A number of studies have investigated the control of aimed movement by requesting subjects to perform a discrete secondary task during the course of movement. The secondary task can be considered as a perturbation, as it may add a processing load to the informational channel controlling the movement. The perturbation may therefore affect the pattern of the primary movement. Alternatively, the performance of the secondary task may be reduced because of the demands of the primary motor task, as in probe reaction time studies [6, 7]. The effect of the perturbation is attributed to interference between the information-processing involved in the two tasks. The locus of the interference could be either a general information-processing channel [8], or a dedicated processing channel specialised for a particular kind of information which is used in both tasks. On the latter view, two tasks will interfere if and only if they both compete for the resources of the same channel at the same time.

This concept can be applied directly to the coordination of hand aperture and hand transport in prehension. Haggard [9] asked subjects to make one of three movements with the right hand according to condition: either a simple pointing movement, or a simple grasping movement involving opening and closing the hand without movement of the arm, or a composite reaching and grasping movement. In each case, the movement of the right hand was repeated continuously with a regular rhythm for 12 seconds. At a random point during the trial an auditory signal was given, following which the subject had to perform one of two discrete secondary tasks with the left limb: either a forward movement of the left arm similar to a reaching movement, or simple flexion of the left digits similar to a grasp. The conditions in which the right hand made a simple reaching or grasping movement thus corresponded to the simple perturbations schematised in the top panel of Fig. 1, while the conditions in which the right hand made a compound reaching and grasping movement corresponded to the complex perturbations schematised in the lower panel of Fig. 1. The results were analysed using phase transition curves [10], which measured the changes of the phase of the primary task rhythm due to performing the secondary task. Unlike probe RT, this method allows direct measurement of the effect of the perturbation on the kinematics of the primary task.

The results showed a pattern of simple and complex task interference which was consistent across four out of five subjects. When subjects made a simple rhythmic grasping movement with the right hand, the rhythm was significantly reset by the requirement to respond to an auditory signal by making a single grasp with the left hand. The same secondary task did not, however, disturb a simple rhythmic reaching movement of the right hand. Nor did the secondary task significantly affect the rhythm of the right hand's grasping when this occurred as part of the composite reach and grasp movement.

These results can be summarised using the processing model shown in Fig. 1, taking the hand aperture component as **A**, the hand transport component as **B**, and the secondary grasping task as **p**. The simple perturbation experiments showed that **A'≠A**, but the complex perturbation experiment showed that **A'=A** and **B'=B**. This suggests that coordination of **A** with **B** in the complex task prevented the perturbation from affecting **A**. That is, component **A** is ***slaved*** by component **B** when they are performed together. Therefore, route **2** does not exist, and route **4** is relatively more important than route **1** in the control of component **A**. The way in which the grasping movements of the right hand were controlled changed when those movements were part of a coordinated prehensile movement, compared to the condition involving grasping alone. This suggests that the control of the digits becomes subordinate to the hand transport component of the movement. Since there is no functional mechanical connection between the digits of the left and right hands, the significant effect of the left grasp perturbation on isolated movement of the right digits presumably arises from interference and competition for computational resources in a neural processing channel. This channel cannot be a general central resource, since the same perturbation does not interfere with all tasks: rather it must be specific for the control of grasping. Further, the pattern of digit movement in preparing hand aperture in a composite prehensile movement seems to be driven directly by the position of the hand approaching the target, and not by an encapsulated neural channel specialised for controlling grasping.

PERCEPTUAL PERTURBATION OF THE MOTOR TASK

In a prehensile movement, the hand transport component moves the hand to the location of the target object, while the hand aperture component involves preparing a suitable hand configuration for grasping the object, and then closing the hand to acquire the object. In an elegant series of perturbation experiments, Jeannerod and colleagues have investigated the motor system's responses to unexpected changes in either of these functions.

Paulignan et al. [11] arranged three pieces of dowel in a dimly lit room, illuminated the central one of the three, and asked subjects to reach out and pick up the illuminated dowel. On a small proportion of trials, the illumination changed as soon as the subject began to move, so that either the left or the right hand dowel became illuminated, instead of the central dowel. This apparent change in the location of the target amounted to a perceptual perturbation of hand transport. Paulignan et al. [11] found that these perturbations of hand transport had extensive effects on both the hand transport and the hand aperture components of the movement. The initial portion of the movement on perturbed trials, was similar to that in control trials directed towards the central dowel. However, after an average 275 ms, the spatial path of the hand, measured by an infra-red marker on the wrist, was seen to curve round towards the new target location. Paulignan et al. [11] also found significantly earlier and lower values for the wrist's peak tangential velocity and acceleration in the perturbed conditions than in the control condition. The lower peak acceleration in perturbed trials occurred some 100 ms after the start of the movement and change in the apparent target location, suggesting very rapid adjustments to hand transport. These patterns suggested that the initial movement directed at the central dowel was interrupted part way through, and replaced by a second submovement towards the location of the new target.

Paulignan et al. [11] also observed changes in the hand aperture component of the movement, which were coordinated with the changes in hand transport. Instead of hand aperture increasing smoothly to a single peak value as in the control trials, they found "paradoxical closing" of the hand in the perturbed trials. In perturbed trials the hand aperture reached a peak after around 217 ms and began to close again briefly. This peak occurred significantly earlier and had a significantly lower amplitude than the peaks in control trials. Some 200 ms later, a second peak in hand aperture was reached, which had a similar amplitude to the peaks in control trials. The characteristic double-peaked pattern of hand aperture on perturbed trials suggested that there were two hand aperture submovements coordinated with the two hand transport submovements. The first peak in hand aperture appeared to be coordinated with the abandoned hand transport submovement, while the second peak in hand aperture appeared to form part of a second submovement directed towards the second new target. In brief, the perturbation of hand transport caused by changing the apparent location of the target had an indirect effect on hand aperture, because of the coordination between these two components. This result corresponds to the second proposition above.

Haggard [12] repeated the experiments of Paulignan et al. [11] to investigate these coordinated responses to perturbation using slightly different methods. A linear actuator was used to displace a single target dowel either to the left or the right, beginning 30 ms after the subject began to move. Thus, the movements of the target were not instantaneous apparent movements, but actual physical movements lasting around 200 ms. Under these conditions, interruptions of hand transport following perturbation were seen only occasionally, perhaps because the subjects could adjust their movements more gradually to these slower perturbations. In contrast with Paulignan et al.'s [11] results, no consistent differences were found in the latency or amplitude of either the forward or tangential peak velocities and accelerations. Interruptions of hand aperture, such as the double-peaked grip pattern seen by Paulignan et al. [11], were also uncommon. Instead, Haggard's data [12] typically showed a single peak in hand aperture on perturbed trials, which was significantly greater than the control value.

To avoid the difficulties of calculating latencies from differentiated data (see above), the latencies of hand transport adjustments were measured directly from the spatial paths, using a new method which does not depend on ***judgement by eye***. First, the X and Y positions of the thumb marker on each subject's control trials were spatially resampled, to show the spatial variability at successive one centimetre ***slices*** along the start-target axis (Fig. 3**A**).

Figure 3 (facing page). Estimating adjustment latencies. ***A****. The spatial path of the thumb in control trials is resampled every centimetre of forward movement.* ***B****. The random lateral deviation appears normally distributed.* ***C****. Mean and confidence intervals for the thumb's spatial path can thus be found.* ***D****. Hand transport adjustments on perturbed trials occur when the lateral deviation exceeds the confidence interval on control trials.* ***E****. The same method applied to the spatial relation of hand aperture with hand transport shows significant increase in hand aperture on perturbed trials.*

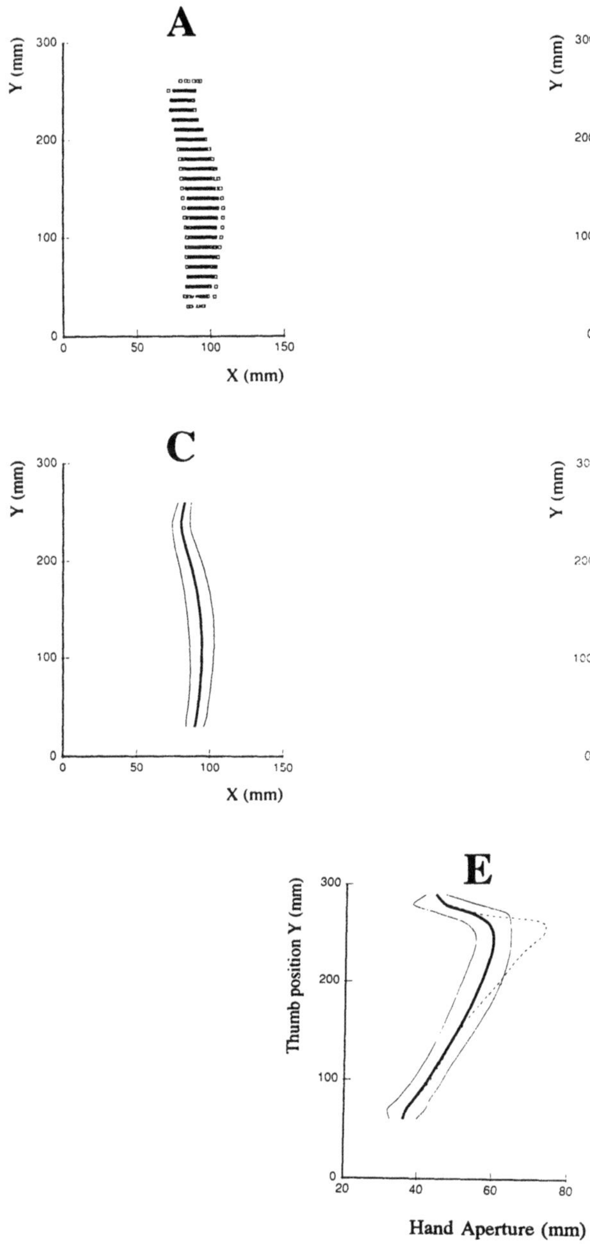
A
Y (mm)
300
200
100
0
0
50
100
150
X (mm)
B
Y (mm)
300
200
100
0
0
50
100
150
X (mm)
C
Y (mm)
300
200
100
0
0
50
100
150
X (mm)
D
Y (mm)
300
200
100
0
0
50
100
150
X (mm)
E
Thumb position Y (mm)
300
200
100
0
20
40
60
80
Hand Aperture (mm)

The X component was approximately normally distributed (Fig. 3**B**). Therefore a mean spatial path of the thumb, and 90% confidence intervals for the lateral variability can be calculated (bold and thin lines in Fig. 3**C**, respectively). Perturbed trials can then be superimposed upon the spatial distribution of the control trials (Fig. 3**D**). When the spatial path of a perturbed trial leaves the confidence intervals for the control trials, the occurrence of a motor adjustment can be inferred, with a known probability of a type I error (0.1 in this case). The latency of the adjustment can then be calculated by returning to the original waveforms. A similar method was used to detect the increased hand aperture following perturbations, plotting the hand aperture as a function of the forward progress of the hand (Fig. 3**E**). This method gives estimates of adjustment latency of 217 ms for hand transport (thumb marker, grand average of 4 subjects), and 206 ms for hand
aperture – some 100 ms longer than those obtained by Paulignan et al. [11]. In fact, this method tends to overestimate the actual spatial variability of control movements, because of the unwarranted assumption that the spatial distributions at successive one centimetre ***slices*** are independent. It thus tends to increase estimates of adjustment latencies. Use of an alternative measure of variability based on Generalised Procrustes Analysis [13], which does not assume independent slices, suggests that this conservative bias contributes only around 25 ms to the latency estimates. Some of this difference between Haggard and Paulignan et al.'s results is probably due to the slower perturbations used in the former study, but further research is required into the latency of such adjustments, and into methods of measuring them. There was no significant correlation between the latencies of the hand transport and hand aperture adjustments on individual trials in Haggard's experiment, despite the similar means, suggesting that the adjustment did not involve a second submovement with a temporal pattern common to both components. Nevertheless, the hand aperture adjustment might be coordinated with other features of hand transport, such as the remaining distance to the new target.

These target displacement experiments have also demonstrated a new aspect of coordination of hand aperture with hand transport. In normal prehension, flexion and extension of the index finger provides most of the modulation of hand aperture with the thumb making a smaller contribution [14]. In perturbed left trials, however, the thumb made a much larger abduction so as to grasp the target object while the index finger remained relatively static. The converse was the case for perturbed right trials. That is, the "ipsiperturbational" effector made a greater contribution to the grasp in the perturbed trials. These results are illustrated in Fig. 4 which shows the typical thumb and finger angles in each condition.

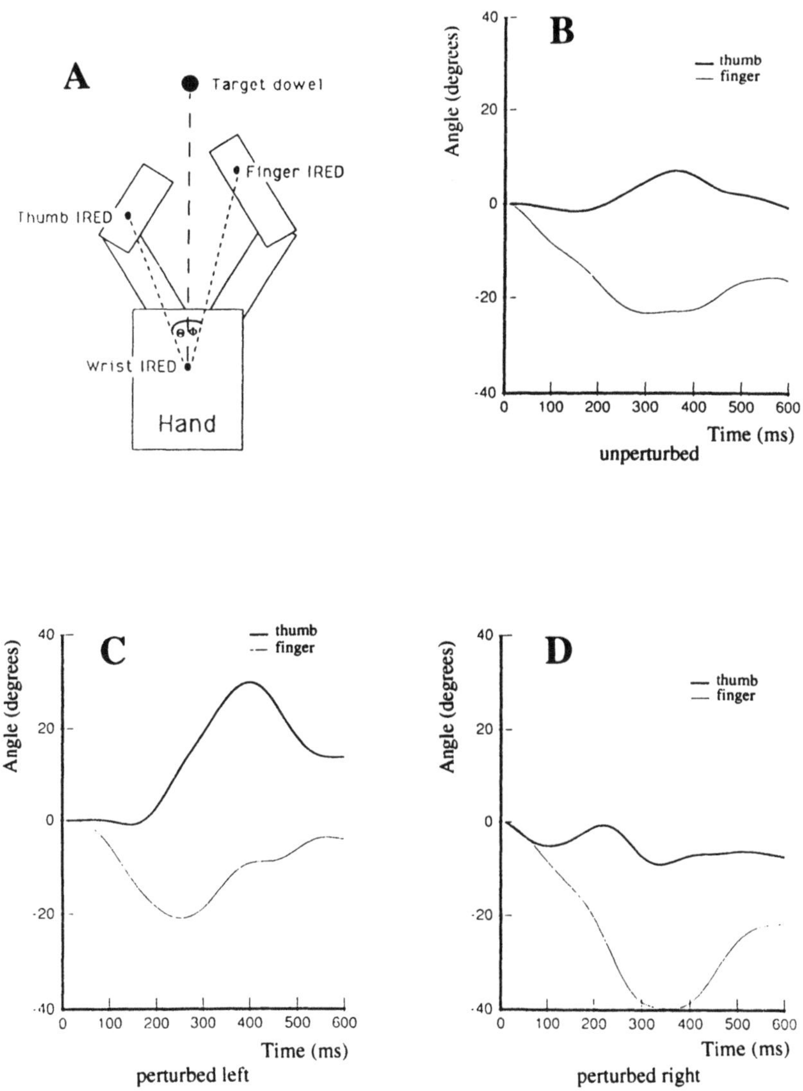

Figure 4. Modulation of grasp configuration with direction of adjustment. ***A****: calculation of finger and thumb angles.* ***B, C, D****: waveforms showing change in finger and thumb angles in typical unperturbed, perturbed left and perturbed right trials respectively. Initial angles have been subtracted in each case.*

Analysis of variance for the mean absolute deviation of finger and thumb from their initial values confirmed this pattern in the index finger. Thumb movement data was more equivocal: although the overall F ratio was significant for each subject, there were variations between subjects in the amount of thumb movement in each condition.

The low adjustment times for the hand transport component reported by Paulignan et al. [11] are particularly interesting, since they are less than some recent estimates of visuomotor reaction time, which suggest a figure of around 200 ms [15]. The latency for hand aperture adjustments coordinated with these changes in hand transport, however, was closer to traditional estimates of RT, at around 200 ms. Paulignan et al.'s estimate for hand aperture was corroborated by Haggard's experiment, despite the different methods and the different form of the hand aperture adjustment observed. In a further study, Castiello et al. [16] investigated the origin of the more rapid adjustments to hand transport. They replicated the estimate of around 100 ms, and also found that subjects took 420 ms to make a vocal response to the same perturbations of target location. Assuming that the vocal response required subjects to have a perceptual awareness of the perturbation, Castiello et al. [16] suggested that the advantage that motor adjustment times show over vocal response times may reflect operation of a subcortical processing pathway for the control of movement whose operation need not enter into subjects' conscious experience. Indeed, their subjects reported perceiving the illumination of the objects to change as they were about to grasp them, although the change actually happened at the very start of the movement. On this argument, the coordination between hand aperture and hand transport must also be implemented at a low level of processing, since the hand aperture adjustment to displacements of the target is also substantially faster than the vocal response.

Two recent experiments by Paulignan et al. [17] and Castiello et al. [18] have investigated the effects of perturbing the size of the object to be grasped. This was achieved by mounting a narrow dowel in the centre of a wide dowel, and interchanging the illumination of the two, as for the perturbation of object location. Both studies found that hand aperture increased following a change in illumination from the small to the large dowel, and decreased following a change in the opposite direction. There was a concurrent interruption of hand transport, which led to a significantly lower latency for the peak deceleration of the hand in perturbed trials. These adjustments occurred after about 300 ms, considerably later than the adjustments following perturbation of object location. However, some aspects of the hand aperture response appear to be much faster. In a more recent study of perturbations of object size, Castiello et al. [18] allowed subjects to change grasp configuration from precision grip to whole-hand prehension as they wished, and also recorded the movement of all the digits. When subjects reached to grasp with the whole hand a large dowel, which was then perturbed to a smaller object, they found that the index finger flexed away from the other digits to form a precision grip after only 174 ms.

It is difficult to draw conclusions from the large range of latency estimates arising from these experiments. In particular, it is difficult to compare the speed of the hand transport and hand aperture pathways because of the different methods to measure latencies in each channel. In particular, analyses of variance of salient kinematic parameters such as peak acceleration, grip aperture and so on, may give different results from methods which detect when the continuous movement waveform exceeds some limiting value calculated from control trials. Future research is needed to devise suitable methods for detecting motor adjustments, whose efficiency can be estimated statistically. However, the present state of the field suggests that perturbations of object location elicit rapid adjustments of hand transport, and coordinated adjustments of grip pattern at comparable latencies. Perturbations of object size appear to elicit adjustments of hand aperture with a greater latency. They also appear to slow down the final phase of hand transport, but the extent to which these two events are coordinated is unclear. Why might adjustments to perturbations of location be faster than adjustments to perturbations of size? It seems likely that the explanation is perceptual, rather than motoric. The proximal muscles involved in redirecting the trajectory of the hand to a new target location may enjoy a slightly lower conduction time from the motor centres in the brain than the digits do, but they also have much greater inertia, because of their greater mass. On the other hand, the visual stimulus provided by a change in object location is very salient, and may be strongly represented in fast subcortical visual pathways, such as those involving superior colliculus, which are known to be involved in producing rapid orienting responses towards changing targets. Changes in object size might involve slower, visual cortical pathways which represent the shape and intrinsic features of visual objects. It would be useful to measure the adjustment latencies of hand transport and hand aperture to these perturbations having removed these visual differences; for example with blindfolded subjects who learn to associate different auditory stimuli with perturbations of location and of size (see Bennett et al. this volume).

MECHANICAL PERTURBATIONS

The final kind of perturbation experiment discussed in this paper involves mechanical perturbations to either the hand transport component or the hand aperture component during the course of movement. These studies involve perturbation at a lower level of control than the studies which alter the perceptual goals of the task. If a mechanical perturbation to one component results in a coordinated adjustment by the second component, then the second component must receive information about the state of the first component as the movement progresses. Coordinated adjustment to mechanical components, then, suggests a continuous closed-loop coordination between the state of the two components.

Rather few studies have used mechanical perturbations to study coordination in reach and grasp movements, perhaps because of the technical difficulties involved. However, Haggard and Wing [19] delivered mechanical perturbations to the hand transport component of movement by attaching a force-servoed actuator to the subject's upper arm. In normal operation, the actuator followed the subject's arm movement, but on some trials it would either pull the subject's arm back towards the start point, or push it towards the target, under computer control. Fig. 5**A** and **B** shows the effect of pull perturbations on two typical trials by different subjects. Note that the perturbation (shown as a black horizontal bar) causes a reversal of hand transport (the solid line), followed by a reversal of the hand aperture trace (dashed line). Spatial plots (Fig. 5**C** and **D**) of the same movements, showing hand aperture as a function of hand transport, have a characteristic loop due to the two reversals. In the four subjects tested, 57 out of a total 139 pull-perturbation trials showed a hand transport reversal. Of these, 34 also showed a hand aperture reversal. Comparison with normal spatial plots (dashed lines) suggests that the loop has the net effect of returning the movement to the same underlying spatial relation as normal trials. Thus, the hand aperture reversals appeared to be coordinated responses to the direct effects of the perturbation on the spatial progress of hand transport. This was supported by a linear regression showing a significant relation between the magnitude of the hand aperture reversals (i.e., the height of the loops) and the hand transport reversals (length of the loops): $t\ (32) = 6.14$, $p < 0.001$. The hand transport reversals occurred on average 124 ms after perturbation onset, and the hand aperture reversals, 198 ms after perturbation onset. A further 40% of pull perturbed trials did not show a loop, but had a characteristic spatial plot like that in Fig. 5**E**. These plots can be treated as a perturbation phase, in which the forward movement of hand transport is impeded, producing a purely vertical section of the plot, and a compensation phase, in which forward movement of hand transport is resumed with little or no increase in hand aperture, producing a more horizontal section of the plot. Again, the net effect of the vertical and horizontal phases is to return the spatial relation between hand aperture and hand transport to its pre-perturbation pattern.

The push-perturbation trials delivered in the same experiment often had no visible effect on the spatial plots of hand aperture against hand transport, partly because they produced hand transport velocities which approached the technical limits of the servo-motor bandwidth. However, a few trials showed a pattern of

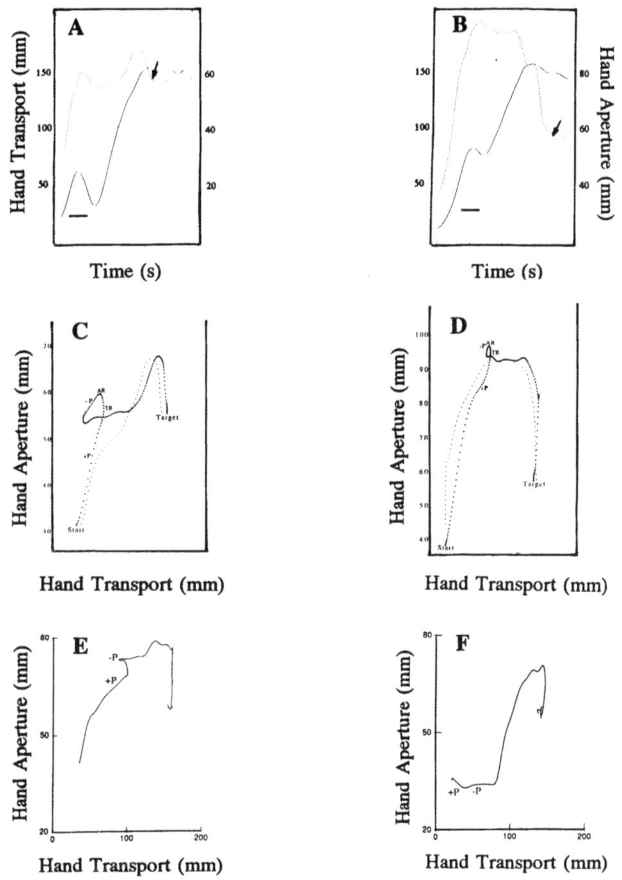

Figure 5. Responses to mechanical perturbations of hand transport. ***A*** *and* ***B****. Waveforms showing hand transport (solid line) and hand aperture (dashed line) in typical* ***loop*** *trials with a pull perturbation (horizontal black bar; abscissa 0-2000 ms). Arrow marks moment of hand's contact with target object.* ***C*** *and* ***D****. Spatial plots of hand aperture against hand transport for the same two trials. Each diamond shows a data point sampled at 100 Hz.* ***+P*** *and* ***-P*** *are onset and offset of perturbation.* ***TR*** *and* ***AR*** *are reversals of hand transport and hand aperture respectively. Thin dashed lines show spatial plot of an unperturbed movement by the same subject, for comparison.* ***E****. A typical pull perturbation trial without loop pattern. Note sequence of a vertical phase followed by a horizontal phase in the spatial plot, returning the movement to the normal pattern.* ***F****. An illustrative push perturbation trial. Note sequence of a horizontal phase followed by a vertical phase, in contrast with* ***E****.*

modulation of the spatial plot which is exactly the converse of that seen for pull perturbations (Fig. 5**F**). In these trials, the perturbation phase involves a rapid forward movement of hand transport with minimal increase in hand aperture, producing a horizontal portion of the plot immediately after the perturbation. This is followed by a more vertical portion, in which hand aperture increases very rapidly during just a few centimetres of forward hand transport. Again, the net effect of these two phases returns the spatial relation between the two components to an underlying spatial plot seen in normal trials.

The mechanical perturbation experiment suggests that hand aperture and hand transport are coordinated so as to maintain a consistent spatial relation between the two components. This evidence contrasts with Jeannerod's original view [20] of independent visuomotor channels for hand transport and hand aperture, which share only a loose temporal coupling. How does this spatial coordination arise? Is it a purely mechanical coupling between proximal and distal musculature, or does it reflect transfer of information about the state of one visuomotor channel to the other, with some active process in the motor system coordinating the two? Two pieces of evidence suggest the spatial relation is computational rather than mechanical in origin. First, the mechanical perturbations have a delayed effect on hand aperture, whereas a mechanical linkage would operate almost instantaneously. Second, a control experiment, in which force pulses were delivered to the static arm without giving subjects any explicit instructions about hand aperture, showed the opposite effect to that seen in the main experiment. That is, pulling the static arm produced small increases in hand aperture, while pushing the static arm produced small decreases in hand aperture. The spatial patterns seen during reaching and grasping thus appear to involve a psychological process which coordinates the spatial state of hand aperture with the spatial state of hand transport during the course of a single movement. The coordination is not instantaneous, but involves a delay comparable to traditional estimates of proprioceptive reaction time.

CONCLUSION

This chapter has described how perturbation experiments have measured people's responses to unexpected changes in the parameters of a coordinated reaching and grasping movement. These experiments have typically perturbed the hand transport component of the movement, and observed compensatory adjustments in the hand aperture component of the movement, suggesting a tight coupling between these two components. The coordination appears to have both temporal aspects (i.e., the two components share a common timing device) and spatial aspects (i.e., the hand aperture component receives information about the state of the hand transport component, and adjusts accordingly). Finally, two issues are likely to be particularly important in future research. The first is to clarify the latency of adjustments in hand transport and in hand aperture, by developing systematic methods of detecting the onset of motor adjustments. The second issue involves devising new methods to perturb the effectors involved in hand aperture, to investigate the interesting possibility that hand transport is coordinated with hand aperture during the final preparation for grasping.

ACKNOWLEDGEMENTS

I am grateful to The Wellcome Trust, whose financial support in the form of a Prize Studentship and a Post-Doctoral Fellowship, made this work possible. I would also like to thank Alan Wing, Ian Nimmo-Smith, David Lee, John Stein, Chris Miall, Keree Bennett, Umberto Castiello and many others for help and advice with this work.

REFERENCES

1 J.A.S. Kelso, B. Tuller, E. Vatikiotis-Bateson, and C. Fowler, J. Exp. Psychol., Hum. Percep. Perf., 10 (1984) 812.
2 J.H. Abbs and C.J. Winstein, in: M. Jeannerod (ed.) Attention and Performance XIII, Erlbaum, Hillsdale, 1984, p. 627.
3 S. Athènes and A.M. Wing, in: S. Wallace (ed.) Perspectives on the Coordination of Movement, North-Holland, Amsterdam, 1988, p. 285.
4 A.M. Wing, in: D.E. Meyer and S. Kornblum (eds), Attention and Performance XIV, Synergies in Experimental Psychology, Artificial Intelligence and Cognitive Neuroscience, MIT Press, Cambridge, 1992, p. 709.
5 H.-L. Teulings and F.J. Maarse, Hum. Mvt. Sci., 3 (1984) 193.
6 M.I. Posner and S.W. Keele, in: Proceedings of the XVIth International Congress of Applied Psychology, Swets and Zeitlinger, Amsterdam, 1968, p. 418.
7 M.I. Posner and S.J. Boies, Psychol. Rev., 78 (1971) 391.
8 D.E. Broadbent, Acta Psychol., 50 (1982) 253.
9 P. Haggard, J. Mot. Behav., 23 (1991) 25.
10 A.T. Winfree, Scientific Am. Mon. Series, 1980.
11 Y. Paulignan, C. MacKenzie, R. Marteniuk, and M. Jeannerod, Exp. Brain Res., 83 (1991) 502.
12 P. Haggard, The Coordination of Human Prehension, Unpublished Doctoral Thesis, University of Cambridge, 1991.
13 P. Haggard and J. Richardson, in preparation.
14 A.M. Wing and C. Fraser, Q. J. Exp. Psychol., 35A (1983) 297.

15 P. McLeod, Perception, 16 (1987) 49.
16 U. Castiello, Y. Paulignan and M. Jeannerod, Brain, 114 (1991) 2639.
17 Y. Paulignan, M. Jeannerod, C. MacKenzie and R. Marteniuk, Exp. Brain Res., 87 (1991) 407.
18 U. Castiello, K.M.B. Bennett and G. Stelmach, Exp. Brain Res., 94 (1993) 163.
19 P. Haggard and A.M. Wing, Neurosci. Lett., 122 (1991) 103.
20 M. Jeannerod, in: J. Long and A.D. Baddeley (eds.), Attention and Performance IX, Erlbaum, Hillsdale, 1981, p. 153.

Insights into the Reach to Grasp Movement
K.M.B. Bennett and U. Castiello (Editors)

CHAPTER 9

VISION AND THE REACH TO GRASP MOVEMENT

K.M.B. BENNETT[1], C. MUCIGNAT[2], C. WATERMAN[3] and U. CASTIELLO[4]

[1] *Dipartimento di Psicologia Generale, Universitá di Padova, Piazza Capitaniato, 3, 35139, Padova, Italy.*
[2] *Istituto di Fisiologia Umana, Universitá di Parma, Via Gramsci 14, 43100, Parma, Italy.*
[3] *Department of Exercise Science and Physical Education, Arizona State University, Tempe, Arizona, U.S.A.*
[4] *Dipartimento di Psicologia, Universitá di Bologna, Viale Berti Pichat, 5, 40127, Bologna, Italy.*

SUMMARY

This chapter reviews kinematic studies which have assessed the patterning of the reach to grasp movement of normally sighted subjects with imposition of visual restrictions. It then presents kinematic assessments of the reach to grasp movement of blind subjects. These latter studies aim to determine how movement patterning is adjusted according to intrinsic and extrinsic object properties under both non-perturbed and perturbed conditions. It is demonstrated that without experience of vision the reach to grasp movement of blind subjects has inherent functional characteristics which are also common to full vision subjects. The benefit of vision lies in giving speed confidence. However, vision is not a necessary agent for such characteristics as coordination between the transport and manipulation components, scaling of grip aperture or suitable responses to perturbations of target size and location. Results for blindfolded subjects suggest that these subjects are acting in an unfamiliar feedback climate - restriction of visual feedback thus assesses their adaptation to this different environment rather than the role of vision for the reach to grasp movement.

INTRODUCTION

In studies of the reach to grasp movement much emphasis is placed upon the importance of vision. According to Jeannerod, "The reaching .. component .. reflects determination by the visual system of the coordinates of a point in a body-centered space. The grasping .. component .. reflects visual computation of shape, size and weight of the object" ([1] p. 235). The pathways responsible for transport and manipulation are each described as individual ***visuomotor*** channels [2]. Similarly, in a hypothetical *coordinated control program* for reaching to grasp an object, Arbib [3] inserts ***visual input*** at the various processing stages. Prior to movement, this input helps to determine location, size and orientation of the target object; during movement execution it continues to influence both the transport and manipulation components.

However, the number of research studies which have investigated the role of vision specifically in relation to the reach to grasp movement is low. In the following section these studies will be described and an outline of how each has contributed to our understanding of this topic will be presented.

Jeannerod [1] had three visual feedback conditions in his prehension study with seven normally sighted subjects. For one (*visual-feedback*), subjects were able to see both movement of the reaching limb and the object to be grasped (except for the first 9 cm or approximately 125 ms in the case of a movement with a velocity of 70 cm/s). In another, (*no-visual-feedback*) the subject was unable to see the reaching limb but could see the object to be grasped (i.e., a reflection of this object). For the third condition (*no-vision*), subjects saw the object to be grasped prior to movement onset. Once movement was initiated the room lights were extinguished and the movement was performed in the dark. In addition to these different visual conditions, object size and the distance of the object from the subject were also varied. To an extent, these latter variations confound the results. For example, comparison between the two subjects who were tested under the *no-vision* condition while reaching either 25, 32 or 40 cm and the two subjects who were tested under the *no-visual-feedback* condition when reaching for objects of different sizes is complicated by the effects of intrinsic and extrinsic object properties upon movement kinematics (see Weir this volume). However, clear results, particularly for the *no-visual-feedback* (all subjects) became evident. For the transport component, movements under this latter condition were often dysmetric: at the end of the movement the hand was one to two centimetres short of the object and final prehension was achieved after a tactile-driven reshaping of finger posture. Movements performed with full vision were of longer duration than those without and they showed a longer low velocity phase (i.e., the phase after peak reach velocity whereby tangential velocity becomes constant). Despite these differences according to visual environment, Jeannerod, nevertheless found that a certain coordination between the manipulation and transport components was maintained: " .. the duration of the finger closure phase .. remained .. correlated to that of the low-velocity phase .." ([1] p. 248). Because of this intercomponent relationship in cases of both full vision and no vision, Jeannerod hypothesized that "the timing of the two components is achieved by a centrally generated pattern and that this pattern remains uninfluenced by feedback control" ([1] p. 250). Visual feedback was thus not implicated in ensuring that the timing of maximum hand opening occurred at a certain important stage of the arm reaching movement. However, feedback of the reaching limb did appear to play a role in the appropriate placement of the hand in relation to the target object. As proposed by Jeannerod, the presence of this feedback may have added to the processing requirements and this would explain the longer movement durations under full visual conditions.

Given that Jeannerod [1] found a form of coupling between the transport and manipulation components, Wing et al. [4] hypothesized that errors of execution in one component should be compensated for by changes in execution of the other. They proposed that one means of forcing errors in the transport component was by excluding visual feedback (eyes closed) during a reach to grasp movement. Indeed, under this condition it was found that the trajectory pathway of the reaching limb showed greater variability when compared to movements performed at normal velocity and with full vision. Compensatory strategies in the manipulation component were suggested by a greater and an earlier maximum hand aperture. These strategies possibly served to increase the chance of contact with the object despite transport variance. However, a number of questions can be raised in relation to this study. It is not clear, for example, why deprivation of visual feedback should exclusively lead to errors in the transport component. The increased hand aperture could reflect an independent effect upon the manipulation component rather than something that countervails increased trajectory variability. Another question relates to the previous visual experience of the subjects. Prior to the fourth block of non-visual trials the subjects performed three blocks of trials with full vision. It is thus difficult to dissociate results due to lack of vision from those due to practice and learning effects. Changes in either component could reflect movement plans based on the memory of immediately preceding task performance with vision.

A third kinematic analysis of the effects of visual feedback on the reach to grasp movement was provided by Jakobson and Goodale [5]. In their study, subjects were required to reach for objects of different sizes which were placed at different reaching distances. For one experiment and using a counterbalanced design, a non-visual condition was compared to a visual condition. Under the former, the room light was extinguished approximately 80 ms after trial initiation. The results confirmed the findings of Wing et al. [4] in that the maximum hand aperture was greater and earlier under the non-visual condition. The finding of this earlier timing, together with longer movement initiation times, prompted the authors to question Jeannerod's conclusions that visual feedback was not influential in the patterning of or in the coordination between the two components. Their view was that vision gave important premovement and on-line information in order both to program the movement and to improve its on-line precision. However, in some respects and as indicated by their results, this position could be reexamined. Firstly, even without vision, the patterning of the manipulation component varied according to intrinsic object properties: maximum grip aperture was positively related to object size. Clearly vision was not necessary for a scaling of pre-grasp aperture. This pattern was also evident in a second experiment by these authors whereby non-visual trials were interspersed and randomized with visual trials. Subjects were thus prevented from predicting when visual feedback would be available. Scaling of hand aperture according to object size was again observed for both visual and non-visual trials. The coding for a suitable adjustment of the grasp-to-object ratio thus appeared to be of central origin.

A second reason as to why their position could be questioned relates to the coordination between the transport and manipulation components. Under the non-visual conditions (and for the unpredictable visual condition) both the maximum height in the trajectory of the wrist (a transport parameter) and the maximum grip aperture (a manipulation parameter) were proportionately earlier than for predictable visual trials. This would suggest that a form of temporal coupling between the two components was indeed maintained despite the absence of vision, although correlation analysis would best confirm this. If coupling is still present this would confirm rather than negate Jeannerod's central origin for intercomponent coordination. Both components show earlier temporal settings for important functional landmarks preceding the phases of final arm approach to and hand closure upon the target. That is, the absence of vision may provoke compensatory mechanisms, such as the larger hand aperture and longer approach/closure times, to increase the chance of successful performance. These mechanisms, too, could be of central origin rather than an indication of the role of visual feedback.

A central origin is also supported by the results of the second experiment of Jakobson and Goodale [5]. This showed, in summary, that when non-visual and visual trials were presented such that neither could be predicted, the patterning of the visual trials resembled that of the non-visual trials. The presence of vision did not lead to later temporal settings either for maximum grip aperture or for the maximum height in the wrist trajectory. This would suggest that the central strategy of a movement without vision had already been 'decided upon' and that visual input was discarded at this programming level and more or less ignored during movement execution. This alternative method of performing the movement would appear to be one that is relatively reproducible when deployed and one than ensures a functional motor equivalence. It could be indicative of a strategy whereby a controller directly monitors disturbances to a system, that is, visual deprivation, "and immediately applies compensatory signals to the controlled system, rather than waiting for feedback on how the disturbances have affected the system." ([3] p. 1466).

Sivak and MacKenzie [6] approached the question of the role of vision from a different viewpoint. They were interested in determining whether feedback from *peripheral* vision had effects upon movement kinematics which differed from those from *central* visual feedback. Peripheral vision was isolated using a contact lens upon which was mounted an opaque lens to block the central ten degrees of visual field. Central vision was isolated with swimming goggles which were painted black except for small apertures at the front to permit vision in the central ten degrees of field. While maintaining a stable head position (chin rest) subjects were required to reach for a small diameter (2.5 cm) target object. In summary, the results indicated differential effects. With only peripheral vision both the transport and manipulation components were affected. For the transport component, movements were of lower velocity and the wrist stopped moving at around 72% of the movement (following a phase of near zero acceleration). This meant that hand closure about the object was tactile driven - a result similar to that found by Jeannerod [1] in his no-visual feedback condition. For the manipulation component, maximum grip aperture was greater, as was the aperture of the hand at the time when the wrist stopped moving. With only central vision, only the transport component showed effects. Subjects consistently underestimated the distance of the dowel and consequently continued the movement with constant deceleration until successful contact. This meant that the time from peak deceleration until the end of the movement was greater (52%) than that under normal visual conditions (39%). The manipulation component in contrast, showed a kinematic patterning which was the same as that found with normal vision conditions.

It thus appeared that both central and peripheral vision gave on-line information which influenced transport component kinematics. This information assured a suitably rapid and metric movement without a final long slow deceleration or zero acceleration of the limb. It can be suggested that, particularly for the approach stages of the transport movement, vision helped to ensure reaching distance accuracy and a non-hesitant deceleration. The arm could complete the movement confidently without having to proceed cautiously until an almost 'perchance' encounter with the target.

Only central vision seemed to contribute to the successful performance of the grasp component. As stated Sivak and MacKenzie ([6] p. 254), " .. subjects based the organization of the grasp component on visual information received from the dowel .. ", having had almost no visual feedback from the moving limb or from the spatial area within which this limb moved. Intrinsic object properties gleaned from central visual feedback were sufficient enough information for appropriate spatial and temporal scaling of the hand opening and closing.

Another interpretation of these results is that a particular compensatory strategy was recruited in the absence of sufficient visual feedback. This was evidenced, for example, by the lack of accommodation effects over the 16 experimental trials with only peripheral vision. The same movement design was executed for each trial. Rather than reflecting a pattern which was determined by the available visual feedback, this may have been an alternative central schema which is discharged in cases of insufficient visual information. With this in mind it is interesting to note that in all the previously mentioned studies, findings of temporal lengthenings to the final arm approach stage and of greater hand apertures under no or restricted vision were almost consistent. Other results from the Sivak and MacKenzie study [6] which could point to a central compensatory strategy are those obtained when subjects were required to reach for and grasp objects of different sizes (1, 2.5 or 5.5 cm diameter) with peripheral vision only. As for the Jakobson and Goodale study [5], maximum grasp aperture was greater for the largest than for the smallest diameter object, suggesting again that a coding for the relationship between object size and spatial characteristics of the grasp component is predetermined. The only hitch to this argument is their finding that the relationship was not as monotonic as that found under the full vision condition. The maximum grasp aperture for the 2.5 cm diameter object showed only a trend to be greater than that for the smallest object and to be less than that for the largest object rather than significant differences.

PREVIOUS VISUAL EXPERIENCE

Common to the experimental design of the aforementioned studies is some form of previous visual experience related to the task. In Jeannerod's experiment [1] subjects either saw the target throughout almost the entire movement (a restricted viewing condition) or saw the target object prior to movement initiation (a no vision condition). In the Wing et al. study [4] subjects performed three full vision blocks of trials prior to the last no-vision block. Under the no-vision condition of the Jakobson and Goodale study [5] subjects saw the target object prior to and for, at least, the first 80 ms of the movement. These visual experiences may have influenced the patterning of the following movement. As an example, Young and Zelaznik [7] found that their results for the accuracy of aimed hand movements without vision of the hand differed from those of Carlton [8]. They attributed this difference to the presence of visual information during the first 50 ms of the movement for their but not for Carlton's study. Applying this to the reach to grasp studies, it could be proposed, for example, that different results may have been found by Jakobson and Goodale [5] if the first 80 ms of the movement had provided no visual information. Similarly, the results for trials with premovement visual information of object size and distance may differ from those for trials with no premovement visual information. Using the same line of reasoning, it is also plausible that immediately preceding rehearsal of the task with visual information could influence subsequent performance without vision.

In addition to these short term impressions on movement patterning are the longer term influences. The reach to grasp movement is one performed under conditions of full visual feedback several times a day. Stating the obvious, sighted subjects have usually had an extensive and a well-rehearsed experience of the reach to grasp movement with full visual information. This experience undoubtedly affects the patterning of the movement. It is thus proposed that care must be taken in determining what is being assessed when a visual deprivation condition is applied to subjects who normally perform movements without this deprivation.

In his studies of motor learning, Proteau [9] found that a specific afferent source, particularly that of vision, plays an increasing role in the control of an ongoing movement as the movement becomes more practiced and rehearsed (see also [10, 11]). This was evidenced by a greater decrease of accuracy upon visual feedback withdrawal after an aiming task had been practised extensively with visual feedback (e.g., 2000 times) than after it had been practised only moderately (e.g., 200 times). This result for adult learning may also apply to the much longer term development of a motor skill. The acquisition of movement obviously involves optimal utilization of available sensory sources. Upon temporary withdrawal of one of these sources (e.g., vision) in adulthood the effect would probably be similar to that following extensive rehearsal, that is, some form of deterioration to movement performance. In other words, the everyday interactions with objects in a visual environment probably increases the dependency upon vision. Therefore, studies which exclude vision are in all likelihood assessing the degree of this dependency. Another way of viewing this is that full vision subjects, more than likely, have visual representations of the reach to grasp movement. In line with the view of Elliott, these representations may " .. provide an adequate substitute for direct visual contact" ([12] p. 41).

The dependency upon vision or the durability of a visual representation would be potentiated if the effect of vision upon motor function was more dominant than other sensory modalities. Researchers from other fields have commented on this idea of visual dominance. For example, Ernest, in a review of imagery and memory of blind subjects, states that "the dominance of the visual modality in the sighted may limit their access to the mnemonic potential of other sensory modalities, such as the auditory and tactual-kinaesthetic" ([13] p. 233). In a defense of the potential of haptic exploration, Landau points out that " .. one cannot reason from the supposed inadequacy of haptic exploration in the sighted to the inadequacy of haptic exploration in general. One can only draw conclusions about how the haptic system functions in the absence of visual feedback guidance" ([14] p. 348). With reference to navigation ability, Loomis et al. proposed that " .. prior visual experience has consequences for tasks performed without vision" ([15] p. 75). If these lines of reasoning are applied to the motor system it can, for example, be hypothesized that the subservience of the motor system to visual feedback could limit the ***motor potential*** contributed by other feedback systems. Similarly, inadequacies in the performance of the reach to grasp movement without the usual visual feedback probably cannot entirely reveal the role of the visual system for this task. Rather they would reflect an adaptation of motor output to the absence of a wonted on-line input.

STUDIES OF BLIND SUBJECTS

Blind subjects have an obvious difference in the sensory means by which they confront the world. However, without ever having had the experience of vision, they develop refined motor skills such as that of reaching to grasp objects. The study of blind subjects thus provides an additional means by which to clarify the role of vision for the reach to grasp movement. Indeed, a number of authors have suggested that data from this subject group is needed. For example, in his book **Movements of the hand**, Charles Phillips asks "Would the initial exploratory location of the target", as performed by blind subjects, "set up a proprioceptive transportation-and-manipulation programme which would operate accurately first time round?" ([16] p. 19). Sivak and MacKenzie quite justly point out that "Congenitally blind infants learn to reach and grasp objects" ([6] p. 256) and raise several research questions which would be pertinent for this group. Although kinematic studies of blind subjects are lacking there have been some attempts to describe the upper limb functioning of this group. In the following paragraphs a short review of these descriptive analyses of blind children and adults is presented.

Blind infants begin reaching to sounding objects at 9 - 12 months of age [17, 18] and use what has been described as a characteristic ***groping*** movement [19]. Reaching proficiency is attained later, sometimes well into the second year of life [17]. Fraiberg [20] describes blind reaching as developing through two stages. In the first stage, the reaches are for sounding objects that are taken from the infants hand. In the second stage, the reaches are for sounding objects presented in the immediate vicinity and for those with which the infant has had no previous contact. The developmental onset of reaching movements to sounding objects by blind children tends to be later than that onset to visual objects by sighted children [17, 20]. However the auditory-manual and visual-manual coordination systems may show a differential development, the former developing later, in any case, also for sighted subjects [18, 21, 22]. According to Schwartz [23] it may be gains in attention span, which enable blind children to process the complexity of a sound and to respond to its location, that mediate the onset of this reaching to sounds.

Blind subjects are obviously practiced in coping with haptic information. As toddlers, they demonstrate an impressive repertoire of upper limb exploratory activities [14]. These activities resemble those of sighted children and are systematic and related to the information extracted. For example, to determine the overall configuration of an object the blind infant uses such manipulations as rotation and whole hand grasp, while for the detection of surface properties the manipulations include scratch and rub. The practiced use of such exploratory activities may contribute to the advantage of blind children over sighted for the haptic modality. This advantage was shown by Millar [24] who found that blind children of less than 16 years of age had shorter response latencies in a yes-no tactile comparison task of three-dimensional nonsense shapes (see also [25]). Similarly, Davidson et al. [26] found that blind adolescents used a more efficient haptic strategy to identify an unfamiliar form. For example, when compared to blindfolded sighted subjects they were better able to select a three-dimensional unfamiliar form from five comparison forms. Blind subjects are also more adept at tactually detecting transformation of a rod from a straight to a curved alignment and the way in which they detect subtle curvatures is characteristic of this subject group [27, 28]. Unlike sighted subjects who scan a rod by sweeping along it with the pad of one finger or by pinching it between the index finger and thumb, blind subjects use a ***gripping*** scanning strategy. With this strategy, three to four fingers are used to simultaneously explore numerous points along the rod. The efficiency of such a global appraisal in the detection of rod curvature was demonstrated by the improvements in sighted subjects' performance when they too used the ***gripping*** technique. Overall, Ernest [13] feels that the available evidence suggests an enhancement rather than an equivalence when comparing the blinds tactually derives memories of shape with those of the sighted. Similarly, Axelrod [29] and Heinricks and Moorhouse [30] feel that blind subjects show a superiority for the sensory modality of touch.

KINEMATIC ANALYSES

Castiello et al. [31] provided a kinematic analysis of the reach to grasp movement of blind subjects. Four congenitally blind subjects (no residual vision; age 25 - 40 years) were compared to age, gender and weight matched control subjects. In one control group (n = 4) subjects were blindfolded before entering the experimental room. In this manner they had no immediate previous visual experience of the experimental setup. The other control group (n = 4) were not subjected to any visual constraints, that is, they performed the reach to grasp tasks under normal room-lit conditions and with full vision both of the upper limb and of the target to be grasped. In summary, the comparison was between three groups of subjects who had all had extensive experience at performing the reach to grasp movement. One group was accustomed to performing the movement without vision, the other two, with vision. It is thus probable that blindfolded subjects would be more *stressed* by the experimental demands of executing a motor output within an unaccustomed visual feedback environment.

One aim of this study was to assess the patterning of the transport and manipulation components with variation of intrinsic and extrinsic object properties. The intrinsic property of size was varied by using either a small diameter (0.7 cm) or a large diameter (6 cm) target object (perspex cylinder). The extrinsic property of distance was varied by placing the target at either 20 cm or 30 cm from the starting position. Blind and blindfolded subjects were given ample opportunity to haptically explore the target objects. In addition, the experimenter passively moved each subject's arm to the target object and passively opened and closed the hand around the object. Prior to each block of trials each subject also performed practice trials. Such extensive training of the experimental task was conducted with the aim of reducing motor learning effects and later analysis, in fact, revealed that there were no such effects (e.g., no difference between first and last trials of a block of ten trials).

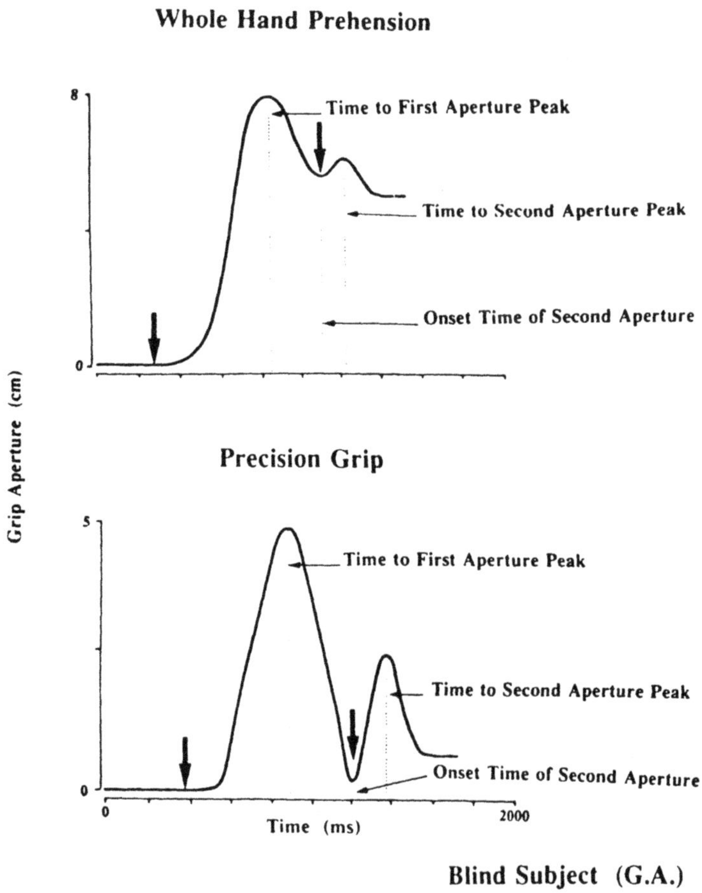

Figure 1. The double manipulation component of a blind subject. Each trace is of a single trial. Measured temporal parameters are indicated.

The kinematic analysis, using an ELITE system [32], revealed one feature which was unique to blind subjects. This was a double opening and closing sequence between the index finger and thumb as illustrated in Fig. 1. That is, the blind subjects opened the hand initially (reaching a peak of maximal aperture at an average of 40% of the total movement duration) began to close it and then opened it again (at an average of 56% of movement duration) to reach a maximal aperture at an average of 70% of the movement duration. Such a pattern was consistent, that is, it was observed for both object sizes, and thus for both precision grip and for whole hand prehension (Fig. 1), and at both reaching distances. Blindfolded and full vision subjects did not demonstrate this double-grip.

The transport component of blind subjects also showed peculiar features although some of these were also evident for blindfolded subjects. With the 30 cm reach of blindfolded subjects but with the 20 cm reach of blind subjects the reach movement showed a low velocity phase (LVP; i.e., a plateau on the falling edge of the arm velocity peak; 85% and 90% of trials respectively). With the 30 cm reach of the blind subjects the velocity profile was doubled - that is, the initial velocity peak was followed by a complete rise to and fall from a second peak. Such features significantly increased the movement time with respect to that of the full vision subjects.

An initial impression of these results is that the experience of vision is related to a singularized reach to grasp movement. However, this conclusion can only be applied to the manipulation component as previous studies have demonstrated both LVPs and submovements under conditions of full visual feedback [1, 33, 34, 35]. In contrast, there have been no previous reports of doubled manipulation under non-perturbed conditions. A remaining query relates to whether or not the double grasp opening and closing is a predetermined feedforward mechanism or one that results from not having on-line visual feedback.

A number of observations suggested the former, that is, that the doubling is inherent and preprogrammed. For example, the first peak of hand aperture was earlier than the single aperture peak of the other subjects (blindfolded and full vision). This anticipation could reflect a strategy for augmenting the time available for the second opening and closing sequence; the relative timing of the second maximum grip aperture (70%) allowing approximately 30% of the movement for the final hand closure phase. Evidence for some form of programming focus on the second opening and closing sequence was also given by the finding of temporal coupling only for the second but not for the first sequence. For the 20 cm reaching distance, temporal correlations were found between the onset of the LVP (transport component) and the second grasp aperture peak (manipulation component). For the 30 cm distance the correlations were between the onset of the second velocity curve and the onset of the second hand opening. Thus, in concurrence with Jeannerod's [1] proposal, visual feedback did not appear to be necessary for coordination between the transport and the manipulation components.

The absence of coordination between the components for the first opening and closing sequence also suggests that this sequence was not simply a mistake, that is, because the first *try* was unsuccessful, a second attempt was implemented. If this were the case, intercomponent coordination would be also expected for the first module. Further, a consistent doubling of both components should be evident. Yet at the 20 cm distance the transport component of the blind subjects showed only a low velocity phase rather than movement doubling. In addition, the onset of the second transport movement at the 30 cm distance was not correlated to the onset of the second hand opening. Finally, if the first movement was a *mistake* it might be expected to show either improvement or inconsistency. There were, however, very few cases of blind singularized grasp movements. Nor were there different grasp patterns such as a very late first opening and closing sequence.

Further evidence of a central origin for the blind reach to grasp movement was provided by findings of similar patterning across subject groups. For example, kinematic parameterization of the blind subjects according to object size resembled that of the full vision subjects. For both object sizes the blind group showed neither a greater proportional opening nor a larger absolute hand opening than the other groups. Maximal hand aperture (first and second peak) was greater for the large than for the small object. The timing of this aperture was earlier for the small than for the large target. For example, the second peak aperture of blind subjects was at around 70% (of movement duration) for the small but at around 80% for the large target. Congruity of patterning in relation to object size was also found for the transport component. The time from the single velocity peak (all subject groups: 20 cm; blindfolded and full vision subjects: 30 cm) or from the second velocity peak (blind subjects: 30 cm) to the end of the movement (i.e., the final deceleration phase) was longer for the small than for the large target. Blind subjects thus demonstrate appropriate mechanisms for altering the duration of the approach phase according to the precision/accuracy requirements of the task.

Kinematic parameterization according to reaching distance was also similar across the three groups. For the transport component, the single (or first - blind subjects: 30 cm) peak of velocity was of greater amplitude for the 30 cm than for the 20 cm reach. The time from this peak until the end of the movement was also greater for the longer distance. Turning to the manipulation component, both the single (or first - blind subjects) and second (blind subjects only) peaks of grasp aperture were later for the longer distance. Once again, blind subjects showed adjustments to parameterization which closely resembled those of full vision subjects.

In summary, the results of this study indicated that the experience of vision was firstly not necessary for the development of an accurate reach to grasp movement. Blind subjects did not show dysmetric reaches or over-scaling of grasp aperture. The two components were activated in parallel and showed appropriate pre-contact variations. Secondly, visual experience is not necessary for intercomponent coordination. The transport and manipulation components of blind subjects were coupled in time. Thirdly, visual experience does not mediate kinematic parameterization according to task requirements such as changes of object size or of object location. It is thus proposed that reaching to grasp is largely an ***innate*** mechanism with many predetermined features. With development, movement and feedback experience then influence the moulding of this mechanism according to user requirements. For blind subjects such shaping results in a form of movement doubling. As proposed by Castiello et al. [31] this doubling may be a calculated functional division of the movement in order to avoid errors and imprecision. For example, the blind subjects obviously glean important task details by haptic exploration and with practice trials. The movement could subsequently be divided into an initial phase which brings the arm and hand to an optimal spatial or temporal stage for the second task completion phase. The information in short-term memory may assist in determining at which point the second phase must be implemented.

For full vision subjects the moulding probably leads to a dependency on vision. This was evidenced by some breakdown of the ***innate*** reach to grasp mechanism when vision was withdrawn. In particular, the blindfolded subjects showed a significant breakdown of temporal coupling between the transport and manipulation components. This was more so for more accurate tasks (e.g., reaching for a small object using a precision grip between the index finger and thumb). The coordination between the reaching and grasping components for full vision subjects thus appeared to be compromised when performing previously tested motor outputs, particularly those requiring more accuracy, in an unfamiliar visual climate. It could thus be hypothesized that temporary deprivation of vision assesses the adaptation of a motor output to an unaccustomed on-line environment. The results of such an assessment thus probably reflect the stilted implementation and coordination of rehearsed but visually feedback-dependent neural channels.

PERTURBATION STUDIES

One role of vision for the reach to grasp movement may lie in the changing of motor output under unexpected conditions. From double-step perturbation studies it has been proposed that comparative visual information about an old and a new target is used for the mediation of rapid and task-related adjustments to the motor requirement. In these studies, perturbation has often been signalled visually, for example, with perturbation of object size, by changing illumination from a small to a large target [36, 37, 38, 39].

Blind subjects can provide informative data as to the importance of vision for the execution of perturbation responses. Thus, in a series of studies we have recently assessed the ability of blind subjects (n = 5) to respond to perturbations of object size and of object location. These responses were compared to those of age and gender matched full vision and blindfolded subjects. None of these subjects had participated in any of our previous experiments. In all cases, the perturbations were signalled with auditory tones. Full vision subjects had no restrictions to visual input while blindfolded subjects were subjected to full masking of vision prior to entering the experimental room. As with previous experiments, the blind and blindfolded subjects were given ample time to haptically explore each target and to practice non-perturbed movements to each target at each location. A description of the specific perturbations and of the preliminary results follows.

Perturbation of object size

Two experiments were conducted in order to assess the effect of a perturbation of object size. In both these, the targets were translucent perspex cylinders; one of small diameter (0.7 cm) and 10 cm high; the other of large diameter (7.5 cm) and 8 cm high. The small cylinder was placed to stand vertically within the vertical centre of the large cylinder (see [36, 39]). The centre of this large cylinder was placed 35 cm from the starting position. Control trials were either to the large (40/100) or to the small (40/100) cylinder, the requirement to begin the movement being signalled by an auditory tone and the frequency of this tone indicating which cylinder was to be grasped. There were two types of perturbed trials: small-to-large (10/100) and large-to-small (10/100). These perturbations were randomly interspersed amoung the control trials and were signalled by a change of tone immediately upon initiation of the movement. In effect, the perturbation required that the subject suddenly change motor output from one to another type of grasp. For the small-to-large perturbation this change was from precision grip to whole hand prehension; for large-to-small, from whole hand prehension to precision grip (see [37, 39] for more details).

Manipulation component. By simply observing the blind subjects reacting to the perturbation, it was clear that they could adeptly and rapidly readjust motor output. This was further confirmed with the results from the kinematic analysis. In fact, many of the trials performed by blind subjects showed a parameterization which largely resembled that of full vision subjects. Rather than a doubling, the perturbed trials of blind subjects often showed only a single peak of grip aperture. Control blind trials showed an incidence of 69% of doubling but this incidence decreased to 27.5% for perturbed trials.

It is not clear why blind subjects often adopted a single movement in response to size perturbation rather than when the movement was not perturbed. The single movement could indicate that blind subjects open the hand early and then wait for contact with the object prior to a tactile driven closure. Against this argument are findings which indicate that the size properties of the object which is finally grasped influenced kinematic parameterization. For the single perturbed large-to-small trials the relative length of the final hand closure phase (from peak grip aperture to contact with the object) was, on average, 43% of the total movement duration. This was significantly longer than that allocated for the final phase of the perturbed small-to-large trials of this subject group (33%) and would suggest that the accuracy requirements of the small cylinder had been accounted for.

One proposition to explain the singled blind trials is that, in general, a greater amount of time can be allocated to final digit closure upon the *new* target. Indeed, a comparison of the relative time allocated to this phase between the single and double trials indicates that this idea is plausible. The mean percentage of time given to the closure phase in double trials was 24% of the movement duration. Using a strategy of a single opening and closing sequence, this mean increased to 38%. A single rather than a double movement to provide a longer closure phase would also presumably decrease the costs both to movement duration and to central processing requirements.

The parameter of manipulation onset (time at which thumb and finger began to open in relation to onset of transport component) also showed changes which were related to the final object to be grasped. For blind control trials to the large target, hand opening began, on average, 94 ms (7% of total movement duration) after the reach onset. With perturbation from large to small target this mean dropped (significantly) to 62 ms (5%), a value comparable to the mean for perturbed trials of full vision subjects (53 ms). For the opposite perturbation the visual and blind subjects (unlike blindfolded subjects) showed the converse finding, that is, the onset of manipulation was later for perturbed small-to-large than for control small trials. This parameter was thus modified according to the end-task requirement. As such, an earlier hand opening is not simply a general non-specific response to perturbation. In addition, the results for this parameter demonstrated that perturbation was immediately 'recognized' by the blind subjects and that appropriate outputs were rapidly mobilized. The signalling of perturbation with an acoustic signal led to blind responses which were comparable in absolute time to those found with visually mediated perturbations of object size [40] and it was clear that vision was not essential for anticipations or delays in the timing of initial hand opening.

The findings for the parameter of peak grip aperture also suggested that blind subjects can more than adequately respond to perturbation - they showed no examples of exaggerated hand opening or of overscaling. For example, mean peak grip aperture between the index finger and thumb of the single large-to-small perturbed trials was 74 mm, a value which was less than that found for the two other groups (full vision: 80 mm; blindfolded: 96 mm). Such precision of scaling was even more evident with the blind double trials, the second peak of grip aperture for the same perturbation showing a mean value of 46 mm. Comparative visual information between the small and the large target or of the changing hand configuration does not appear to be necessary for appropriate scaling of hand aperture in response to size perturbation.

The most obvious difference between blind and full vision subjects related, once again, to movement speed. Both non-perturbed and perturbed trials of blind subjects were of longer duration than those of full vision subjects. The absolute temporal settings of almost all manipulation component parameters were later for blind than for full vision subjects. As an example, the peak of grip aperture for the single perturbed small-to-large trials of blind subjects showed an average value of 831 ms; for the full vision subjects it was much earlier (567 ms). Vision would thus appear to play an important role in contributing to the speed of movement performance.

Transport Component. Given the functional/temporal/spatial coupling between the manipulation and transport components, a perturbation primarily directed at disturbing the manipulation component should, and as has already been demonstrated [36, 37, 38, 39] also affect the transport component. Perhaps the experience of vision and visual feedback is used to assist in this coupling. If so, it would be expected that blind subjects would not demonstrate changes to the transport component with perturbation of object size. For this subject group, perhaps the perturbation would force greater independence of the manipulation and the transport components. However, this was not the case: the transport component was clearly affected by perturbation of object size and it showed suitable parameterization according to the final object to be grasped.

As was found for the manipulation component, blind subjects showed examples of both singularized (one velocity peak) and doubled transport movements, however the incidence of singularized movements did not increase with perturbation. With perturbation from small-to-large the changes in both the single and double movement were late - a result also found for the full vision subjects. For example, with the single movements both blind and full vision subjects showed a trend for the peak of arm deceleration to be relatively earlier for perturbed than for control trials. Similarly, with the blind doubled trials the changes were primarily to the second rather than to the first movement, the second deceleration peak being relatively earlier for perturbed than for control trials.

A look at the (first) acceleration peak indicated that blind subjects could also execute **early** transport changes: this peak was of lower amplitude for perturbed than for control trials. As an example, with large-to-small perturbed trials the mean amplitude was 2754 mm/s^2 while with control large trials the value was 2910 mm/s^2. It must be restated, however, that the absolute temporal settings of this parameter were generally later for blind than for other subject groups. Hence it is difficult to establish whether blind subjects are capable of adjustments to the transport component which are as early as those found for full vision subjects (see also [37] for a visually mediated perturbation). Nevertheless, from these results it can be concluded that visual feedback is not necessary for acting upon the parameterization of the acceleration peak in cases of size perturbation.

Perturbation of object location

Three experiments were conducted in order to assess the effect of a perturbation of object location. In all cases the targets were three translucent perspex cylinders each 10 cm high and of 1.5 cm diameter. Each cylinder was placed to stand vertically 35 cm from the starting position, one immediately in front of the subject, and the others respectively 10° to the left and right of this central dowel. In one experiment, each subject was required to reach to grasp the central cylinder (40/50 trials) upon hearing a start auditory tone. Randomly interspersed perturbed left trials (10/50) were signalled by a change of tone at movement onset and required that the subject grasp the left rather than the central cylinder. In effect, this meant that the subject had to unexpectedly change the direction of the reaching movement. In another experiment, the control trials were the same (central cylinder; 40/50) but the perturbed trials (10/50) required that the right cylinder was grasped. These two experiments will be referred to as ***simple*** perturbations. A further experiment increased the complexity by combining perturbed left and perturbed right trials; it will be referred to as the ***complex*** perturbation. Each subject once again performed control trials to the central cylinder (80/100). Perturbation was random and was signalled at movement onset by a change of auditory tone; one tone indicating that the left cylinder should be grasped (10/100), another, that the right cylinder should be grasped (10/100) (see [41, 42] for further details).

Transport component. Again, the blind subjects showed very similar results to the full vision subjects. All groups showed a doubling of the transport component with perturbation; that is, the velocity profile of the reaching arm showed two distinct peaks. From visual perturbation studies, this doubling has been taken to indicate that the first movement to the central cylinder has been arrested and that a second to the lateral cylinder has been mobilized [41, 42].

With the ***simple*** experiments, a comparison between the single control trials to the central cylinder and the perturbed trials to either the right or left cylinder showed remarkably similar results across subject groups. In summary, all groups showed a shortening of the first movement - a strategy which was presumably undertaken in order to accommodate execution of the second movement. The similarity of results for the blind subjects once again indicated that vision is not essential for the appropriate patterning of perturbation responses in the transport component.

From the results for these ***simple*** experiments it was not possible to determine whether blind subjects could elicit early changes in the transport component. For example, the peak of acceleration was relatively (i.e., when expressed as a percentage of movement duration) earlier for perturbed than for control trials but showed no difference of absolute value. However, with the ***complex*** perturbation experiment, such early responses became manifest; the results, in fact, mirroring those found previously for visually mediated perturbations of object location with full vision subjects [41, 42]. Anticipation of the **absolute** value of the (first) peak of acceleration was found when comparing the doubled perturbed to the single control trials (except in the case of blindfolded subjects). The complexity generated by adding two possible perturbations appeared to force earlier changes in the transport component than were evident with the ***simple*** perturbation experiments, and blind subjects were just as adept in executing these changes as full vision subjects. The only obvious difference lay, once again, in the later absolute temporal setting of this parameter. For example, the first peak of acceleration for the perturbed left trials of the blind subjects was at an average of 278 ms, a value significantly greater than the average for the full vision subjects (190 ms). Hence with this experimental paradigm it can be stated that blind subjects show the same patterning in response to a perturbation of object location but it is not possible to determine whether they can execute transport component changes which are as rapid in absolute terms as those of full vision subjects.

Manipulation component. By suddenly changing the location of an object the primary disturbance is to the transport component - the arm must change the direction of its reach. However, visually signalled studies have indicated that the manipulation component also shows changes in response to a perturbation of object location [41, 42, 43]. The results of the current study indicated that these changes are also evident with a location perturbation which is signalled by different auditory tones and that blind subjects show very similar adjustments to the manipulation component as full vision subjects. Once again, this would suggest that vision is not a prerequisite for intercomponent coordination under conditions of perturbation.

Examples of a double opening and closing sequence of the hand were observed for the perturbed trials of all groups. For full vision subjects the incidence of these doubled trials (results for simple and complex experiments combined) was 38%. For blind subjects the incidence of doubled perturbed trials (49%) was considerably less than the incidence of doubled control trials (84%). Hence the effect upon the manipulation component was similar whether the perturbation was of the reaching movement or of the grasping movement, that is, in both cases there was an increase in the incidence of trials with a single hand opening and closing sequence. The reason for this greater incidence may once again relate to the length of the final hand closure: blind subjects showed a greater allocation of time to this phase for perturbed single trials (30-33%) than for perturbed double trials (19-20%).

Whether a single or a double movement was adopted blind subjects showed appropriate temporal and amplitude modification of the peak(s) of grip aperture. For example, a comparison between the single control trials to the central cylinder and the perturbed trials to either of the lateral targets showed that the first (single) peak of grip aperture was earlier for the perturbed trials. Similarly, in the spatial domain, the second peak of grip aperture, as an example, was similar for blind (mean = 49 mm) and full vision subjects (44 mm). Clearly, the parameterization of the adjusted manipulation component enabled a successful and accurate grasp of the end-target.

CONCLUSION

In themselves, the results for the blind subjects, and particularly those of the manipulation component, are very interesting and raise many questions worthy of future investigation. When reaching to grasp an object and under conditions of no perturbation, this subject group more often use a doubled opening and closing sequence of the hand. This would suggest that there is some form of functional advantage for the adoption of this mode. The basis of this advantage can, however, only be speculated upon. Perhaps a movement division enables a planning focus on end-task requirements. By using haptic memories of the object's intrinsic and extrinsic properties, blind subjects must determine what these requirements are and then calculate what needs to be done towards the end of the movement in order to fulfill these task demands. Viewed in this light, the first hand movement could serve to ***set*** the temporal and spatial starting point of the final movement.

Under perturbed conditions, and this is irrespective of whether the manipulation or the transport component is perturbed, the blind subject more often adopts a single opening and closing sequence of the hand. Presumably advantages also exist for this operating mode under perturbed conditions and one that became apparent from the results was a lengthening of the final hand closure phase. This lengthening would, for example, have the effect of increasing the time available for proprioceptive information to mediate corrective mechanisms. Another reason why this single movement is adopted could relate to the decreased processing demands: compiling and mobilizing one output pattern is presumably less taxing than the same processing for two patterns. Executing only one pattern could also decrease the costs to the speed of the movement and hence to its duration.

Whatever the reason for adopting a single movement under perturbed conditions, this mode would appear to be of less functional advantage than the doubled mode. If not, it would surely be always used under non-perturbed conditions. The only indications as to why the doubled movement was preferentially used were subtle and inconsistent and this is something that we are now trying to further elucidate. Early results suggest that, taken as a whole, parameterization of the single movement is less related to the intrinsic and extrinsic properties of the target than parameterization of the double movement.

Apart from speculating as to why the blind subject uses certain movement strategies it is also interesting to discuss how these results contribute to our understanding of the role of vision for the reach to grasp movement. As mentioned several times throughout the chapter there appear to be many functions that do not necessarily require the experience of vision or on-line visual feedback:

A) Matching of appropriate transport and manipulation component parameterization to the intrinsic and extrinsic properties of the target to be grasped. The transport component of blind subjects is modified with changes to object location; the manipulation component, with changes to object size.
B) Intercomponent coordination. Parameters measured from the transport component of blind subjects show temporal correlations with parameters measured from the manipulation component.
C) Appropriate scaling of hand aperture to target size. Blind subjects did not show examples of overscaling, their mean peak grip apertures largely resembling those of full vision subjects.
D) The rapid and suitable mediation of perturbation responses. Blind subjects quickly adapt the transport component in response to a perturbation of target location. They equally rapidly adjust the manipulation component in response to a perturbation of target size. Perturbation of one component also influences parameterization of the other component. This indicates that comparative information between old and new end-task requirements can be provided by other sensory modalities.

The only obvious advantage provided by vision was a movement 'confidence'. The velocity of blind subjects was consistently lower and movement durations were consistently longer. Some evidence indicated that this was a strategy or a habit for slowness rather than a dysfunction in the ability to move quickly. For example, under perturbation conditions the blind subjects were clearly able to mobilize the manipulation component as rapidly as full vision subjects. Most probably the slower movements reflect a safety tactic - the lower the speed of arm movement, the less physical damage in case of contact with an unexpected obstacle. However, in terms of contributing to task performance, a lower speed of arm movement would, as mentioned previously, increase the time available for proprioceptive information to contribute to corrective mechanisms.

Vision is obviously of greater importance for those who are accustomed to using it. There were several examples of kinematic parameterization which suggested that blindfolded subjects were acting in an unfamiliar feedback environment. The grip aperture was often exaggerated, intercomponent coordination disturbed and the patterning of the movement did not always match end-task requirements. In contrast to the work of

Jeannerod [1] and to that of Wing et al. [4] the movements of our restricted vision subjects were always of lower velocity than the movements of the full vision subjects. This would suggest that the immediately preceding visual experience of the reach to grasp movement at normal or increased speeds influences the velocity of the subsequent task performance. Similarly, even minimal premovement or on-line visual information about the task can give different results. In our studies, the blindfolded subjects, having been masked prior to entering the experimental room, had no specific experience of the experimental setup but obviously had more than ample prior visual occasions of performing the movement under non-experimental conditions. It can thus be proposed that using a paradigm which restricts vision during the performance of the reach to grasp movement assesses the ability of subjects to cope without vision rather than assessing the role of vision for this movement.

In conclusion, the patterning and execution of a motor pattern is similar for subjects who have no vision to those who have vision. Despite differing sensory influences, the output is similar and retains task-related characteristics. Overall, such results support the very early propositions of Jeannerod, that is, that functional characteristics of the reach to grasp movement are largely predetermined at the central processing level and that they are little influenced by visual feedback.

REFERENCES

1 M. Jeannerod, J. Mot. Behav., 16 (1984) 235.

2 M. Jeannerod and R.G. Marteniuk, in: L. Proteau and D. Elliott (eds.), Vision and Motor Control, North-Holland, Amsterdam, 1992, p. 197.

3 M.A. Arbib, in: V.B. Brooks (ed.), Handbook of Physiology, The Nervous System II, Amer. Physiol. Soc., Bethesda, 1981, p. 1449.

4 A. Wing, A. Turton and C. Fraser, J. Mot. Behav., 18 (1986) 245.

5 L.S. Jakobson and M.A. Goodale, Exp. Brain Res., 86 (1991) 199.

6 B. Sivak and C.L. MacKenzie, in: L. Proteau and D. Elliott (eds.), Vision and Motor Control, North-Holland, Amsterdam, 1992, p. 233.

7 R.P. Young and H.N. Zelaznik, Acta Psychol., 79 (1992) 59.

8 L.G. Carlton, Q. J. Exp Psychol., 33A (1981) 87.

9 L. Proteau, in: L. Proteau and D. Elliott (eds.), Vision and Motor Control, North-Holland, Amsterdam, 1992, p. 67.

10 J.A. Adams, E.T. Goetz and P.H. Marshall, J. Exp. Psychol., 92 (1972) 391.

11 J.A. Adams, D. Gopher and G. Lintern, J. Mot. Behav., 9 (1977) 11.

12 D. Elliott, in: L. Proteau and D. Elliott (eds.), Vision and Motor Control, North-Holland, Amsterdam, 1992, p. 33.
13 C.H. Ernest, in: M.A. McDaniel and M. Pressley (eds.), Imagery and Related Mnemonic Processes. Theories, Individual Differences and Applications, Springer-Verlag, New York, 1987, p. 218.
14 B. Landau, in: J. Stiles-Davis, M. Kritchevsky and U. Bellugi (eds.), Spatial Cognition: Brain Bases and Development, Erlbaum, Hillsdale, 1988, p. 343.
15 J.M. Loomis, R.L. Klatzky, R.G. Golledge, J.G. Cicinelli, J.W. Pellegrino and P.H. Fry, J. Exp. Psychol., Gen., 122 (1993) 73.
16 C.G. Phillips, Movements of the Hand, Liverpool University Press, 1986.
17 E. Adelson and S. Fraiberg, Child Devel., 45 (1974) 114.
18 A.E. Bigelow, Brit. J. Devel. Psychol., 4 (1986) 355.
19 D.W. Muir, G.K. Humphrey, P.C. Dodwell and D.E. Humphrey, in: D.H. Warren and E.R. Strelow (eds.), Electronic Spatial Sensing for the Blind, Martinus Nijhoff, The Hague, 1985, p. 299.
20 S. Fraiberg, Insights from the Blind: Comparative Studies of Blind and Sighted Infants, Basic Books, New York, 1977.
21 F.F. Perris and R.K. Clifton, Infant Behav. Devel., 11 (1988) 473.
22 D.M. Stack, D.W. Muir, F. Sherriff and J. Roman, Perception, 18 (1989) 69.
23 M. Schwartz, in: L. Lipsitt and C.K. Rovee-Collier (eds.) Advances in Infancy Research, Ablex, Norwood, 1984, p. 23.
24 S. Millar, Brit. J. Psychol., 65 (1974) 253.
25 P.J. Hampson and C. Duffy, Can. J. Psychol., 38 (1984) 411.
26 P.W. Davidson, J.K. Barnes and G. Mullen, Neuropsychologia, 12 (1974) 395.
27 P.W. Davidson, J. Exp. Psychol., 93 (1972) 43.
28 P.W. Davidson and T.T. Whitson, J. Exp. Psychol., 102 (1974) 687.
29 S. Axelrod, Effects of Early Blindness: Performance of Blind and Sighted Children on Tactile and Auditory Tasks, American Foundation for the Blind, New York, 1959.
30 R.W. Heinrichs and J.A. Moorhouse, New England J. Med., 280 (1969) 72.
31 U. Castiello, K.M.B. Bennett and C. Mucignat, Exp. Brain Res., 96 (1993) 152.
32 G. Ferrigno and A. Pedotti, I.E.E.E. Trans. Biomed. Eng. B.M.E., 32 (1985) 943.
33 D.E. Meyer, R.A. Abrams, S. Kornblum, C.E. Wright and J.E.K. Smith, Psychol. Rev., 95 (1988) 340.

34 D.E. Meyer, J.E.K. Smith, S. Kornblum, R.A. Abrams and C.E. Wright, in: M. Jeannerod (ed.), Attention and Performance XIII, Erlbaum, Hillsdale, 1990, p. 173.
35 M. Gentilucci, U. Castiello, M.L. Corradini, M. Scarpa, C. Umiltá and G. Rizzolatti, Neuropsychologia, 5 (1991) 361.
36 Y. Paulignan, M. Jeannerod, C. MacKenzie and R. Marteniuk, Exp. Brain Res., 79 (1990) 431.
37 U. Castiello, K.M.B. Bennett and G.E. Stelmach, Exp. Brain Res., 94 (1993) 163.
38 U. Castiello and M. Jeannerod, Neuroreport, 2 (1991) 797.
39 U. Castiello, K.M.B. Bennett and Y. Paulignan, Behav. Brain Res., 50 (1992) 7.
40 U. Castiello, K.M.B. Bennett and G.E. Stelmach, submitted
41 Y. Paulignan, C. MacKenzie, R. Marteniuk and M. Jeannerod, Exp. Brain Res., 83 (1991) 502.
42 U. Castiello, Y. Paulignan and M. Jeannerod, Brain, 114 (1991) 2639.
43 M. Gentilucci, S. Chieffi, M. Scarpa and U. Castiello, Behav. Brain Res., 82 (1992) 47.

Section Four: Clinical and Rehabilitation Studies

Insights into the Reach to Grasp Movement
K.M.B. Bennett and U. Castiello (Editors)

CHAPTER 10

THE NEURAL SUBSTRATES OF VISUALLY GUIDED PREHENSION: THE EFFECTS OF FOCAL BRAIN DAMAGE

L.S. JAKOBSON[1] and M.A. GOODALE[2]

[1] *Department of Psychology, Queen's University, Kingston, Ontario, Canada, K7L 3N6.*
[2] *Department of Psychology, University of Western Ontario, London, Ontario, Canada, N6A 5C2.*

SUMMARY

In this chapter, we will describe some of the various disturbances in the programming and execution of visually guided reaching and grasping movements that are associated with focal damage to different regions of the human cerebral cortex. It will be argued that the visual control of prehension depends on transformations of visual information within the dorsal stream of projections connecting primary visual cortex with posterior parietal cortex and, ultimately, with areas in the frontal lobe. In developing this argument, we will discuss the effects of damage at different points along this pathway. Finally, we will suggest that the visuomotor transformations underlying the production of visually guided prehension are functionally and neurally separated from the mechanisms underlying what is commonly described as ***visual perception****.*

INTRODUCTION

One of the most important adaptations in primate evolution was the appearance of hands capable of some degree of prehensile function. Manual prehension is a complex, skilled behaviour. In executing a grasping movement, the agent must precisely localize the goal object in viewer-centered coordinates; he must compute information about the object's size and shape, and about its orientation relative to the reaching limb. But more than this, to ensure accuracy and postural stability during the execution of the coordinated movement, there must be constant cross-referencing of information about the position of the eyes, head, limb, hand and trunk (c.f. [1]). In addition, along with visual information, haptic information about intrinsic object properties (e.g., weight, friction coefficients, etc.) is used to calibrate grip and load forces during the manipulatory phase of the movement [2-5].

Given the complexity of this action system, it should not be surprising that damage to a number of different brain areas (both cortical and subcortical) can interfere with prehension in a variety of ways. In the following chapter, we will describe various deficits associated with cortical damage that influence the unfolding of a reach to grasp movement. Our main focus will be on the effects of damage to the posterior parietal cortex (PPC) in humans. In interpreting these deficits, we will draw on electrophysiological, anatomical, and behavioural studies in the monkey which suggest that this region of the primate brain plays a pivotal role in carrying out the computations required to achieve the behavioural goal of grasping an object. As we will see below, extrastriate visual areas which project to this region appear to process information about intrinsic and extrinsic characteristics of goal-objects and information about self- and object-motion, all of which is needed for the programming and on-line control of goal-directed grasping. The pattern of connectivity within the parietal lobe allows the PPC to combine this incoming visual information with information provided through other sensory modalities. The PPC is also intimately connected with areas in the frontal lobe which have been implicated in the control of prehension. These interconnections presumably permit the PPC to transform sensory information into useful motor acts through the construction and application of various egocentrically-defined frames of reference (head-centered, shoulder-centered, etc.; [6]).

In sighted humans vision clearly provides most of our distal information about the disposition of objects we wish to grasp. Thus, vision provides the optimal input to assist with the programming and on-line control of prehension prior to contact, particularly when that action is directed at a novel object. For this reason, we will begin by briefly outlining some major features of the organization of the cortical visual system in the primate (for a more detailed review, see [7]). In focusing on vision, we do not wish to minimize the important contributions of the other sensory systems, especially somatosensation, to the control of a reach to grasp movement [8, 9]. Unfortunately, space does not permit us to explore this important topic.

TWO CORTICAL VISUAL SYSTEMS

In recent years much has been learned about the perceptual analysis which underlies our phenomenological visual experience of the world and the objects within it. This field of research was revolutionized by a proposal made by Ungerleider and Mishkin [10] over a decade ago. They described a series of experiments showing that monkeys with lesions of inferotemporal cortex were severely impaired on an object discrimination test but not a spatial discrimination test, while those with lesions of the PPC showed the opposite pattern. Ungerleider and Mishkin used these behavioural observations, along with complementary findings from electrophysiological and anatomical studies, to argue that the cortically-based visual system in the monkey could be divided into two functionally and anatomically separate streams of processing. Specifically, their work implicated a ventral projection stream connecting primary visual cortex (V1) with inferotemporal cortex in object recognition (the ***What*** system), and a dorsal stream connecting V1 with the PPC in spatial perception (the ***Where*** system).

Although the ***What*** *versus* ***Where*** dichotomy is appealing, accumulating evidence is challenging this simple account of visual processing, particularly at the functional level. Recently, a new interpretation of the division of labour between the two streams has been proposed, the details of which have been elaborated in a series of papers [11-13]. Briefly, according to this new account, functional specialization within the primate visual system evolved to support two broad categories of visual abilities (see Fig. 1). One set of abilities, which appears to depend on computations carried out in the ventral stream emanating from V1, supports our capacity to parse the visual array into recognizable objects and events to which we can attach meaning and significance. This capacity for visual recognition and visual learning underlies long-range planning, problem solving, and the establishment and maintenance of social discourse. Indeed, this capacity is probably coincident with what we call ***visual perception***. (Note that in using the term **perception** in this way we are referring to high-level processes used in representing objects and events within the world rather than to the low-level sensory analysis that might precede these and other processes).

The second broad category of abilities which depend on vision in primates includes the large number of visually-guided behaviours which we engage in on a daily basis. Of course we share many of these abilities with other vertebrates, and the ancient evolutionary history of these abilities is reflected in the fact that many of them depend on dedicated subcortical systems (e.g., the control of saccadic eye movements by the superior colliculus, a prominent midbrain structure). As noted earlier, however, primates also

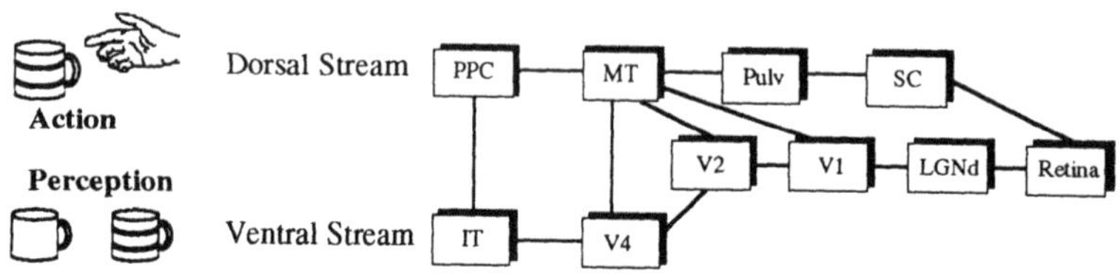

Figure 1. Schematic diagram illustrating some of the main components of the ***dorsal*** *and* ***ventral*** *streams of visual processing in the primate cerebral cortex. The dorsal stream is specialized for the control of visually guided actions (e.g., grasping a mug), while the ventral stream is specialized for object identification and recognition (e.g., perceiving the mugs shown here and knowing which belongs to you).* ***LGNd*** *dorsal lateral geniculate nucleus.* ***V1*** *primary visual cortex.* ***V2*** *visual area 2.* ***V4*** *visual area 4.* ***IT*** *inferotemporal cortex.* ***MT*** *middle temporal area.* ***PPC*** *posterior parietal cortex.* ***SC*** *superior colliculus.* ***pulv*** *pulvinar. Modified from [11].*

display a remarkable set of motor abilities involving precise control of the hands and fingers - abilities that have dramatically expanded the range of behaviours they can engage in. According to Goodale and Milner [12, 13], the visual control of prehension depends on cortical elaborations of basic subcortical systems. Specifically, these elaborated systems are thought to depend on networks in the dorsal stream of processing connecting V1 to the PPC and, ultimately, to areas in the frontal lobe. In the following section we will present some of the evidence supporting this proposal, with special emphasis given to observations of human patients with damage to the dorsal stream including its target area, the PPC.

ACTION SYSTEMS IN THE DORSAL STREAM

Optic ataxia: defining the nature of the deficit

Damage to the PPC in humans produces a constellation of deficits which affect the execution of visually-guided movements. One of the most striking of these is ***optic ataxia***, a disorder characterized by the inability to reach accurately to objects that are clearly seen. This sign is associated with damage to the posterior region of the superior parietal lobule [14, 15], in an area dorsal to that associated with hemispatial neglect [16]. It is important to note that these two signs, optic ataxia and hemispatial neglect, often occur independently; they should not be confused. Optic ataxia can be demonstrated by having the patient fixate centrally, presenting a visual target in the periphery, and asking the patient to reach toward it. Depending on the nature of the damage, affected patients will misreach with one or both hands into the visual field opposite the side of the lesion. As we will see below, this deficit cannot be accounted for simply on the basis of motor or sensory deficits. An analogous deficit is seen in parietally damaged monkeys (e.g., [17, 18]). Typically, however, experimental lesions of the PPC in monkeys have been rather large, presumably interrupting many different visual processes. Thus, the observed deficits in reaching and grasping must be viewed in the context of a broad range of possible impairments. The same interpretational problem plagues many studies of human patients with damage to the PPC; thus, it is often difficult to establish whether the observed disturbance in visuomotor control reflects a primary deficit or is secondary to other problems.

In the past, researchers have sometimes interpreted misreaching, or trouble avoiding obstacles when walking, to visuospatial perceptual failure (e.g., to a disturbance in size or distance estimation, or to a deficit in stereoscopic vision [19, 20]). The bias to do so has been strengthened in recent years with the widespread acceptance of Ungerleider and Mishkin's [10] proposal that the dorsal visual pathway is responsible for "spatial vision". But the assessment of a visuospatial deficit in these cases is sometimes complicated by the co-occurrence of another deficit known as simultanagnosia (e.g., [21, 22]). Simultanagnosia refers to a disturbance of attention which makes it difficult for affected individuals to ***see*** more than one object at a time. In these cases it may be the attentional problem, rather than a deficit in spatial vision, which underlies the difficulties these patients experience when required to make *relative* size or distance judgments, and to navigate around obstacles [19, 21]. Simultanagnosia might also be expected to interfere with the on-line control of visually-guided behaviours, such as prehension, by making it impossible for the patient to monitor simultaneously the goal object (which is typically foveated) and the reaching limb (as it moves through the visual periphery).

Although in some patients perceptual and/or attentional deficits may be producing (or exacerbating) problems with the visual guidance of movement, in others it is not clear that this is the case. Denes et al. [23], for example, described a patient with bilateral damage to the PPC whose difficulty in executing visually guided movements could not be reduced to a disorder of spatial perception, broadly defined; this patient could reliably indicate the absolute or relative positions of objects in space, and their absolute or relative sizes. Neither simultanagnosia nor neglect were described in the case report. An even more striking example of optic ataxia was reported by Levine et al. [15]. This patient's right superior parietal lesion produced no significant deficits in visual sensation, visual attention, taction, proprioception, motility, praxis, or visuospatial performance. This patient could accurately indicate the longer of two lines "even when the differences in their length became so small (i.e., 6 cm versus 6.15 cm) that control subjects found the task difficult" (p. 559). Moreover, when shown two lines differing in orientation, he had no difficulty selecting the orientation corresponding to a sample line shown previously. He also had no difficulty finding his way about, describing routes, locating cities or states on a map, or performing tests of constructional praxis. Yet, he produced "moderately" to "grossly inaccurate" reaching movements when using the left limb and/or when reaching into the left visual field under a variety of experimental conditions, although movements directed towards objects in central vision were accurate.

The fact that misreaching can occur in only a certain region of space when using one hand but not the other, as was the case in the patient just described, challenges the notion that optic ataxia can be reduced to a perceptual or attentional problem. (For other accounts of field- and hand-dependent deficits, see [16, 24, 25, 26]). Additional support for this conclusion comes from studies showing that optic ataxia can affect pointing movements but not eye movements directed toward objects in a single visual field [27]. Clearly, it is difficult to interpret such effector-dependent deficits in terms of a general failure of ***spatial vision*** or some other superordinate perceptual function. In order to do so, one would have to account for the fact that such a failure can come and go, depending on the response that is required.

We believe that optic ataxia is perhaps best characterized, as Perenin and Vighetto [16] have suggested, as a primary deficit in visuomotor control. This proposal is certainly consistent with what is known about the response properties of cells in the PPC of nonhuman primates. Early electrophysiological studies identified six classes of cells in the PPC, all of which might be expected to play a role in visually-guided prehension; these include saccade, fixation, pursuit, reach, manipulation, and light-sensitive cells [28]. The response properties of these cells, unlike those of cells in the ventral stream, are greatly dependent on the concurrent behaviour of the animal with respect to the visual stimulus. Indeed, Andersen [29] has suggested that their operation can only be understood by taking into account both sensory- and motor-related responses. In other words, he considers them to be visuomotor in nature. It is important to emphasize that these various cell types are segregated in different regions of the dorsal stream. This modular organization is what one might expect if specialized computations were being carried out in different dedicated systems. Presumably, these dedicated systems are recruited in different combinations depending on the specific act the organism intends to engage in.

The distal deficit in optic ataxia

As noted above, one of the striking features of optic ataxia is the large directional errors which are observed in aiming movements (see [27] for additional details). But it is important to note that a deficit in proximal limb control is not the only feature of the disorder. Patients with optic ataxia also display a profound disruption in grip formation ([8, 16, 24, 30]; Fig. 2).

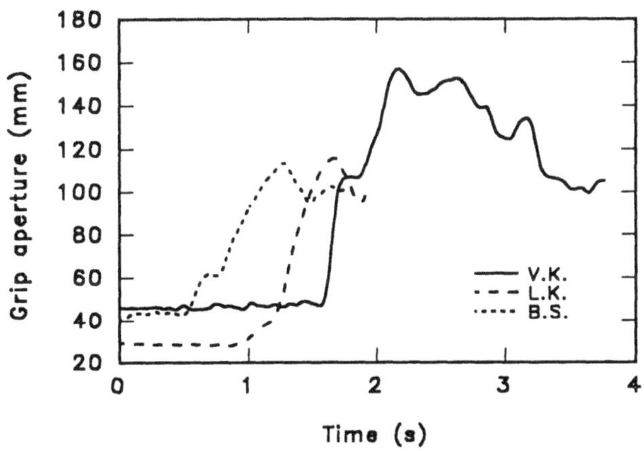

*Figure 2. Representative grip aperture profiles from individual reaching trials executed by a patient with optic ataxia (**VK**) and by two age-matched control subjects (**LK** and **BS**). Movements were tracked with an optoelectronic recording system (WATSMART) as the subjects reached out to grasp oblong blocks of various sizes (for full details see [30]). Movement onset time (indicated by the flat portion at the beginning of each trace) and movement duration were both longer in the patient than in the controls. More importantly, the patient opened her hand much more than controls overall and showed frequent reposturing of her fingers during movement execution. In addition, maximum grip aperture in the patient was poorly correlated with object size. Modified from [30].*

Perenin & Vighetto [16], for example, showed that patients with lesions of the PPC not only misdirected their reaching movements, but were also unable to orient their hand with respect to a slot into which they were to insert their fingers. They were similarly unable to extract a small object from within the slot using a precision grip (for similar observations in the monkey see [17]). Kinematic analyses of grasping movements in two other patients with optic ataxia have shown a failure to calibrate the grasp with respect to object size [30, 31], and a failure to use information about object shape effectively when positioning the fingers on the perimeter of the goal object ([31]; we will return to this latter observation in a subsequent section). Here, then, we have evidence that damage to the dorsal stream can interfere with the ability to use information about the size, shape and local orientation of a visual stimulus to control the distal musculature during grasping, in addition to interfering with accurate placement of the limb. It should be emphasized, however, that these patients, like those described earlier, remained perceptually aware of these same object features; that is, they remained quite capable of using these features to discriminate between pairs of objects and to make other perceptual judgements about them. This finding suggests that these cognitive or perceptual abilities are mediated by other regions of the brain. This is a topic to which we will return later.

It is interesting to note that recent studies of the PPC in the monkey [32] show that some of the manipulation cells mentioned in the previous section not only fire when the monkey manipulates an object, but are also sensitive to the intrinsic features of that object, such as its size and orientation - features that determine the posture of the hand and fingers during a grasping movement. These electrophysiological observations from monkeys taken together with the neuropsychological findings just described underscore our argument that the function of the dorsal stream is best interpreted in terms of the control of skilled actions, rather than in terms of ***spatial vision*** or other broad, perceptual concepts.

Other deficits in prehension associated with damage to the dorsal stream

Typically when one reaches out to grasp an object, the head and eyes move so as to bring the object into central vision, and movements of the hand and arm are tracked in peripheral vision [33]. If the object is moving and must be caught, then the trajectory of that object must be extrapolated in order to time the interceptive act. For these and other reasons, the neural systems involved in the control of prehension must utilize information about visual movement. One of the major components of the dorsal stream, the middle temporal area (MT, see Fig. 1), appears to be intimately involved in processing this type of information. Electrophysiological studies in monkeys have implicated cells in this region in coding both the speed and direction of motion [34]. This area also contains an expanded representation of the lower visual field, a feature which some have argued makes it well-suited to play a role in eye-hand coordination [34, 35]. In the adjacent middle superior temporal area (MST), which receives input from MT, there are many movement-selective cells whose response amplitude is modulated by the animal's behavioural response to the stimulus. For example, many of the cells in MST show optimal firing when the movements of the visual stimulus are tracked by movements of the head and eyes [36, 37] and some show sensitivity for moment-to-moment changes in the disposition of objects in three-dimensional space [38]. A recent kinematic study of grasping in humans [39] has shown that timing of hand closure is controlled by the rate of expansion of the target image on the retina, the same sorts of changes that excite the cells in MST.

A recent description of the visuomotor performance of a patient with damage to the presumptive human homologue of MT, and perhaps MST, demonstrates the importance of motion processing in the control of grasping. This patient was able to achieve accurate pointing to a stationary target only by markedly slowing his movements and continuously looking back and forth between his finger and the target [40]. These behavioural observations suggest an impairment in the ability to track the moving limb in peripheral vision. While precise homologies between the monkey and human parietal cortex are not yet well established, these results are in good agreement with the electrophysiological findings discussed above.

Coordination between visuomotor channels

When a given response requires the coordination of multiple visuomotor systems, damage to any one of these systems would be expected to interfere with the achievement of the overall movement goal. It is well known that bilateral damage to the PPC can produce marked impairments in the ability to fixate a stationary object, or to track a moving target visually using smooth pursuit [41-44]. Botez and colleagues [41-43], for example, report the presence of large, slow deflections in the electrooculographic (EOG) records of patients with such damage as they attempted to fixate a particular locus. It should be clear that damage to the dorsal stream system(s) dedicated to oculomotor control might indirectly affect the visual guidance of grasping movements. Indeed, it has been demonstrated that in patients with oculomotor control problems associated with bilateral damage to the PPC, failures to track a moving target *manually* coincide in time with abrupt deflections in the EOG record [41-43].

We are still left with the very difficult conceptual problem of understanding how the movements of different effectors (eyes, arms, hands, trunk, etc.) are coordinated during the act of prehension. This question has generated considerable interest in recent years. In this regard, Stein [6] has recently proposed that the PPC is in an excellent position to play a pivotal role in this coordination. In his review of the literature, Stein notes that neurons in the PPC receive somaesthetic, proprioceptive, vestibular, auditory and visual sensory inputs together with information about eye, head, limb and locomotor movement. He has characterized this region as "a neural network that implements algorithms for converting one set of vectors (e.g., retinal) into another set of vectors (e.g., oculomotor or arm-centered)". In this way, he suggests, the egocentric spatial relationships needed to guide a particular action are represented. In short, he proposes that the brain does not use a single "map of space" to control action; instead, it relies on a flexible system in the parietal lobes, capable of selecting and combining information from different frames of reference (retinotopic, shoulder-centred, etc.) to achieve particular behavioural goals. Theoretical accounts such as Stein's are not inconsistent with the idea developed here that the PPC is best understood as a network of visuomotor modules. Such accounts, however, continue to emphasize only the *spatial* frames of reference in which different actions are played out and do not deal directly with the problem of transforming visual information about intrinsic object properties, such as size, shape, and local orientation, into the motor coordinates specifying the grasp.

DISSOCIATIONS BETWEEN PERCEPTION AND ACTION

We suggested earlier that the deficits in reaching and grasping which accompany damage to the dorsal stream cannot be reduced to a deficit in ***spatial vision*** [10]. Indeed, as noted above, there are now several reports of patients with damage in PPC who retain the ability to make perceptual judgements about the size, shape, and orientation of objects, and their location in space, but who nonetheless appear unable to access information about these object features to guide their motor responses [15, 16, 23, 31]. In many of these same patients, perception of the world and the objects and events within it remains remarkably intact (e.g., [30, 31]). Thus, the suggestion [45] that grip formation errors result from faulty processing within a recognition or representational system is unlikely to be true.

We recently had an opportunity to undertake a strong test of these ideas by examining both the perceptual and visuomotor abilities of a patient (DF) with ***visual form agnosia***. DF survived an accidental carbon monoxide poisoning which affected ventrolateral regions of extrastriate cortex, but left V1 largely intact [46]. Elementary visual functions are well preserved in DF, yet she has great difficulty identifying objects visually, failing on even the simplest tests of visual form discrimination [46, 47]. But amazingly, although perceptual tests reveal a complete *insensitivity* to the size or orientation of objects placed in front of her, the visuomotor system controlling skilled manual prehension in DF appears to have full access to this information [47]. Thus, when she reaches out to pick up an object, DF's hand adapts in-flight to the size and orientation of that object! Additional experiments have shown that although DF is unable to discriminate the outline shape of objects, she uses information about contour to position her fingers on points which maximize the stability of precision grip ([48]; see Fig. 3). Thus, in essence, DF is the ***mirror image*** of the optic ataxic patients we described earlier.

A similar dissociation between the ability to ***perceive*** visual stimuli and the ability to direct spatially accurate motor responses to them is seen in cases of so-called ***blindsight*** resulting from damage to the occipital lobe. Damage to V1 produces subjective blindness in the affected part of

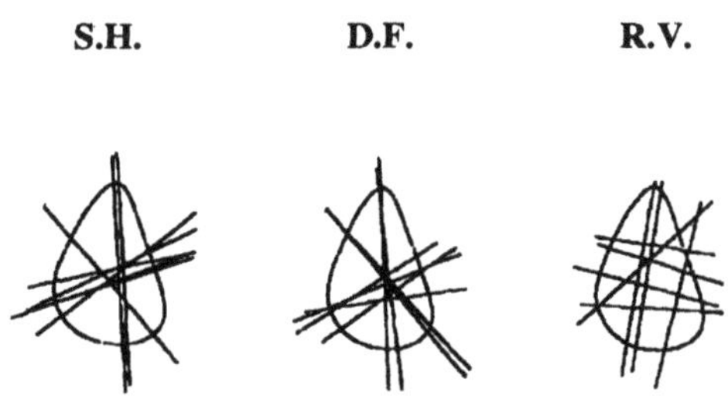

Figure 3. Representative ***grasping*** *axes for one of several irregularly shaped objects grasped by a normal control subject (****SH****), a patient with visual form agnosia (****DF****), and a patient with optic ataxia (****RV****). Each line passes through the points on the perimeter of the shape where the index finger and thumb first made contact with the object on a single trial.* ***DF*** *(like the control subject) places her finger and thumb on appropriately opposed points on the shape (i.e., on points that allow for a stable lift); the patient with optic ataxia (****RV****) does not. This is true despite the fact that at a perceptual level* ***DF*** *cannot use contour information to discriminate between shapes like this, while* ***RV*** *can. Based on [31].*

the visual field. Yet patients with lesions of V1 have been shown to be able to look towards stimuli presented within their scotomata and to localize them by pointing (see [49] for a recent review); some are even said to have shown " ... better than chance preparatory adjustments of the wrist, fingers, and arm in reaching for objects of differing shape, orientation, size, location, and distance in the blind field" (see [50], p. 276). Although there is not complete agreement in this regard, these residual visual abilities are widely believed to be mediated by a separate visual system arising from the retina which has its primary synapse in the superior colliculus, rather than in the lateral geniculate ([49]; see Fig. 1). This phylogenetically older pathway to the superior colliculus, which has been linked to the control of orientation movements of the head and eyes, has direct connections with motor nuclei in the brainstem. But, in addition, it projects (via the thalamus) to extrastriate regions within the dorsal (but not the ventral) stream and is capable of activating cells in these areas independent of V1 (for review, see [51]).

These and other observations support the argument recently advanced by Goodale and Milner [12, 13] that the visual control of motor skills like prehension depends on the dorsal stream, while object recognition and ***conscious*** visual perception rely primarily on the ventral stream. The functional separation of the two cortical visual systems advocated here, one which emphasizes the relative independence of computations underlying perception and action, differs in several important respects from the popular ***What*** *versus* ***Where*** theory of Ungerleider and Mishkin [10]. Most importantly, according to this new account, both streams process information about object orientation and shape, and about spatial relationships including depth. Each stream, however, uses this visual information in different ways. Transformations carried out in the ventral stream permit the formation of perceptual and cognitive representations of objects and their spatial relations with each other; those carried out in the dorsal stream, which utilize the instantaneous and egocentric features of objects, mediate the control of goal-directed actions.

CONCLUSIONS

In this chapter, we have argued that the deficits in reaching and grasping seen after damage to the dorsal stream cannot be reduced to a failure of ***spatial vision***. Indeed, some patients who are unable to direct accurate goal-directed movements nonetheless perform normally on tests of perceptual functioning which demonstrate unequivocally that they ***know*** where the object is located. A similar dissociation between perception and visuomotor control is seen in patients with damage to the dorsal stream who show exquisite sensitivity to intrinsic object features in perceptual tests and yet remain unable to use this kind of information to preshape their hand in grasping. These observations demonstrate that action systems in the dorsal stream utilize information about intrinsic as well as extrinsic object features. In short, the dorsal stream clearly computes more than simply ***Where*** an object is.

We would characterize the dorsal stream as a hierarchically organized and distributed processing network containing a number of dedicated visuomotor systems which can be recruited alone or in various combinations, depending on task demands. According to this proposal, deficits in prehension could arise primarily from input failure within one or more of these dedicated systems, from damage to the systems themselves, from a failure to coordinate different systems, or from a failure to transmit the outputs of this stream of processing to other brain regions. This latter point is one of considerable importance. In focusing on the dorsal stream in the present chapter, we do not wish to suggest that it is *solely* responsible for carrying out all of the steps involved in the visual control of manual prehension. It is important to remember that this visual system is intimately and reciprocally interconnected with the numerous areas in the frontal lobes [52], and with visuomotor structures in the midbrain [51] and pons [53]. Thus, we believe that any of the visuomotor modules to which we alluded earlier extend well beyond the PPC and depend on these reciprocal connections. A full understanding of the deficits arising from damage to the PPC must, then, take into account its position in this complex neural network.

ACKNOWLEDGEMENTS

This work was supported by a grants from the Natural Sciences and Engineering Research Council of Canada (to L.S.J.) and the Medical Research Council of Canada (to M.A.G.).

REFERENCES

1 C. Trevarthen, in: M.A. Goodale (ed.), Vision and Action: The Control of Reaching and Grasping, Ablex, Norwood, 1990, p. 181.

2 A.M. Gordon, H. Forssberg, R.S. Johansson and G. Westling, Exp. Brain Res., 83 (1991) 477.

3 S.J. Lederman and R. Klatzky, in: M.A. Goodale (ed.), Vision and Action: The Control of Reaching and Grasping, Ablex, Norwood, 1990, p. 98.

4 A.M. Smith, in M.A. Goodale (ed.), Vision and Action: The Control of Reaching and Grasping, Ablex, Norwood, 1990, p. 128.
5 C.J. Winstein, J.H. Abbs and D. Petashnick, Exp. Brain Res., 87 (1991) 465.
6 J.F. Stein, Behav. Brain Sci., 15 (1992) 691.
7 D.J. Felleman and D.C. Van Essen, Cerebral Cortex, 1 (1991) 1.
8 M. Jeannerod, Behav. Brain Res., 19 (1986) 99.
9 M. Jeannerod, F. Michel and C. Prablanc, Brain, 107 (1984) 899.
10 L.G. Ungerleider and M. Mishkin, in: D.J. Ingle, M.A. Goodale and R.J.W. Mansfield (eds.), The Analysis of Visual Behavior, MIT Press, Cambridge, 1982, p. 549.
11 M.A. Goodale, Curr. Opin. Biol., in press.
12 M.A. Goodale and A.D. Milner, TINs, 15 (1992) 20.
13 A.D. Milner and M.A. Goodale, in: T.P. Hicks, S. Molotchnikoff and T. Ono (eds.), The Visually Responsive Neuron: From Basic Neurophysiology to Behavior, Prog. Brain Res., Elsevier, Amsterdam, 95 (1993) 317.
14 A.R. Damasio and A.L. Benton, Neurology, 29 (1979) 170.
15 D.N. Levine, K.J. Kaufman and J.P. Mohr, Neurology, 28 (1978) 556.
16 M.T. Perenin and A. Vighetto, Brain, 111 (1988) 643.
17 R. Haaxma and H.G.J.M. Kuypers, Brain, 98 (1975) 239.
18 R.H. Lamotte and C. Acuña, Brain Res., 139 (1978) 309.
19 G. Holmes, British J. Ophthal., 2 (1918) 449.
20 G. Holmes and G. Horrax, Arch. Neurol. Psychiat., 1 (1919) 385.
21 H. Hécaen and J. de Ajuriaguerra, Brain, 77 (1954) 373.
22 J. Montero, J. Peña, D. Genis, R. Rubio, J. Peres-Serra and L. Barraquer-Bordas, Acta Neurol. Belg., 82 (1982) 270.
23 G. Denes, F. Caviezel and C. Semenza, Cortex, 18 (1982) 165.
24 M.T. Perenin and A. Vighetto, in: A. Hein and M. Jeannerod (eds.), Spatially Oriented Behavior, Springer-Verlag, New York, 1983, p. 305.
25 A. Tzavaras and M.C. Masure, Lyon Medical, 236 (1976) 673 (in French).
26 P. Rondot, J. de Recondo and J.L. Ribadeau-Dumas, Brain, 100 (1977) 355.
27 G. Ratcliff and G.A.B. Davies-Jones, Brain, 95 (1972) 49.
28 V.B. Mountcastle, J.C. Lynch, A. Georgopoulos, H. Sakata and C. Acuña, J. Neurophysiol., 38 (1975) 871.
29 R.A. Andersen, in: F. Plum and V.B. Mountcastle (eds.), Handbook of Physiology, Am. Physiol. Soc., Rockville, 1987, p. 483.

30 L.S. Jakobson, Y.M. Archibald, D.P. Carey and M.A. Goodale, Neuropsychologia, 29 (1991) 803.
31 M.A. Goodale, K. Murphy, J.P. Meenan, C. Racicot and D. Nicolle, Soc. Neurosci. Abstr., in press.
32 M. Taira, S. Mine, A.P. Georgopoulos, A. Murata and H. Sakata, Exp. Brain Res., 83 (1990) 29.
33 J. Paillard, in: D.J. Ingle, M.A. Goodale and R.J.W. Mansfield (eds.), The Analysis of Visual Behavior, MIT Press, Cambridge, 1982, p. 367.
34 J.H.R. Maunsell and D.C.J. Van Essen, Neurophysiol., 49 (1983) 1127.
35 J.H.R. Maunsell and D.C.J. Van Essen, J. Comp. Neurol., 266 (1987) 535.
36 W.T. Newsome, R.H. Wurtz and H. Komatsu, J. Neurophysiol., 60 (1988) 604.
37 P. Thier and R.G. Erickson, Eur. J. Neurosci., 4 (1992) 539.
38 H. Saito, M. Yukie, K. Tanaka, D. Hikosaka, Y. Fukada and E. Iwai, J. Neurosci., 6 (1986) 145.
39 G.J.P. Savelsbergh, H.T.A. Whiting and R.J. Bootsma, J. Exp. Psych., Hum. Percep. Perf., 17 (1991) 315.
40 M. Rizzo, D. Rotella and W. Darling, Neuropsychologia, 30 (1992) 711.
41 M.I. Botez, Eur. Neurol., 13 (1975) 101.
42 M.I. Botez and T. Serbanescu, J. Neurol. Sci., 4 (1967) 289.
43 M.I. Botez, T. Serbanescu and I. Vernea, Neurology, 14 (1964) 1101.
44 F. Girotti, C. Milanese, M. Casazza, A. Allegranza, F. Corridori and G. Avanzini, Cortex, 18 (1982) 603.
45 M. Jeannerod and J. Decety, in: M. A. Goodale (ed.), Vision and Action: The Control of Grasping, Ablex, Norwood, 1990, p. 33.
46 A.D. Milner, D.I. Perrett, R.S. Johnston, P.J. Benson, T.R. Jordan, D.W. Heeley, D. Bettucci, F. Mortara, R. Mutani, E. Terazzi and D.L.W. Davidson, Brain, 114 (1991) 405.
47 M.A. Goodale, A.D. Milner, L.S. Jakobson and D.P. Carey, Nature, 349 (1991) 154.
48 J.P. Meenan, M.A. Goodale and H. Bülthoff, Invest. Ophthal. Vis. Sci., 34 (1993) 1131.
49 L. Weiskrantz, Proc. Royal Soc. Lond., 239 (1990) 247.
50 A.J. Marcel, Cog. Psych., 15 (1983) 238.
51 C.G. Gross, Neuropsychologia, 29 (1991) 497.

52 C. Cavada and P.S. Goldman-Rakic, in: T.P. Hicks, S. Molotchnikoff, T. Ono (eds.), The Visually Responsive Neuron: From Basic Neurophysiology to Behavior. Prog. Brain Res., Elsevier, Amsterdam, 95 (1993) 123.
53 J.F. Stein and M. Glickstein, Physiol. Rev., 72 (1992) 967.

Insights into the Reach to Grasp Movement
K.M.B. Bennett and U. Castiello (Editors)

CHAPTER 11

THE REACH TO GRASP MOVEMENT OF PARKINSON'S DISEASE SUBJECTS

U. CASTIELLO[1], K.M.B. BENNETT[2] and M. SCARPA[3]

[1] *Dipartimento di Psicologia, Universitá di Bologna, Viale Berti Pichat, 5, 40127, Bologna, Italy.*
[2] *Dipartimento di Psicologia Generale, Universitá di Padova, Piazza Capitaniato, 3, 35139, Padova, Italy.*
[3] *Clinica Neurologica, Universitá di Modena, Via del Pozzo, 71, 41100, Modena, Italy.*

SUMMARY

The limited number of studies which have assessed the reach to grasp movement of PD subjects have demonstrated appropriate patterning of movement kinematics according to reaching distance and object size. However, under predictive movement conditions the onset of the manipulation component is delayed. Double-step perturbation paradigms have also shown a suitable rearrangement of movement patterning in more responsive situations. Dysfunction is only evident at the changeover from one to another movement. PD subjects are thus able to perform both preprogrammed and feedback based movements but show deficits in the closing and in the opening of motor subsets. Neuroanatomical and neurophysiological mechanisms which could explain this dysfunction are discussed.

INTRODUCTION

In humans, the study of motor behaviour finds one of its foundations in the differences between healthy subjects and those with cerebral lesions. In particular, with reference to voluntary motor acts, it is important to examine patients with deficits of motor areas such as those of the basal nuclei. A classic example of a basal nuclei deficit is that of Parkinson's disease (PD).

The following sections give a brief overview of the clinical picture of PD and of its underlying pathophysiological and neurochemical mechanisms. This review was included in order to give those readers who have had little contact with PD subjects some background knowledge.

PD is a progressive disease of largely unknown aetiology. It is characterized by four cardinal signs: tremour, rigidity, akinesia and loss of postural reflexes. The age of onset is variable (17 - 89 yrs.) but in 60 - 70% of cases it becomes clinically obvious during the sixth and seventh life decades. The incidence in the United States and in Europe is one in every 1000 and 12 in 1000 within higher risk areas. The disease appears with one or more of the classic symptoms and the syndrome is usually completed within a few years. The clinical picture may also include signs such as dysphonia, dysarthria, dysphagia and vegetative dysfunction. In addition, neuropsychological deficits of variable degree have been reported and, in some cases, these progress to advancing dementia. Patients with PD are thus often not self-sufficient and assistance from the family and from social and health care structures is frequently needed.

PATHOPHYSIOLOGY

From the pathophysiological point of view, PD is characterized by neural loss of the substantia nigra at the mesencephalic (pars compacta) level and consequent changes to neural conduction along the nigrostriatal pathway. The lesion mainly involves the central and caudal portions of the pars compacta. It is bilateral and generally symmetric and can be associated with moderate gliosis [1]. Neuronal degeneration is also often evident at the level of the locus coeruleus and within the motor dorsal nucleus of the vagus, the basal nucleus of Meynert and within other subcortical grey matter formations [2]. In 80% of cases Lewy bodies may be observed at both cortical and spinal cord levels. These bodies are not specifically indicative of PD but appear during early stages of the disease.

NEUROCHEMISTRY

Neurochemically (for review see [3]), PD is characterized by prevalent but not exclusive lesions of the central dopaminergic routes. There is generally a significant reduction of dopamine in the mesencephalon and in all the telencephalic structures containing the terminals of the mesostriatal, mesocortical and mesolimbic routes [4-6]. The loss of the dopaminergic innervation leads to compensatory changes to the residual dopaminergic neurons and to their post-synaptic receptors. These residual neurons become *overloaded* in their attempt to augment dopamine turnover [7]. In addition, D2 post-synaptic receptors become hypersensitive and increase in number [8, 9]. Modifications also occur to the level of neurotransmitters such as serotonin, glutamic acid decarboxylase and noradrenaline in the locus coeruleus and such as acetylcholine in the pontine nuclei. These modifications have been tentatively related to secondary effects of lesions to the dopaminergic routes [10].

NEUROANATOMY AND NEUROPHYSIOLOGY

Study of the neuroanatomical and neurophysiological nature of the connections between the basal nuclei and the motor cortical regions has greatly contributed to the current understanding of both the underlying mechanisms of Parkinson's disease and the control of voluntary actions.

The basal ganglia do not project directly to the primary motor cortex. A major portion of their output from the globus pallidus and from substantia nigra zona reticulata is directed, via the thalamus, to the supplementary motor area (SMA) and to the lateral premotor cortical region [11]. Basal ganglia projections are thus upstream to those neurons which project to the primary motor cortex and thence to corticospinal tracts [12]. This higher order sequencing has been used to suggest that basal ganglia output is likely to be influential in movement organization. This is further supported by projections to the basal ganglia from area 46 [13, 14]. A role in cognitive functioning has also been proposed from those basal ganglia pathways which are directed, via other thalamic relay nuclei, to the prefrontal cortex. However, the extent to which this basal ganglia output is concerned with cognitive rather than movement processing is uncertain.

Evidence for the role of the basal ganglia for motor behaviour has been obtained largely from lesion studies and from single unit recordings in behaving animals [15-18]. Overall, the results from neurophysiological studies have indicated that basal ganglia neurons show discharge activity in relation to such movement parameters as amplitude, velocity, direction and load. For example, in animals trained to perform a step-tracking task in which the amplitude, speed and direction of arm movements were varied, significant relationships were found between globus pallidus and substantia nigra neuronal discharge and all three of these parameters [19-22]. If we relate this to the reach to grasp movement, this would suggest that these basal ganglia neurons would show discharge in relation to the direction of the reaching arm, to the amplitude of this reach and possibly also to the amplitude of the grasp aperture, and to the velocity of the arm and hand movements.

Current evidence points more to a coding by the basal ganglia for higher order movement parameters, in particular, direction, than for lower level components such as joint angle or muscle activity. This was evidenced in a study by Crutcher and DeLong [23] whereby monkeys were trained to perform a visuomotor task requiring elbow flexion and extension. By using assisting and opposing loads this task enabled a dissociation between direction of arm movement and pattern of muscular activity. It was found that a greater percentage of putamen neurons showed a discharge pattern which was related to direction (50%) than to the activity of any individual muscle active during the task (13%). This gave an indication that the basal ganglia are more involved in the parameterization of movement than in the specification of which muscles should be activated.

The pattern of discharge activity at the level of the basal ganglia shows a semblance to that at higher cortical levels to which the basal ganglia is connected. This is most obvious when looking at directional effects. These effects within the globus pallidus [17] correspond to those within the putamen and subthalamic nucleus [22, 23], regions which both project to the globus pallidus. In turn, the directional features within these latter nuclei resemble the discharge patterns of neurons in the precentral and parietal cortical areas which project to the putamen and the subthalamic nucleus [24-26]. Similarly, neurons of the basal ganglia show a discharge relationship to steady and dynamic load which mirrors that of the motor cortex, although this is less obvious than the directional semblance (see [23] for review). Hence the neuronal relationship to movement parameters in the basal ganglia may reflect the nature of the inputs to this structure from the motor, premotor and somatosensory cortices. On this basis it has been suggested that the basal ganglia serve as a ***funnel*** from association areas to the motor cortex [21, 27, 28]. Overall, it appears that influences from the sensorimotor and premotor cortices are largely directed, via the caudate, pallidum and thalamus, to premotor areas while influences from association areas are directed, via a separate caudate-pallidum-thalamus pathway, to the prefrontal cortex. These segregated parallel cortico-subcortical loops probably subserve different functions in the planning and execution of a motor task [17].

MOVEMENT DEFICITS IN PARKINSON'S DISEASE

Apart from the obvious clinical signs, two deficits of movement performance are mainly reported for PD subjects. One is an impaired ability to perform simultaneous or sequential movements. The other is more debated but relates to how PD subjects use advance information for movement planning.

Quantitative evaluation of the first dysfunction has been sparse. Schwab et al. [29] found that the performance of a squeezing task with one hand was severely disrupted both in the temporal and spatial domains when the contralateral hand performed a non-homologous task such as drawing a triangle. As if unable to perform these movements simultaneously, some PD subjects adopted a sequential mode of performance, drawing, for example, one line of the triangle and then squeezing the rubber bulb. A later study by Talland and Schwab [30] not only confirmed this difficulty with the simultaneous performance of a complex task with one hand (picking up beads with tweezers) and a more simple tally counter press with the other hand, but reported that five PD subjects were unable to successfully perform the task. In addition, it was found that when performing a sequence of movements (marking specified alphabetic letters) PD subjects had " .. difficulties in switching from one criterion to another" ([30] p. 51).

The simultaneous performance of the *simple* but separate motor patterns of elbow flexion and isometric opposition between the index finger and thumb in one limb [31] demonstrated that movement time for this combination movement was well above that time to perform either task in isolation. A similar result was found for the performance of the same tasks but in sequence [32]: For PD subjects, not only was movement time of this sequential pattern longer than that which would be expected by simply adding the individual movement times of each task when performed in isolation, but the interval between termination of the first pattern and activation of the second was prolonged. Harrington and Haaland [33], in a comparison of the sequential performance of one hand task to the sequential performance of distinct hand postures (e.g., from pressing a button with the tip of the index finger to grasping a handle), not only showed longer inter-response times for the PD subjects with the latter task but a greater incidence of error.

A second reported dysfunction relates to aspects of task predictability in relation to motor performance. Flowers [34, 35] suggested that PD subjects were unable to utilize ***internal models*** [36] for the spontaneous generation of movements which could be predicted. He found that the performance of PD subjects on a sinusoidal or ramp tracking task differed more from that of controls than did a task whereby the subjects were expected to track *noise* [34]. Further, when unexpectedly faced with visual gaps in the waveform, PD subjects were less able to anticipate the correct tracking pathway [35]. Various investigators have since demonstrated that PD subjects conditionally use advance information for accurate motor performance. Bloxham et al.

[37] found little differences between control and PD subjects when the task required tracking of a circular pathway. They attributed this contrast with Flower's works to the absence of the requirement for sharp directional changes and thus of the initiation of different movement segments in their task. Day et al. [38] showed that PD subjects, after having performed a series of apparently random task sequences, could improve tracking lag of either the elbow or wrist joints when the pattern of required responses was known. The finding that PD subjects did not show such a marked concurrent decrease in the number of positioning errors, prompted the suggestion that use of a feed-forward motor strategy resulted in less accuracy of movement. This latter idea was supported for the oculomotor system [39]: PD subjects adopted a feed-forward mode for alternating saccadic eye movements when visual information of target position was unexpectedly and intermittently removed. However, accuracy with these anticipatory movements was diminished.

WHY STUDY PREHENSION MOVEMENTS IN PD SUBJECTS?

Recently both the issue of simultaneous/sequential movement performance and that of using predictive movement information have been studied in relation to the reach to grasp movement of Parkinson subjects. This prehension movement has been chosen for a number of reasons. It is a functional well-rehearsed task. When performed without visual or other constraint, the task is thus already known to the subject and may reflect the recruitment of an ***internal model***. This is in contrast to other previously studied movements which must be learnt for the experiment - the result being an assessment of this learning process and thus of the manifestation of neural mechanisms which are not fully rehearsed. It is proposed that the study of a **functional** movement would shed a different light on the mechanisms that subserve the implementation of complex actions.

In addition, and as reflected in this book, numerous studies have characterized the various aspects of this movement. It is thus now used as a basic experimental task, procedures and results on normal subjects being well-established. This makes the comparison between PD subjects and non-PD subjects more valid.

Extrapolating from the results of kinematic studies, this prehension movement consists of two principal components: transport (reach) and manipulation (grasp). These are activated in parallel but are thought to reflect the operation of separate neural channels which are loosely coupled in time [40, 41]. The dysfunction in the simultaneous/sequential activation of movements in PD has traditionally been described for those movements which are temporally and functionally distinct and which are subserved by completely different motor programs. Study of the reach to grasp movement can reveal whether this dysfunction also applies to functionally and temporally coupled movement components. Although, when activated, the two components run in parallel, the manipulation component is often activated at the same time as or shortly after the transport component (< 50 ms; [42]). If PD subjects have difficulty with simultaneous or sequential activation of separate movements, and if this dysfunction also applies to initiation of the components (or *segments*, c.f. [37]) of a coordinated movement, the prediction is for a longer delay of manipulation component onset for PD subjects. In turn, this delay may affect the kinematic patterning of each component and the inter-component coordination.

Study of this prehension movement with PD subjects can thus assist in determining the role of the basal ganglia in the activation, patterning and coordination of a functional well-learnt task.

SIZE AND DISTANCE EFFECTS

Following an initial study by Castiello and Scarpa ([43] - see PERTURBATION SECTION this chapter), Castiello et al. [44] studied the kinematics of the reach to grasp movement of eight Parkinson and eight matched control subjects. The reach was of either 15, 27.5 or 40 cm. The grasp was of either a small (0.7 cm) or a large diameter (8 cm) cylindrical dowel.

Results for the PD subjects confirmed those originally found by Castiello and Scarpa [43]. Performance was slow but there were no deficits in the ability to modify the spatiotemporal characteristics of the prehension pattern in response to experimentally imposed changes in either the distance of the object from the subject or the size of the object [43, 45]. Transport time of the arm, and the timing and amplitude of the velocity and acceleration/deceleration peaks of this movement all increased with reaching distance. There were no additional bursts of speed at the end of the movement (in contrast to the findings of Hallett and Koshbin [46]). These results are in accordance with those from non-PD subjects when reaching different distances [42]. In addition, the PD subjects of this study showed changes of the transport component according to the type of grasp adopted. Thus, the time from peak velocity to the end of the movement was longer when a precision grip (the grasp naturally adopted for the small object) rather than a whole hand prehension (the grasp naturally adopted for the large object) was performed (see Fig. 1). This supports previous results, from studies of non-PD subjects, of a prolongation of the approach phase for

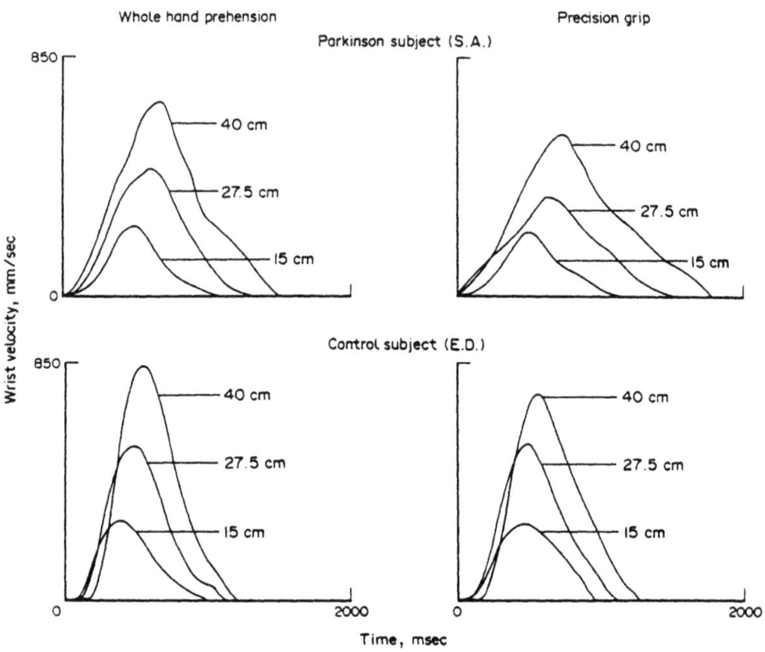

Figure 1. Wrist velocity profile when reaching to dowels placed at 15, 27.5 and 40 cm. From [44].

movements requiring greater accuracy such as when reaching to grasp more fragile [47] or smaller objects [42, 48, 49].

PD subjects were thus able to correctly regulate movement parameters. They exhibited no inability to activate the required and appropriate motor components. Further, this study illustrated that PD subjects showed no abnormalities in the modification of the timing and amplification of hand opening in relation to the reaching distance or to the size of the object to be grasped. As with control subjects, the peak of hand opening was relatively earlier when reaching to grasp closer and/or smaller objects. That the timing of the peak of hand opening changed as a function of movement duration demonstrates how aspects of one component are sensitive to changes in the other [40-42, 48-50]. PD subjects show no dysfunction in this sensitivity.

The overall form of the motor program of PD subjects thus appears to be maintained. The selection of muscles and the timing of their activation enables the correct relative timing of all movement parameters of the reach to grasp movement. A suitable number of neuronal sets is mobilized and the temporal arrangement of these sets is maintained.

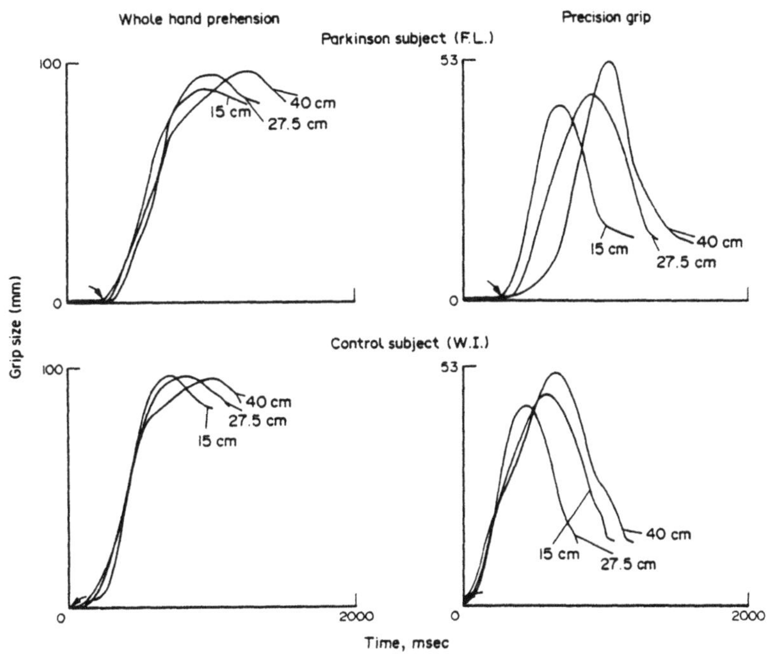

Figure 2. Grip size profile when reaching to dowels placed at 15, 27.5 and 40 cm. For the PD subject (above) the onset of the manipulation component (see arrows) occurs well after that of the control subject. From [44].

For PD subjects, it was the relative activation of the two components which revealed abnormalities: the onset of the manipulation component was delayed with respect to the onset of the transport component (PD: 80 ms; Control: 31 ms; Fig. 2; [43, 45]). PD subjects were not able to initiate the two components in a near-simultaneous manner. This result was confirmed when the onset time of the manipulation component was expressed as a percentage of the total movement time - this *relative* onset was also later for PD subjects. Further, in order to test that the delayed onset was not simply related to the slower PD movements, a regression analysis was performed between the duration of this delay and movement time. The absence of significant correlations indicated that the delay of manipulation onset was not a function of movement duration.

These results confirmed that the reach to grasp movement may indeed be directed by two distinct motor programs which are typically executed almost simultaneously. It appears to be this near concurrent activation which is desynchronized by the specific impairment in the management of synchronous motor programs in Parkinson's disease.

These findings were further confirmed in subsequent studies by Bennett et al. [51] and Castiello et al. [52] in a subject with hemiParkinson's disease. The increased delay of onset for the manipulation component of the hemiParkinson subject was not confined to the affected side but was also observed for the unaffected limb. The dysfunction is thus more global than is shown both by the unilateral clinical signs and by measures of several other kinematic parameters. These subtle findings are in accordance with evidence that the pathology tends to be bilateral even when it is markedly asymmetric [53]. Clearly impairments in programming goal-directed prehension movements are not only related to damage of cortical areas [54, 55] but also to damage of the deep nuclei and the pathways running through them [56, 57].

PERTURBATION STUDIES

The administration of a perturbation paradigm to PD subjects assesses their ability to rapidly change motor output. If it is assumed that PD subjects have a dysfunction in the sequential activation of motor programs, abnormalities should be easily evidenced at the changeover from one to the following program. For example, if the perturbation requires a change from precision grip to whole hand prehension the expectation would be for some form of kinematic disturbance at the transition phase between the two grips.

Perturbation paradigms, as the name suggests, consist of an unexpected change in task requirement. In double-step paradigms this change can be introduced at the onset of the movement [49, 58-62]. By ensuring that these perturbed trials are very infrequent and random, these trials thus cannot be predicted. As such, this paradigm can be additionally utilized to test the question of how PD subjects perform in a more responsive situation. For example, it could assess whether the activation of the two movement components or the patterning of each show changes with the more *uncertain* condition (perturbed trials).

Perturbation of object location

With eight PD and eight control subjects, Castiello and Scarpa [43] applied a double-step paradigm [60], perturbing the reach to grasp movement at its onset. Subjects performed two sessions. In one session no perturbations were applied (*blocked* trials). Subjects were required to reach and grasp a spherical transparent object (4 cm diameter) which contained three light emitting diodes. One such sphere was located in the mid-sagittal plane at 15 cm from the starting position. The two other spheres were located at 27 and 40 cm, respectively. The subject began the movement immediately upon illumination of one of the spheres and then reached, grasped and brought the target sphere back to the resting position. The requirement was for a movement speed which allowed an error-free performance. In short, the results from these blocked trials exactly reproduced those found by the previously reported studies for different reaching distances. Again, too, a delay was found for the onset of the manipulation component.

The other session consisted of 60 experimental trials. The go-signal was always the illumination of the sphere located at 15 cm. For 40/60 trials, this same target remained illuminated until it had been grasped (*control* trials). For 20/60 trials, a perturbation was applied immediately upon movement onset (*perturbed* trials). For ten of these trials, illumination shifted to the sphere at 27 cm; for ten, it shifted to the sphere at 40 cm. The subject was thus required to reach and grasp the second target, which remained illuminated until it was grasped. The perturbed and control trials were randomly distributed.

In summary, the results indicated that PD subjects showed little dysfunction in the ability to appropriately respond to a spatial perturbation. As was found for control subjects, the first peak of acceleration for the perturbed trials was earlier than the single peak of the control trials. The same was found when comparing the first peak of velocity for the perturbed trials to the single peak of the control trials. These earlier temporal settings of peak acceleration and velocity indicated that PD subjects, like controls, had perceived the perturbation and had suitably and quickly reacted by allowing more time for the second task. In keeping with previous interpretations [58, 62], the first movement appeared to be curtailed as part of a rapid, on-going, motor reorganization to mobilize the second movement. The only obvious difference for PD subjects was the time at which the transition occurred. For perturbed trials of this subject group peak acceleration occurred within a range of 150 - 270 ms after movement onset. For control subjects, this peak was much closer to the minimum estimated time for visual feedback correction (that is, at around 100 ms). The nature of this later change in PD subjects could, on the one hand be related to their overall slowness of movement. On the other, it could, once again, be an indication of a difficulty with sequential movements, that is, with the ability to close (or semiclose) one movement and to open the succeeding action. Despite this dysfunction the overall integrity of responsive behaviour in Parkinson's disease is maintained [34, 35, 63-65]. When presented with an unexpected visual stimulus, PD subjects can recall and execute the correct motor program to reach and grasp the new target.

Object size perturbation

In non-Parkinsonian subjects, it has previously been found [48, 49] that the natural response to a perturbation from small to large target stimulus is a change of the distal patterning from one type of grasp (e.g., precision grip between the index finger and thumb) to another (whole hand prehension). Kinematic changes to the transport component are also evident.

Castiello et al. [66] performed a kinematic assessment of the reach to grasp movement and its adaptive response to a perturbation of object size in 15 PD and 15 control subjects. For blocked trials, subjects were instructed to reach 35 cm to grasp and lift an illuminated small (0.7 cm) or large (8 cm) diameter cylinder. Looking first at this simple assessment, it was once again evident that the patterning of movement kinematics for PD subjects largely resembled that for non-PD subjects. The grasp adopted for each cylinder was appropriate, that is, a precision grip for the small and a whole hand prehension for the large cylinder. A comparison of parameters across the two conditions showed that PD subjects demonstrated appropriate adjustment according to accuracy requirements of the task. Thus, movement initiation time (so called because no emphasis was placed on reacting as quickly as possible to the illumination stimulus) and the duration of the movement were longer for trials to the small than for the large cylinder. Likewise for this small, as opposed to the large cylinder, deceleration time (from peak arm velocity to the end of the movement) was prolonged and the time of maximum grip aperture was earlier. These are all indications, that mechanisms for the recognition of end-task requirements and for the execution of suitable motor patterns are not obviously affected by Parkinson's disease.

The perturbed trials of this experiment consisted of a change in illumination at movement onset; this onset being the release of the starting switch by the reaching arm. The change was either from the small to the large cylinder or vice versa. Again, such perturbed trials were infrequent (10/100 for each type of perturbation) and randomized with trials whereby illumination did not change.

PD subjects showed appropriate responsive behaviour with perturbed trials. They quickly adapted the grasp to successfully complete the movement. Thus, with perturbations from large to small cylinder, the end-grasp was a precision grip (PG) despite an initial execution of whole hand prehension (WHP). Suitable kinematic adjustments were also evident in the transport component and reproduced those results found with studies of non-PD subjects. For example, the timing of peak deceleration was earlier for perturbed than for non-perturbed trials (see Haggard this volume).

Differences between PD and control subjects were at transitional stages of the movement. This was particularly evident for perturbations requiring a change from precision grip to whole hand prehension. The profile of grasp

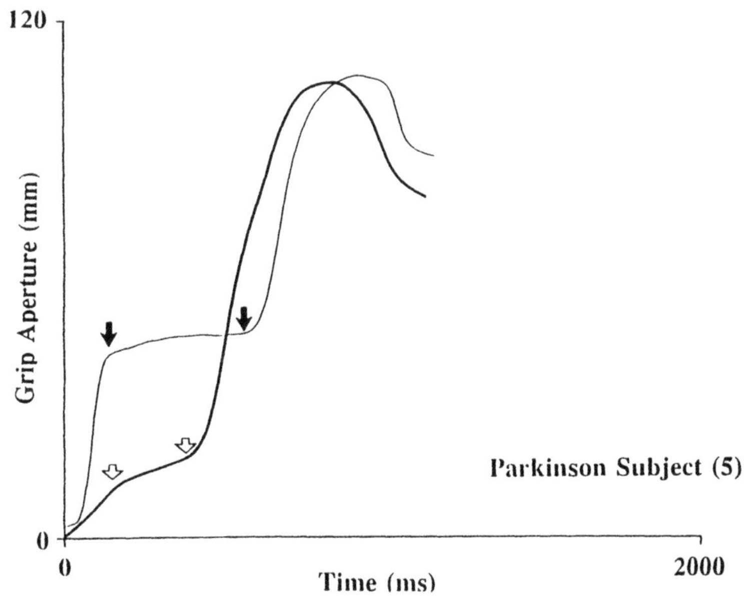

Figure 3. Grip aperture profile for perturbed trials from small to large cylinder for a Parkinson subject. Light line: grasp change. The filled arrows indicate the times of inflection (left) and the onset time for whole hand prehension. Dark line: aperture change. The open arrows indicate the times of inflection (left) and the onset time for the larger aperture.

aperture over time showed a plateau between the initial opening for PG and that following for WHP. This showed an average duration of 400 ms and an amplitude appropriate for the small cylinder (Fig. 3). It was as if the first pattern (PG) perseverated, or was not easily closed, before the second pattern (WHP) could be expressed. Such difficulty in ***releasing*** the first motor output is in accordance with the neurophysiological findings of Brotchie et al. [67, 68]. These investigators found a proportion of neurons in the anterior globus pallidus which showed phasic discharge in relation to the end of a wrist movement. It was proposed that this mechanism could operate both to terminate sustained activity in the supplementary motor area for an existing movement and to prepare for execution of the forthcoming movement. With basal ganglia dysfunction, such results would predict deficits/delays when the motor requirement is for cancellation of already activated channels and for the setting of cortical excitability for an upcoming movement. The absence or malfunctioning of this end-of-movement phasic influence could explain the long transition between closure of the precision grip channels and activation of those for whole hand prehension.

Subjects also performed trials which perturbed only grasp aperture [49]. For one set of these perturbed trials, subjects were asked to always perform a precision grip and thus, with perturbation from small to large target object the demand was for a change from a small to a large PG aperture. That is, the type of grasp did not change. The other set of perturbed trials assessed change of the WHP aperture. In contrast to the perturbed trials requiring a change of grasp type, for the perturbation of grasp aperture a prolonged transition was not found. From the small to the large aperture of either PG or WHP the shift was smooth, being barely evident on the aperture profile (Fig. 3). PD subjects, thus, did not show a dysfunction with the amplification of an existingly activated program. Again, this is in accordance with results of Brotchie et al. [67, 68]. They found that pallidal neurons show little relationship to movement parameters such as the amplitude of angular wrist movement or the amount of torque production. This supports the contention that the basal ganglia are not directly involved in the execution of a movement pattern once it is in operation.

The rapid and smooth aperture change also indicates that the long transition from PG to WHP cannot be attributed to mechanical factors. For example, the well-known rigidity of PD subjects does not limit the speed with which the fingers can open for a larger grip. Central neural mechanisms for the transition are more supported.

A further finding of interest was the conditional delay of manipulation component onset. For blocked trials, this component showed the same delay reported earlier (see SIZE AND DISTANCE EFFECTS). For control and perturbed trials this delay was not evident, the manipulation component being activated at the same time as, or shortly after (around 50 ms), the transport component. This result further indicates that the delay is not simply an adjunct of the slower PD movement. Secondly, it shows that there is a difference of motor output according to the predictability of the task. In blocked trials there was no uncertainty - the subject knew that ten trials were dedicated solely to one task. Premovement planning reliably corresponded to subsequent movement execution, that is, this type of trial was predictable. Given the proposal that PD subjects have a dysfunction with predictive behaviour, this could explain the later onset of the manipulation component. In the control and perturbed trials there were a number of premovement uncertainties such as whether or not the small or the large cylinder would be illuminated or whether the trial would be perturbed or not. These lessened task predictability and this could explain why the delay was not present for these trials.

FUNCTIONAL TASK

Recently a further kinematic study was conducted [69], again with PD and control subjects, in order to assess a functional action which includes the reach to grasp movement, that is, a drinking action. In this case, the reach to grasp is usually a necessary requirement for bringing a glass to the lips. With such an assessment it is possible to observe movement kinematics of the reach to grasp when this movement is performed within an example of its usual context. In this study, the first arm movement phase consisted of a reach to grasp. The second phase consisted of a movement in the opposite direction to bring the receptacle to the lips. Overall, the action thus consists of both simultaneous component activation (transport and manipulation in the first phase) and sequential motor activation (first followed by second phase). The aim of this experiment was thus to further test the deficit of simultaneous/sequential component/movement activation when performed in a fully functional manner. Subjects were required to reach 28 cm, to grasp a cup filled with 180-200 ml of water, and to bring the cup to the lips in order to take a sip.

For seven of the nine PD subjects at least one of the ten reach-grasp-drink trials (38% of the total number of trials for this group) showed a transition period between the end of the ***reach to grasp*** movement and the onset of the ***take to lips*** movement. The duration of this phase was, on average, 337 ms and comprised 7% of total movement time. The control subjects (bar for one trial of one subject) showed no transition phase between the two movements, the time at which the ***reach to grasp*** movement finished corresponding with the time of onset of the ***take to lips*** movement. The transition phase can thus be regarded as a sign which is almost exclusive to PD subjects but one that is not characteristic of all these subjects. Its presence was not related to age, gender or severity of disease.

CONCLUSION

The findings presented in this chapter confirm that PD subjects have a dysfunction with the simultaneous or sequential activation of motor programs [29-33]. The abnormality is more evident when the activation is of motor patterns which are subserved by different neural substrates. In the case of prehension movements the control channels for transport are most probably distinct from those of manipulation [70-73]. Similarly those for precision grip may differ from those for whole hand prehension [71, 74, 75]. For PD subjects, Benecke et al. [31, 32] described particular deficits with the activation of separate motor programs, that is, those which showed no evidence of being controlled by a single complex motor program (c.f. [76]). Thus, the size perturbation study shows a delay in the transition from one grasp (precision grip) to another (whole hand prehension). So, too, the drinking study shows a transition period between grasping the cup and moving it to the lips. However, from our studies on prehension, a disturbance of the near-simultaneous or sequential activation of two coordinated movement **components** also emerged. Under predictive conditions the manipulation component is activated obviously later than the transport component. The deficit of PD subjects in the activation of programs thus not only applies to those motor programs which are completely independent but also to those which are largely independent but which show functional coordination.

Two main but distinct arguments can be advanced to explain these delays/transitions. One is that the Parkinson subject places greater emphasis upon the utilization of movement related feedback. The second argument is that of a central delay in the activation and probably also in the suppression of motor programs.

Previous studies have demonstrated that PD subjects become more dependent upon visual feedback to guide movement [77, 78]. As proposed by Goldberg [79], this could reflect greater reliance upon the responsive, feedback dependent, lateral, premotor system involving the arcuate premotor area and the cerebellum. In the studies presented, feedback has not been withdrawn and thus we cannot make any definitive conclusions as to the degree to which it is used by PD subjects. The delayed onset of manipulation for blocked trials could, however, represent processing times of the visual and proprioceptive information resulting from arm transport. Such rapid processing of movement related information for on-line correction has been shown for the visual [59, 62, 80], proprioceptive [81, 82] and kinaesthetic [83] modalities. This feedback, together with premovement cues, could be utilized to trigger onset of the manipulation component. Similarly, on-line movement feedback could be utilized for the change from one grasp to another in the size perturbation experiment. Thus, for the perturbed trials from precision grip to whole hand prehension the transition took about 400 ms - more than sufficient time to allow for the processing of feedback and activation of whole hand prehension. The transition period of 337 ms (average) between the two phases of the drinking task was likewise of adequate duration for operating upon afferent input and mobilizing the second phase.

It has been suggested that a greater utilization of feedback may avoid the errors associated with the non-feedback movement performance by PD subjects [38, 39]. In the prehension experiments reported, the incidence or type of errors was not determined. Nevertheless, it was clear that PD subjects adapted quickly to a perturbation and successfully performed the end-task [43, 52, 66]. Further, despite transition periods from one grasp to another in the size perturbation experiment, coordination between arm transport and hand opening was maintained. For perturbed trials, PD subjects continued to show temporal coordination between the point of maximum arm deceleration and that of maximum grip aperture. Feedback during the transition period could thus assist in ascertaining the current status of each component for activation of the second grasp pattern and for intercomponent recalibration.

Together with the supplementary motor area, the basal ganglia are thought to form a medial system which operates largely in a feed-forward mode (see [79] for review). The explanation that the delays of movement activation are reflective of a greater dependence upon feedback processing implies that this medial system has been bypassed and that PD subjects turn " .. to remaining functions of the relatively spared lateral premotor system to attempt to substitute for those lost through medial system impairment" ([79] p. 582). An alternative argument is that the medial system **has** been activated: the delays of movement activation would then directly reflect dysfunction of the basal ganglia rather than rerouting strategies. If activation of this system is assumed, it can further be proposed that particular characteristics of the task compel use of the medial system. These characteristics are probably best described as ***certainties*** during both the premovement and execution phases of the task. Thus for blocked trials, all task demands are known in the premovement phase and for the perturbed trials, no further unexpected elements are presented following perturbation. The predictable features of these trials may thus dictate neural involvement of the ***feed-forward*** motor circuit [78, 84].

The motor circuit (medial system) is thought to consist of multiple cortico-striato-nigro-thalamocortical circuits arranged in a parallel and topographical manner [85, 86]. With the loss of striatal influence resulting from dopamine depletion, an increased inhibition of the thalamo-cortical pathway has been proposed [87, 88]. Areas within the supplementary motor area and motor cortex would thus be less responsive to activation - the pattern of ***readiness*** to triggers from sources other than that of the basal ganglia, having not been set. If this lack of responsiveness was confined to a specific neural channel (e.g., transport or manipulation; whole hand prehension or precision grip) this would explain why a movement shows a delay of activation.

Why, however, is it the manipulation component, in particular, that shows the later activation? The neural mechanisms subserving this component may have a degree of neuroanatomical complexity which is greater than that of the transport component. With normal functioning of the basal ganglia the ground-work of appropriate cortical responsiveness normally hides this complexity such that the manipulation component is activated at the same time or shortly after the transport component (see also [42, 89, 90]). With inadequate cortical preparation rapid activation of a complex movement component may become more difficult.

There are a number of remaining issues for future research of the reach to grasp movement in PD subjects. One which we are currently addressing is the issue of whether or not the bradykinesia of PD subjects determines the delays and transition phases. To this end we are testing non-PD subjects to perform prehension with durations equivalent to those of PD subjects. If, as we believe, the deficit in sequential-simultaneous movements is of central origin, the prediction is for an absence of delays/transitions in the non-PD group only. Conversely if PD subjects were to be instructed to perform the movement as rapidly as possible a central origin for the deficit would predict that the delays/transitions would nevertheless be still evident.

We are also currently investigating the performance of bimanual reach to grasp movements by PD subjects. An example of such a movement is when one reaches to both grasp a can of soft drink and pull open the tab. In this example, one arm performs a reach and whole hand prehension movement; the other arm performs a reach and precision grip movement. In total, four components (two transport and two manipulation) need to be activated almost simultaneously at the beginning of this movement. In a study of non-PD subjects [91] it has been demonstrated that with this movement both arms show similar reaching durations yet each has a kinematic parameterization which is matched to its specific end-task. Thus the arm which performs a precision grip shows an earlier peak of acceleration than the contralateral limb - this anticipation allows a longer deceleration phase and reflects the greater accuracy requirements of the precision grip task. As yet, the results for the bimanual study of PD subjects are not available but several questions will be addressed. Given the problem of simultaneous\sequential component activation how will four components be activated in a near-simultaneous manner? Will movement of one arm begin after movement of the other arm, for example? How will the accuracy requirements of the task influence the sequence of component activation? Will, for example, the arm performing the precision grip show a later activation of the manipulation component than the arm performing the whole hand prehension? Finally, how will the kinematics of the movement be organized if there has been delays in component activation?

ACKNOWLEDGEMENTS

Our gratitude is extended to Alberto Bazzini for his technical work upon which many of the experiments are based. Carl Waterman is gratefully acknowledged for technical and programming assistance. Members of the Arizona chapter of the American Parkinson's Association are thanked for their participation in these experiments.

REFERENCES

1 K. Jellinger, in: C.D. Marsden and S. Fahn (eds.), Movement Disorders 2, Butterworth, London, 1987, p. 124.
2 F. Gray, Rev. Neurol., 144 (1988) 229.
3 Y. Agid, M. Ruberg, R. Raisman, E. Hirsch and F. Javoy-Agid, in: G. Stern (ed.), Parkinson's Disease, Chapman and Hall Medical, Cambridge, 1990, p. 99.
4 O. Hornykiewicz, Fed. Proc., 32 (1973) 183.
5 I.J. Farley, K.S. Price and O. Hornykiewicz, Adv. Biochem. Psychopharmacol., 16 (1977) 57.
6 K.S. Price, I.J. Farley and O. Hornykiewicz, Adv. Biochem. Psychopharmacol., 19 (1978) 293.
7 H. Bernheimer and O. Hornykiewicz, Klin. Wochenschrift, 43 (1965) 711.
8 T. Lee, P. Seeman, A. Rajput, I.J. Farley and O. Hornykiewicz, Nature, 273 (1978) 59.

9 U.K. Rinne, P. Loennberg and V. Koskinen, J. Neural. Trans., 51 (1981) 97.
10 O. Hornykiewicz and S.J. Kish, in: M.D. Yahr and K.J. Bergmann (eds.), Parkinson's Disease, Advances in Neurology, No. 45, Raven Press, New York, 1986, p. 19.
11 E.G. Jones, in: Motor Areas of the Cerebral Cortex, Ciba Foundation Symposium, No. 132, Wiley, Chichester, 1987, p. 21.
12 M. Wiesendanger, H. Hummelsheim, M. Bianchetti, D.F. Chen, B. Hyland, V. Maier and R. Wiesendanger, in: Motor Areas of the Cerebral Cortex, Ciba Foundation Symposium, No. 132, Wiley, Chichester, 1987, p. 40.
13 P.S. Goldman-Rakic and L.J. Porrino, J. Comp. Neurol., 242 (1985) 535.
14 I.A. Ilinsky, M.L. Jouandet and P.S. Goldman-Rakic, J. Comp. Neurol., 236 (1985) 315.
15 G.E. Alexander, M.R. DeLong and P.L. Strick, Ann. Rev. Neurosci., 9 (1986) 357.
16 M.R. DeLong, M.D. Crutcher and A.P. Georgopoulos, J. Neurosci., 3 (1983) 1599.
17 M.R. DeLong, A.P. Georgopoulos, M.D. Crutcher, S.J. Mitchell, R.T. Richardson and G.E. Alexander, in: Functions of the Basal Ganglia, Ciba Foundation Symposium, No. 107, Pitman, London, 1984, p. 82.
18 M.R. DeLong, G.E. Alexander, W.C. Miller and M.D. Crutcher, in: Motor Areas of the Cerebral Cortex, Ciba Foundation Symposium, No. 132, Wiley, Chichester, 1987, p. 21.
19 M.R. DeLong, J. Neurophysiol., 34 (1971) 414.
20 M.R. DeLong and P.L. Strick, Brain Res., 71 (1974) 327.
21 M.R. DeLong and A.P. Georgopoulos, in: J.M. Brookhart et al. (eds.), Handbook of Physiology, The Nervous System, Amer. Physiol. Soc., Bethesda, 1981, p. 1017.
22 A.P. Georgopoulos, M.R. DeLong and M.D. Crutcher, J. Neurosci., 3 (1983) 1586.
23 M.D. Crutcher and M.R. DeLong, Exp. Brain Res., 53 (1984) 244.
24 E.V. Evarts, J. Neurophysiol., 29 (1966) 1011.
25 A.P. Georgopoulos, J.F. Kalaska, R. Caminiti and J.T. Massey, J. Neurosci., 2 (1982) 1527.
26 J.F. Kalaska, R. Caminiti and A.P. Georgopoulos, Exp. Brain Res., 51 (1983) 247.
27 E.V. Evarts and W.T. Thach, Ann. Rev. Physiol., 31 (1969) 451.
28 J.M. Kemp and T.P.S. Powell, Brain, 93 (1970) 525.

29 R.S. Schwab, M.E. Chafetz and S. Walker, Arch. Neurol. Psychiat., 72 (1954) 591.
30 G.A. Talland and R.S. Schwab, Neuropsychologia, 2 (1964) 45.
31 R. Benecke, J.C. Rothwell, J.P.R. Dick, B.L. Day and C.D. Marsden, Brain, 109 (1986) 739.
32 R. Benecke, J.C. Rothwell, J.P.R. Dick, B.L. Day and C.D. Marsden, Brain, 110 (1987) 361.
33 D.L. Harrington and K.Y. Haaland, Brain, 114 (1991) 99.
34 K. Flowers, Brain, 101 (1978) 19.
35 K. Flowers, Brain, 101 (1978) 35.
36 R.L. Gregory, Ergonomics, 13 (1970) 25.
37 C.A. Bloxham, T.A. Mindel and C.D. Frith, Brain, 107 (1984) 371.
38 B.L. Day, J.P.R. Dick and C.D. Marsden, J. Neurol. Neurosurg. Psychiat., 47 (1984) 1299.
39 T. Crawford, S. Goodrich, L. Henderson and C. Kennard, J. Neurol. Neurosurg. Psychiat., 52 (1989) 1033.
40 M. Jeannerod, in: J. Long and A. Baddeley, Attention and Performance IX, Erlbaum, Hillsdale, 1981, p. 153.
41 M. Jeannerod, J. Mot. Behav., 16 (1984) 235.
42 M. Gentilucci, U. Castiello, M. Scarpa, M.L. Corradini, C. Umiltá and G. Rizzolatti, Neuropsychologia, 29 (1991) 361.
43 U. Castiello and M. Scarpa, submitted.
44 U. Castiello, G.E. Stelmach and A.N. Lieberman, Neuropsychologia, 31 (1993) 395.
45 T.W. Zakon, Unpublished doctoral thesis, McGill, University, 1989.
46 M. Hallett and S. Khoshbin, Brain, 103 (1980) 301.
47 R.G. Marteniuk, C.L. MacKenzie, M. Jeannerod, S. Athenes and C. Dugas, Can. J. Psychol., 41 (1987) 365.
48 U. Castiello, K.M.B. Bennett and Y. Paulignan, Behav. Brain Res., 50 (1992) 7.
49 U. Castiello, K.M.B. Bennett, G.E. Stelmach, Exp. Brain Res., 94 (1993) 165.
50 L.S. Jakobson and M.A. Goodale, Exp. Brain Res., 86 (1991) 199.
51 K.M.B. Bennett, C. Adler, G.E. Stelmach and U. Castiello, Neuropsychologia, 31 (1993) 709.
52 U. Castiello, K.M.B. Bennett, C. Adler and G.E. Stelmach, Neuropsychologia, 31 (1993) 717.
53 A.J. Martinez and R.A. Utterbach, Neurology, 23 (1973) 170.
54 L.S. Jakobson, Y.M. Archibald, D.P. Carey and M.A. Goodale, Neuropsychologia, 29 (1991) 803.
55 M. Jeannerod, Neuropsychologia, 24 (1986) 41.

56 E. Agostoni, A. Coletti, G. Orlando and G. Tredici, J. Neurol. Neurosurg. Psychiat., 46 (1983) 804.
57 E. DeRenzi, P. Faglioni, M. Scarpa and G. Crisi, J. Neurol. Neurosurg. Psychiat., 49 (1986) 1030.
58 Y. Paulignan, M. Jeannerod, C.L. MacKenzie and R.G. Marteniuk, Exp. Brain Res., 87 (1991) 407.
59 Y. Paulignan, C.L. MacKenzie, R.G. Marteniuk and M. Jeannerod, Exp. Brain Res., 83 (1991) 502.
60 M. Gentilucci, S. Chieffi, M. Scarpa and U. Castiello, Behav. Brain Res., 82 (1992) 47.
61 U. Castiello and M. Jeannerod, Neuroreport, 2 (1991) 797.
62 U. Castiello, Y. Paulignan and M. Jeannerod, Brain, 114 (1991) 2639.
63 K.A. Flowers, Neurology, 99 (1975) 413.
64 K.A. Flowers, Brain, 99 (1976) 269.
65 A. Baroni, F. Benvenuti, L. Pantaleo and F. Urbani, Neurology, 34 (1984) 868.
66 U. Castiello, K.M.B. Bennett and G.E. Stelmach, submitted
67 P. Brotchie, R. Iansek, M.K. Horne, Brain, 114 (1991) 1667.
68 P. Brotchie, R. Iansek, M.K. Horne, Brain, 114 (1991) 1685.
69 K.M.B. Bennett, C. Waterman and U. Castiello, submitted.
70 J. Brinkman and H.G.J.M. Kuypers, Brain, 96 (1973) 653.
71 G. Rizzolatti, R. Camarda, L. Fogassi, M. Gentilucci, G. Luppino and M. Matelli, Exp. Brain Res., 71 (1988) 491.
72 S.P. Wise and R. Desimone, Science, 242 (1988) 736.
73 B. Alstermack, A. Lundberg, N. Norrsell and E. Sybirska, Exp. Brain Res., 42 (1981) 299.
74 R.B. Muir and R.N. Lemon, Brain Res., 261 (1983) 312.
75 R.B. Muir, Exp. Brain Res. Suppl., 10 (1985) 155.
76 M.C. Carter and D.C. Shapiro, J. Neurophysiol., 52 (1984) 787.
77 J.D. Cooke, J.D. Brown and V.B. Brooks, Can. J. Neurol. Sci., 5 (1978) 413.
78 Y. Stern, R. Mayeux, J. Rosen and J. Ilson, J. Neurol. Neurosurg. Psychiat., 46 (1983) 145.
79 G. Goldberg, Behav. Brain Sci., 8 (1985) 567.
80 R.P. Young and H.N. Zelaznik, Acta Psychol., 79 (1992) 59.
81 R.G. Lee and R.G. Tatton, Can. J. Neurol. Sci., 2 (1975) 285.
82 E.V. Evarts and E.J. Vaughn, in: J.E. Desmedt (ed.), Cerebral Motor Control in Man: Long-loop Mechanisms, Progress in Clinical Neurophysiology, Karger, Basel, 1978, p. 178.
83 R.S. Johansson and G. Westling, Exp. Brain Res., 66 (1987) 141.
84 C.D. Marsden, Neurology, 32 (1982) 514.

85 A. Parent, T.I.N.s, 13 (1990) 254.
86 G.E. Alexander and M.D. Crutcher, T.I.N.s, 13 (1990) 266.
87 G. Chevalier and J.M. Deniau, T.I.N.s, 13 (1990) 277.
88 M.R. DeLong, T.I.N.s, 13 (1990) 281.
89 R. Porter and M.M.C. Lewis, Brain Res., 98 (1975) 21.
90 J.T. Murphy, Y.C. Wong and A.C. Kwan, J. Neurophysiol., 53 (1985) 435.
91 U. Castiello, K.M.B. Bennett and G.E. Stelmach, Behav. Brain Res., 56 (1993) 43.

Insights into the Reach to Grasp Movement
K.M.B. Bennett and U. Castiello (Editors)

CHAPTER 12

TASK-SPECIFIC TRAINING OF REACHING AND MANIPULATION

L. ADA, C.G. CANNING, J.H. CARR, S.L. KILBREATH and R.B. SHEPHERD

School of Physiotherapy, Faculty of Health Sciences, The University of Sydney, P.O. Box 170, Lidcombe, N.S.W., Australia, 2141.

SUMMARY

Full upper limb function following brain damage is frequently not achieved despite therapeutic intervention. A new approach to treatment is described in which strategies based on research in the movement sciences are employed to aid the recovery or relearning of functional upper limb movements. The chapter begins by describing some aspects of the dyscontrol characteristics resulting from central nervous system lesions. This is followed by an account of commonly observed adaptive motor behaviors resulting from the effects of the lesion. Finally, examples of task-specific clinical intervention derived from research in relevant movement sciences are given.

INTRODUCTION

The major functional use of the upper extremities is in reaching for and manipulating objects, the arm acting to place the hand in the appropriate position and orientation in space to interact with the environment. Recent interest in the study of reaching and grasping reflects a desire to understand not only the neuromotor control mechanisms of these actions but also the relationship between the observable movements and their interaction with neuromotor processes. For example, some investigators (e.g., [1-4]) have studied the coordination of arm movement with hand movement in reaching and grasping actions, and others (e.g., [5, 6]) have used this kind of analysis to investigate the development of prehension in infants and children. In addition, other investigators (e.g., [7-9]) have described how the physical characteristics of objects dictate the hand postures assumed and the orientation of the hand in preparing for grasping.

This chapter reviews some common movement problems which emerge in adults and children following central nervous system (CNS) damage and explores the scientific basis for clinical intervention in the analysis and training of reaching and grasping. Some clinical examples of task-specific training are given to illustrate the process of incorporating scientific findings related to the control of reaching and grasping into rehabilitation. The concept of developing task-specific rehabilitation intervention from the sciences related to human movement (biomechanics, motor learning, human ecology, neural mechanisms, muscle biology) is described in detail elsewhere [10-13].

Outcome of rehabilitation of upper limb function has been investigated following stroke more than any other neurological condition. An early study [14] reported that, 7 months following stroke, 43% of patients had regained at least three quarters of the normal range of active motion at the wrist, elbow and shoulder but the authors did not report on the performance of functional tasks. More recent studies have documented outcome in terms of functional ability (e.g., [15-19]). Outcome varies from that reported by Gowland [15] in which 5% of the 223 patients studied regained functional use of their arm and hand to that of Dean and Mackey [18] in which 52% of the 70 patients studied were able to perform tasks such as combing the hair at the back of the head with the affected hand by the time of discharge from rehabilitation. It is difficult to draw conclusions from outcome studies, however, because different inclusion criteria, duration of therapy, and measurement instruments were used. Furthermore, there is seldom any detail of the type of physiotherapy given (Dean and Mackey's study [18] is one exception). Type of intervention, for example, facilitation or movement-science based, would be expected to affect outcome, since external events have been shown to affect the process of brain reorganization (e.g., [20]).

Even the studies with the best reported outcomes [16, 18] suggest that up to 50% of patients never regain full function of their affected arm. In addition, those patients who achieve the maximum score on functional testing do not necessarily feel that their upper limb has fully recovered and may not use the affected arm to perform normal bimanual activities [17].

These findings suggest that there is a need to continue developing and refining intervention which will assist people with brain damage to regain their optimum function using the upper extremity. Since factors such as location and severity of the lesion can be expected to play a part in recovery, it may be possible in the future, with improved diagnostic techniques, to identify individual types of lesions that best respond to rehabilitation.

DYSCONTROL CHARACTERISTICS FOLLOWING BRAIN DAMAGE

In this section, the most common features of brain lesions which produce motor impairment will be discussed. These features include weakness, spasticity and sensory impairment associated with upper motor neuron lesions (UMNL), and dyssynergia and dysmetria associated with cerebellar lesions. Cognitive deficits such as spatial neglect, poor attention span, memory loss and dysphasia also interfere with motor behavior but will not be described in this section.

The neurologist Hughlings Jackson, in the late 19th century, observed that the motor problems resulting from lesions of the CNS can be categorised as negative or positive. More recently, Burke [21] has described the negative features as those that represent a loss of function previously present, such as loss of muscle strength and dexterity, while positive features are additional, such as abnormal proprioceptive and cutaneous reflexes.

Negative features of upper motor neuron lesions

Burke [21] suggests that the major impairment to motor control following UMNL is more associated with the negative rather than the positive features. Weakness or paralysis, that is, an inability or decreased ability to produce appropriate amounts of force in muscles contralateral to the side of the lesion, is typically the major negative feature that affects reaching, grasping and manipulation following CNS damage. Systematic investigations of deficits resulting from discrete lesions following stroke shed some light onto both the pattern and physiological basis of weakness. Historically, extensor muscles have been reported to be more affected than flexor muscles, and distal muscles of the hand more affected than the more proximal muscles of the shoulder and elbow [22-24]. However, recent studies do not substantiate these clinical observations (e.g., [25-27]). Extensor muscles are not typically more affected than flexor muscles, and while there is a general tendency for the hand muscles to be weaker than the shoulder muscles, there is no stereotypical pattern of muscle weakness. Thus, both the ability to open and to close the hand are likely to be particularly affected.

The severity with which the hand muscles are affected may be related to whether or not the pyramidal tract is interrupted. The pyramidal tract comprises, in part, monosynaptic corticomotoneuronal fibres. There are more of these monosynaptic fibres to the distal hand muscles than to the more proximal muscles of the arm [28, 29]. Furthermore, the cortical cells from which these monosynaptic fibres arise demonstrate greater activity during finely coordinated hand movements [30] than during an activity which requires less skill but greater force. It is not surprising, therefore, that the interruption of these cortical projections in humans has been associated with profound impairment in the performance of motor tasks requiring manipulation (e.g., [31]).

Although the major impairment following unilateral cerebrovascular accident (CVA) is on the contralateral side, the ipsilateral side is also affected (e.g., [27, 32]). On the ipsilateral side, the proximal muscles are more affected than the distal muscles, unlike the contralateral side where the hand muscles tend to be weaker (e.g. [27, 32]). It has been suggested that this impairment in strength on the ipsilateral side is more likely to be a result of uncrossed pyramidal tract fibres (e.g., [33]) rather than simply disuse [27].

Positive features of upper motor neuron lesions

Of the dyscontrol characteristics that have been labelled as positive, spasticity is probably the most common feature following CNS damage. A widely accepted view of spasticity describes it as a motor disorder resulting from hyperexcitability of the stretch reflex, which is characterised by exaggerated tendon jerks and velocity-dependent increases in tonic stretch reflexes [34]. Hyperexcitability of the stretch reflex, especially of the tonic component, produces the hypertonia that clinicians recognise as increased resistance to passive movement.

The clinical assumption that spasticity is the major cause of dysfunction following CNS damage has been challenged in recent years. Using EMG recordings, Sahrmann and Norton [35] showed that spasticity in a muscle could not explain the slowness of movement in its antagonist. Tang and Rymer [36] have confirmed abnormality of the EMG output of paretic muscles. Other studies in which spasticity has been reduced either by voluntary control [37], or by extraneous measures such as drugs [38, 39] have demonstrated that while it was possible to eliminate (or at least decrease) hyperreflexia, this did not necessarily lead to an improvement in function.

As a general rule, spasticity develops slowly, although fast developing spasticity may occur immediately following high brainstem lesions [40]. The slow onset of spasticity has led to the suggestion that spasticity is an adaptation to neural damage rather than a direct result of the lesion [21, 40]. It is also likely that the adaptive neural processes may be exacerbated by peripheral changes happening in concert, such as the development of muscle contracture resulting from the immobility imposed by muscle weakness or paralysis. The mechanism by which spasticity and contracture are related is probably due to the facilitatory influence of muscle length on the stretch reflex. For example, the gain of the tonic stretch reflex is increased if it is elicited with the muscle in the lengthened position, that is, when it is stretched. When a muscle is shortened due to contracture, the stretch reflex will be elicited earlier in range because, for a given joint angle, the muscle will be stretched more than normal. This suggests that muscle shortening potentiates the stretch reflex and a vicious cycle is set up whereby immobility leads to muscle contracture which in turn increases the hyperreflexia. In other words, spasticity is implicated in movement dysfunction through its relationship to secondary musculoskeletal complications. Dietz and colleagues [41, 42], for example, have shown that the familiar problem of ankle stiffness following CNS damage appears to result more from the increased stiffness that accompanies muscle contracture rather than the abnormal reflex activity.

Sensory impairment

Loss of kinaesthesia, that is, loss of afference arising from cutaneous, joint and muscle receptors is another feature which may occur following brain damage. As with muscle weakness, the hand is often more affected than the upper arm [43]. As well as providing information about the environment (e.g., objects) and limb position, afferent information signals that a motor action has occurred. In addition, the cutaneous afferents facilitate the motoneuron pools of relevant hand muscles peripherally [44]. Centrally, cutaneous afferents provide information necessary for sequencing components of multiphasic tasks such as grip and load forces [45] and will have direct consequences on motor output.

Features of cerebellar dysfunction

Unlike lesions involving the pyramidal tract, lesions of the cerebellum produce errors in planning and execution of movements. For example, the structure of the inputs and outputs to specific regions within the cerebellum equip it to compare internal signals of the intended movement with external signals of the actual movement. The cerebellum, therefore, generates corrective signals that operate on the descending motor pathways (e.g. [46, 47, 48]). Thus, not only may the cerebellum be important during motor learning [49, 50] but also in learned movements because everyday motor tasks require constant adjustments to changing conditions [51].

Individuals with cerebellar dysfunction have difficulty executing coordinated movements, broadly referred to as ataxia. This includes distinctive dyscontrol characteristics such as delay in initiating movement, dysmetria which is error in the range and force of movement, as well as dysdiadochokinesia which is error in the rate and rhythm of movement. Error in the relative timing of the components of complex movements is referred to as dyssynergia or decomposition of movement which is due, at least in part, to an inability to coordinate agonist and antagonist muscles. As descending pathways originating in the cerebellum are uncrossed, discrete unilateral lesions of the cerebellum and its connections will produce deficits ipsilateral to the side of the lesion, although in practice deficits are usually bilateral.

The negative and positive features following brain damage as well as sensory impairment and cerebellar dysfunction produce unique dyscontrol characteristics. These dyscontrol characteristics, in turn, force the system to adapt to its new circumstances.

ADAPTIVE MOTOR BEHAVIOR

Following CNS damage, adaptive or compensatory motor behavior spontaneously emerges as soon as individuals attempt to accomplish a particular goal. That is, an action is performed in the most biomechanically effective manner possible given the dyscontrol characteristics of the lesion, the nature of the musculoskeletal linkage and the environmental context. Movement emerges, therefore, as a result of both internal and external factors. Adaptive motor behavior illustrates the ability of the lesioned system to put together an action out of what remains of the various sub-systems and the state of the musculoskeletal system, with the emergent movement isomorphic with the demands of the environment. For example, normally the actions of reaching, grasping and manipulation emerge as a result of such factors as segmental alignment in sitting, the size, shape and orientation of the object to be grasped, what is to be done with the object, and the strength and control of the upper limb muscles. These same factors also influence the emergence of movement following a brain lesion. Given the effects of the

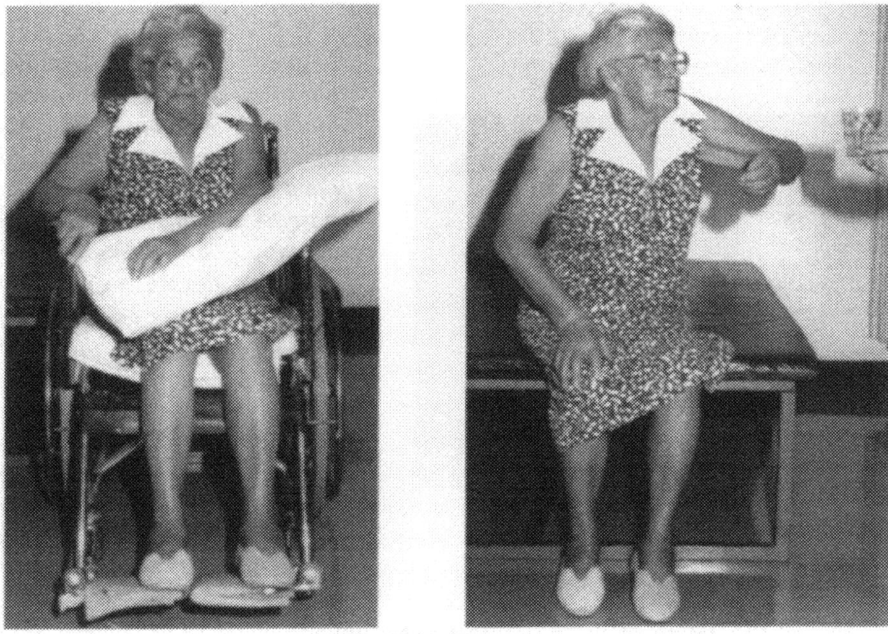

*Figure 1. **A**. This woman's arm following CVA rests on a pillow in the lap. Such a position, in time, predisposes specific muscles to adaptive shortening. **B**. The effect of poor motor control, muscle weakness and adaptive shortening is illustrated when she attempts to reach out for a glass.*

Figure 2. When asked to point to the picture, this man, following a CVA, tilts his body to the right side, elevates the left shoulder girdle, and abducts and internally rotates the glenohumeral joint in his attempt to accomplish the goal. The adaptive movement typically results from inability to contract or to control the muscles which, in reaching, flex and externally rotate the shoulder and extend the elbow.

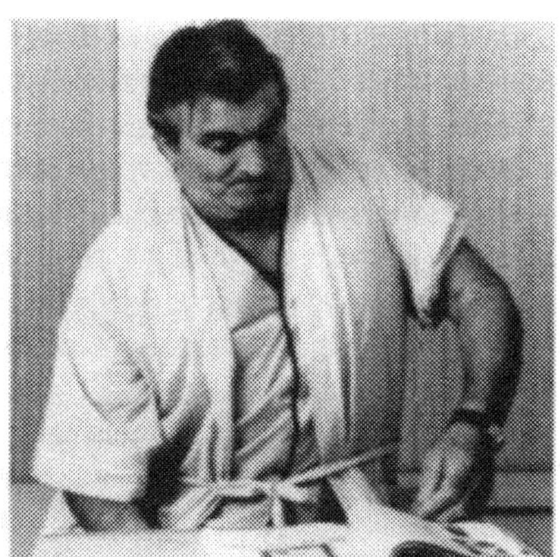

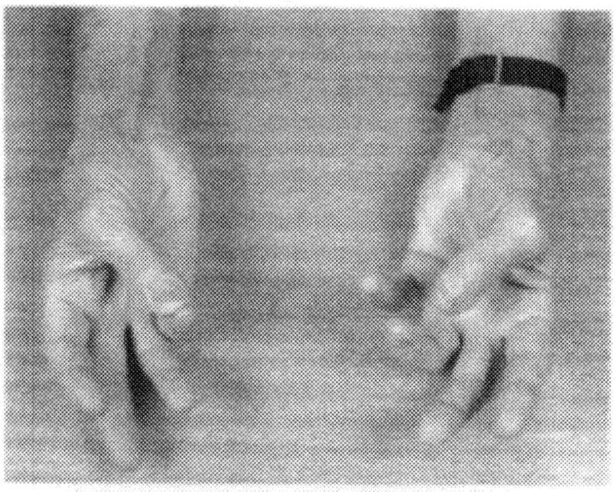

Figure 3. This man, following a CVA, cannot touch his thumb to his little finger with the left hand as he does with the normal right hand. He attempts to achieve the goal by flexing excessively at the metacarpophalangeal (MCP) joints of the thumb and little finger. These adaptive movements result from inability to abduct and rotate the carpometacarpal (CMC) joint of the thumb and rotate the fifth metacarpal.

lesion and the nature of the musculoskeletal linkage, there are probably only a certain number of possibilities for the brain-impaired individual to adapt the alignment of body parts in order to achieve a particular goal. These possibilities themselves depend upon the integrity of the musculoskeletal system.

Muscle weakness or paralysis effectively immobilize the upper limb, contributing to the development of secondary soft tissue contracture and sometimes to musculoskeletal deformity. Enforced immobility following a lesion of the CNS will cause length-associated changes in muscles. Persistent positioning of the upper extremity, as in Fig. 1**A**, may result in contracture of such muscles as internal rotators and adductors of the shoulder, flexors of the elbow, pronators of the forearm, flexors of the wrist and long flexors of the fingers and thumb. Contracture, if allowed to occur, will seriously interfere with the ability to reach out toward and preshape the hand appropriately for objects (Fig. 1**B**).

Figs. 2 - 8 illustrate some examples of commonly observed adaptive movements.

An important factor in reaching and manipulation is the ability to balance in sitting or standing since adaptive motor behavior takes place, not only in terms of the focal action (e.g., the reaching), but also in the associated postural adjustments.

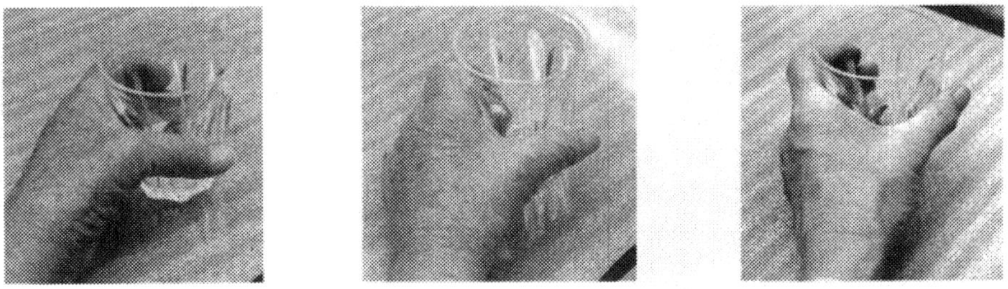

Figure 4. ***A**. The hand on the left illustrates the pattern of grasp used to hold a large glass when the web space between the thumb and index finger has adaptively shortened. Insufficient abduction of thumb CMC joint forces adaptive MCP flexion and abduction. Note that abduction of the MCP joint of the thumb can only occur if there is flexion at the joint.* ***B**. The malalignment of the hand, with the little finger side not in contact with the glass, is due to excessive pronation of the forearm and lack of wrist extension. Note also the flexion of the MCP joint of the thumb.* ***C**. The hand on the right shows the normal alignment of wrist, forearm and thumb in grasping the same glass.*

Figure 5. In order to adapt to poor control over the terminal phase of reaching, the hand is stabilised on the table during approach. Inappropriate pre-shaping has resulted in an excessive grasp aperture for the size of the object. These are typical adaptive behaviors seen in individuals with cerebellar dysfunction.

Normally we adapt our motor performance to take into account the characteristics of an object and the intention. For example, Iberall and colleagues [8] have illustrated how we orient our hand in manipulating an object, such as a cylinder, according to whether we are to shake it or place it on a table.

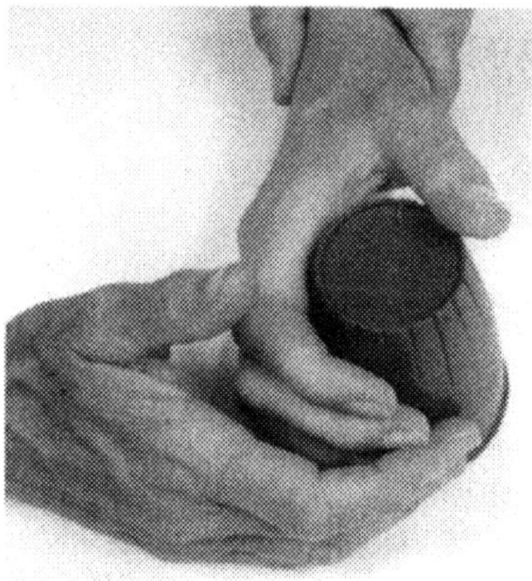

Figure 6. When attempting to release an object, if the thumb cannot be abducted and rotated at the CMC joint, release can be achieved by extension at the CMC and MCP joints.

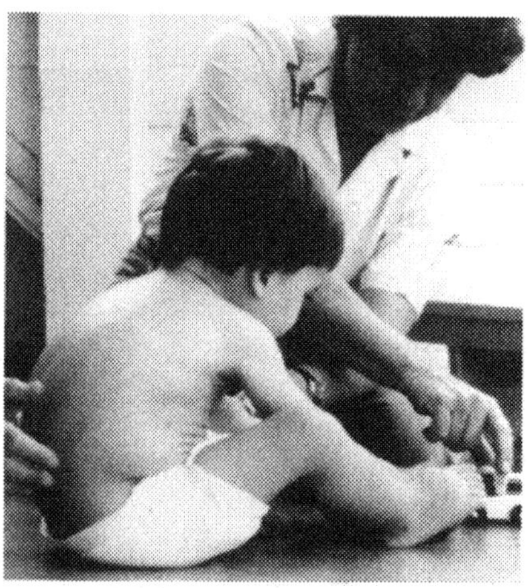

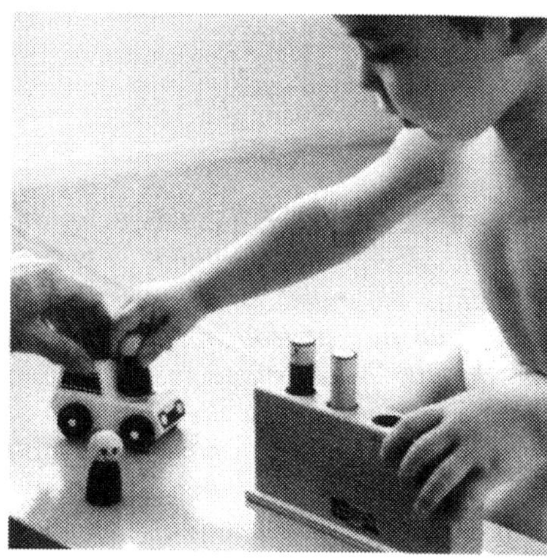

*Figure 7. **A**. This child with cerebral palsy is having difficulty reaching and playing with his toys because of his inability to balance while sitting on the floor. He needs his hands for support. Contracture of his leg muscles also makes sitting on the floor difficult. **B**. Sitting on a small chair at a table, however, with his feet on the floor enables him to use his hands in play.*

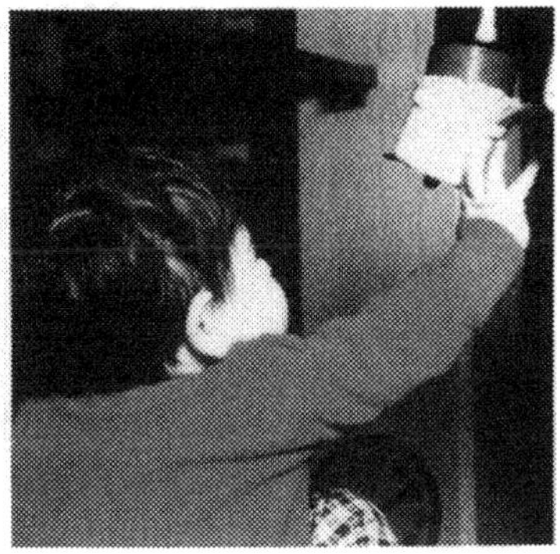

Figure 8. The object presented to this child with left hemiplegia is too big for him to grasp and the orientation of his hand is therefore unimportant. He simply bats at it to make it swing.

Muscle weakness or paralysis may also lead to secondary behavioral adaptation as a result of what Taub [52] has called "learned non-use". He demonstrated this adaptation in several studies of monkeys following surgical interruption of the pyramidal tract. However, in studies in which the monkey's intact limb was restrained and the affected limb was specifically trained with feedback reward, the affected upper limb became functional (e.g., [52, 53]). Similarly, restraint of the intact side in individuals following CNS damage, although not commonly utilized in rehabilitation, has been reported to result in improved function in manipulative tasks in both adults and children [54, 55]. More investigation is needed to establish the optimum timing and method for such intervention.

The adaptive motor behaviors of the brain-lesioned individual may be effective in a limited way, enabling the individual to ***get by*** given that assistance will be provided by others and that the rehabilitation environment will be less demanding than usual [12]. Frequent repetition of such adaptive behaviors, however, is likely to generate strong neural connections ensuring that it is these adaptive or compensatory movements which become ***learned*** or more stable, rather than more effective and efficient movements. In 1980, Le Vere [56] suggested that compensation is incompatible with recovery of function, that is, if compensation occurs then there will be no stimulus to the partially damaged neuronal system to recover. It is interesting that many rehabilitation settings actively encourage the learning of these adaptive substitution behaviors by providing, for example, special implements which have been redesigned so that they can be used with one hand with the aim of providing early independence.

If, as seems likely, motor behavior emerges as a result of both internal and external factors, what occurs in rehabilitation, that is, the process of rehabilitation, may be very important. Autopsy or computerized tomography (CT) scanning some time after CNS damage suggests that recovery often occurs in spite of the continued existence of the original lesion (e.g., [57]). Recently, new technology such as positron emission tomography (PET; see Decety this volume) is shedding light on the functional reorganisation of the brain following a lesion which may underlie the recovery of motor function (e.g., [58, 59]). The aim of training reaching, grasping and manipulation in children and adults with brain damage is, therefore, to take advantage of the potential for and actually guide the functional reorganisation of the brain by intervening early to facilitate this recovery. Such early intervention should utilize strategies to optimize the performance of motor tasks and prevent unnecessary adaptive behavior. The nature and type of clinical intervention and the environment in which it takes place is very likely to affect outcome.

In addition to motor deficits, sensory and neuropsychological impairment such as inattention, perceptual deficits and memory problems may affect recovery. These problems may benefit from specific retraining (e.g., [60]). However, implicit in the concept of task-specific motor training is the use of objects with different characteristics (e.g., shapes, sizes, textures) for a variety of different purposes and goals. Such training provides the stimulus to increase attention span and select the visual and other sensory information specific to the task as a means of improving skill in movement. In this way, training provides the possibility of improving the ability to select, attend to and respond to relevant sensory inputs and to control the muscles and limb segments. The environment is set up to enable practice to occur. Objects and tasks are modified to facilitate the required motor pattern and tasks practised are challenging but not impossible. As part of the application of motor learning principles (e.g., [61]), individuals with movement dysfunction are encouraged to become active in searching for optimal motor solutions to motor problems, to use Bernstein's phrase [62]. In the next section, we suggest some training strategies derived from literature aimed at overcoming the motor deficits affecting the performance of reaching and manipulation tasks.

TRAINING REACHING, GRASPING AND MANIPULATION

One of the difficulties with the analysis and training of hand function is the complex and varied nature of the tasks we perform either with one hand or both hands. However, relatively recent investigations of the development of reaching and grasping and of the performance characteristics of adults provide information of considerable value to both pediatric and adult clinical practice. Infants who need to develop control over the linked segments of the upper limb and adults who need to reestablish control must gain the ability to match their motor performance to the characteristics of objects within their environment. They must learn to judge whether or not an object should be picked up in one or both hands; how the hand should be oriented to match the characteristics (e.g., size and shape of the object) and what is to be done with it; and whether all or only some of the fingers are needed in order to grasp the object. In reaching, the motor-disabled infant or adult has to learn to control for direction and distance. They have to gain the ability to judge the distance over which they can successfully reach, which means knowing the length of the arm and the distance over which movements of other body segments will extend the arm's reach. Of particular importance, therefore, is gaining or regaining the ability to use vision to guide hand movement, since vision provides information about the environment with which the person must interact as well as feedback about movement performance.

Therapy for motor-impaired infants and children has typically addressed upper limb function in terms of fine motor skills (by which is meant grasping and manipulation) and support (e.g., through the hands or forearms in prone or sitting). Reaching as such is not often discussed in the clinical literature although the therapist is very likely in practice to include reaching in a child's therapy programme (e.g., [63, 64]). Similarly, methods of testing upper limb function test manipulation and support functions and not reaching (e.g., [65]).

However, given the evidence from studies of reaching development (e.g., [5, 66]), it can be inferred that, in infants with motor impairment, training of active reaching to grasp can commence earlier than has typically been the case in clinical practice. With motor-disabled infants with cerebral palsy, for example, it has often been assumed that reaching and grasping cannot be expected to occur until the infant develops control of the head and sitting balance. It has further been believed that development of the ability to support the body through the arms (in the prone position, for example) is a necessary pre-requisite for reaching. It generally follows that therapists may concentrate principally on an infant gaining head control and the ability to bear weight through the arms and very little on reaching toward objects.

The benefits of early practice of reaching include the opportunity to develop eye-hand and object-hand coordination, oculomotor control and an awareness of the possibilities offered by interactions with the environment. In normal infants, it has been shown that supporting the head and trunk enhances the reaching capability [67, 68]. The absence of head control and of sitting balance need not, therefore, be a deterrent to early training. Von Hofsten [66] has shown that even neonates are able to attempt reaching toward an object if certain conditions are met. His methodology provides a guide as to how to optimise reaching practice in infants and young children. For example, the infant can be supported in a semi-reclined position and the object should be attractive, graspable and within visual range. In addition, the young infant may see an object more clearly when it moves across the visual field rather than being stationary. With a motor-disabled infant, additional support may need to be given in order that reaching will be possible and the angular displacements and hand path will be optimal. For example, when the infant is supported in a seat, the head may need to be supported so it does not fall to one side; the back rest supporting the trunk may need to wrap around the shoulders to encourage shoulder flexion rather than abduction. Similarly, in adults, early practice of reaching will be optimised by positioning the individual to enable even small amounts of muscle activity to move the limb (see Fig. 15).

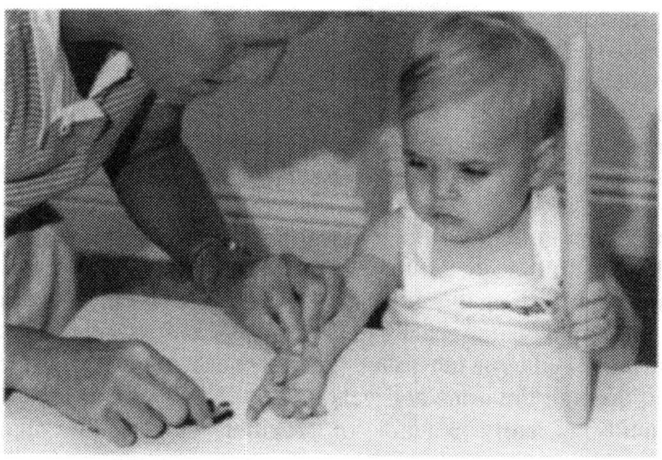

Figure 9. In this game, the child must supinate the forearm to be given a toy. Some assistance with supination is necessary initially. From [84].

A

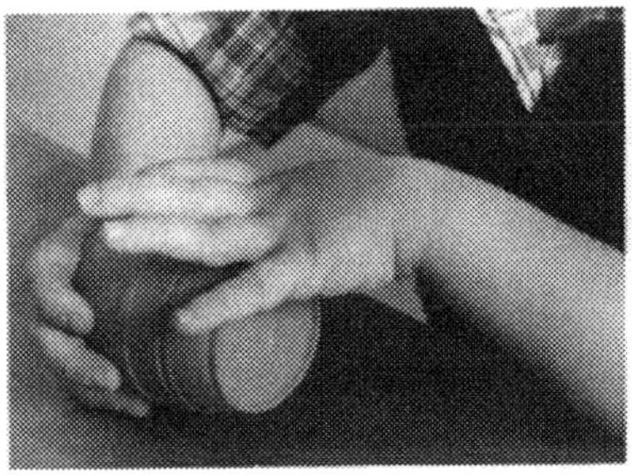

B

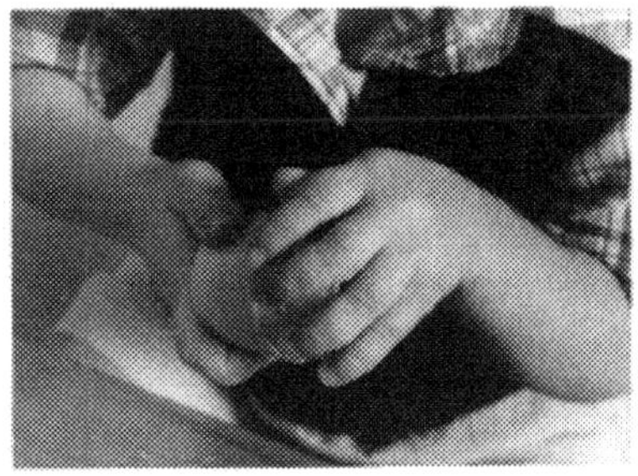

C

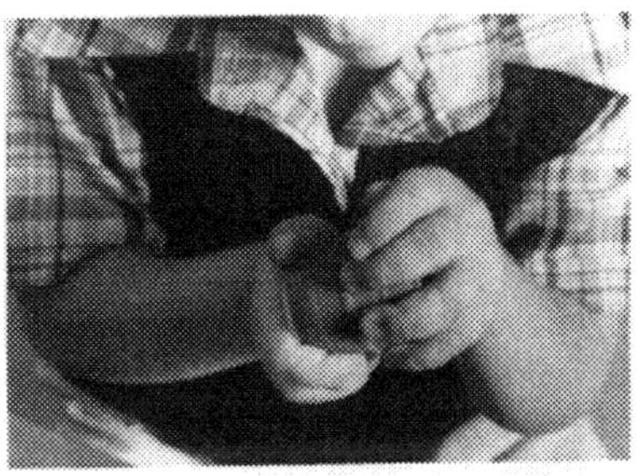

Figure 10. ***A***. *This child demonstrates poor pre-shaping of the left hand, with wrist flexion and excessive finger extension at MCP and IP joints.* ***B***. *He is having difficulty with the application of force through the fingers.* ***C***. *He is about to succeed in unscrewing the barrel. He may have decreased the need to control shoulder and elbow muscles by holding his arms stiffly into his side.*

It is now recognised that motor performance is governed to a considerable extent by objects and their ***affordances***, a term coined by Gibson [69]. Objects, by their properties, offer certain possibilities for interaction. As with motor training in adults, objects should be chosen, therefore, not only for their inherent interest and usability but also for the options they offer for hand orientation. Given that objects and tasks require different grasps, it is probably necessary that training of hand function involves practice with a variety of objects to perform different tasks [11, 12]. If a child is offered toys which can be grasped in many different ways or which are too large to be grasped, the movement pattern used by the child will reflect the options provided by the lesioned system and will probably be the ***easiest*** one (Fig. 8).

When a child has difficulty controlling the orientation of the hand and limb and reaches persistently with the shoulder internally rotated and the forearm pronated (see Fig. 8), an object or a task should be chosen that demands a relatively externally rotated and supinated approach (Fig. 9). In addition, objects can be chosen and games played that actively encourage the action with which the infant has difficulty (Fig. 10). In training young disabled infants, it is useful to consider that normal infants appear to be more likely to reach when a task is challenging than when it does not capture their attention and interest [70, 71]. This point also applies to adults of course.

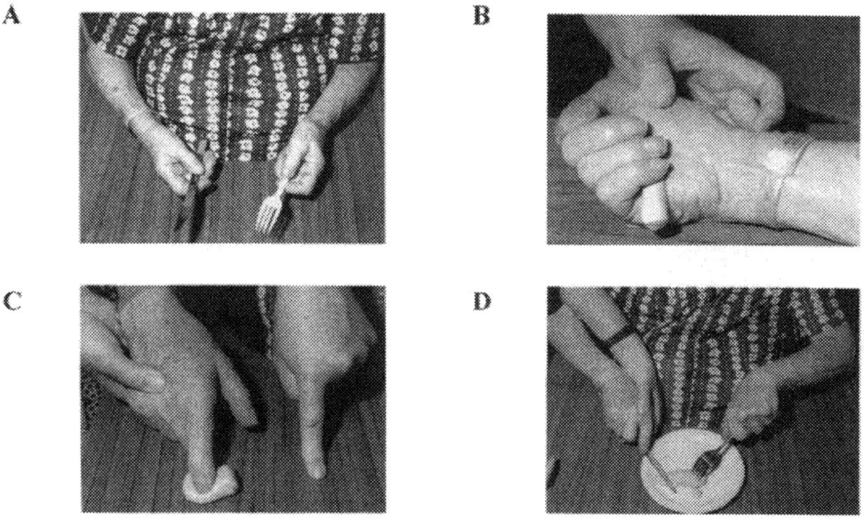

Figure 11. ***A****. This woman following CVA has difficulty holding the knife.* ***B****. She practices flexing her three ulnar fingers.* ***C****. She then concentrates on extending her index finger while maintaining flexion of the three ulnar fingers.* ***D****. She now has some ability to grasp the knife with assistance.*

There have been several attempts over the years to categorise grasps (e.g., [72]) or oppositions [8] in order to facilitate analysis of hand function. Iberall and colleagues [8] have suggested that fingers are grouped as functional units called "virtual fingers". Each functional group performs a component of a task in a cooperative way. For example, when picking up a mug the hand is organised into three functional groupings - the thumb, the fingers that go through the handle and the fingers that go underneath the handle. The way in which the fingers are incorporated into the three virtual fingers is, however, determined by the size of the handle in relation to the hand.

Cooperation between the functional groups produces appropriate levels of opposing forces. For example, in order to use a knife and fork (Fig. 11) the implements are held firmly within the palm of the hand while pressure is applied through the index finger to pierce or cut the food. The three ulnar fingers secure the handle in the palm of the hand while opposing forces produced by the thumb prevent the handle slipping sideways.

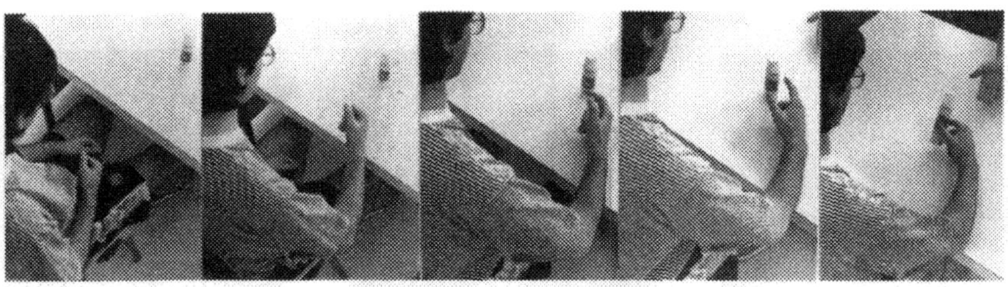

Figure 12. This man, after a head injury, is shown reaching for a small diameter bottle. Not only is the transport phase completed before he begins to open his hand but the overall movement time is at least three times that of normal. Given that he can perform both the transport and grasp components of this task, the aim of training would be to promote a more normal relationship between the two. Getting him to complete the movement faster using a stop watch to provide feedback and having him practice ***getting the idea*** *of commencing the transport and grasp components together should encourage improved performance.*

The kinematic relationship between the transport and the grasp components of reaching has been investigated by Jeannerod in a series of experiments [1, 2]. The findings indicate that the grasp component is linked to the transport component such that both begin at the same time and that maximum aperture of the hand coincides with the beginning of deceleration of the hand towards the object. These characteristics are tuned depending on the nature of the object and the purpose of the task. For example, if the object is large and is located close to the person, the hand opens faster than for smaller objects located further away. Furthermore, if greater precision is required, the aperture will be more exact and the overall movement time, in particular the deceleration phase, will be longer [73]. Following brain damage, although individuals may be able to reach for and grasp objects, the normal timing relationships are often disrupted as illustrated in Fig. 12.

During the deceleration phase of the transport component, Wing and colleagues [4, 74] have found that once maximum aperture is achieved, thumb position is relatively invariant while movement of the fingers completes the grasp. The position of the thumb is probably, therefore, of primary importance for the alignment of the hand in training reaching for objects (Fig. 13).

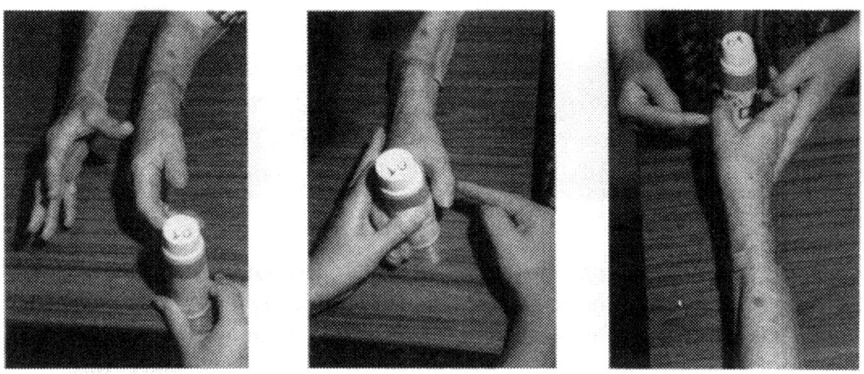

*Figure 13. **A**. This woman following a CVA cannot open her hand to grasp the bottle. One component of the task requiring practice is abduction of the thumb. **B**. She practices thumb abduction with the therapist's finger providing the goal. **C**. When she has some control, she practices reaching for the bottle with assistance.*

In individuals with problems activating and controlling hand musculature, the thumb typically lies in an adducted position and web space soft tissues rapidly become short (see Fig. 2). In training infants, for example, it may be necessary to support the thumb in a small splint so that the carpometacarpal joint is maintained in abduction and opposition. The splint will serve to maintain passively the extent of the grasp aperture so that, in an immobile infant, the soft tissues between thumb metacarpal and index metacarpal will not shorten. If the splint is designed so that active abduction of the thumb away from the splint is possible, the splint may, when worn during training sessions, promote activity in the thumb abductor and extensor muscles. A single case study [75] showed that when a four-year old child with cerebral palsy wore an opponens splint for six hours a day and all night for a four-week period, a significant improvement in active range of movement and in some functional tasks occurred.

Reaching to grasp and manipulate objects involves not only the focal upper limb muscles but also muscles involved in postural adjustments. For example, studies of fast arm raising in standing have indicated that muscle activity occurs in the postural leg muscles before the focal arm muscle and before the arm is moved (e.g., [76, 77]). Similar muscle activations have also been found in reaching in sitting [78, 79]. In training, the necessary postural adjustments have to be incorporated into reaching and manipulation in upright positions such as sitting and standing so that it is possible to reach beyond arm's length without losing balance (e.g., Figs. 7**A**, 14 and 15**C**).

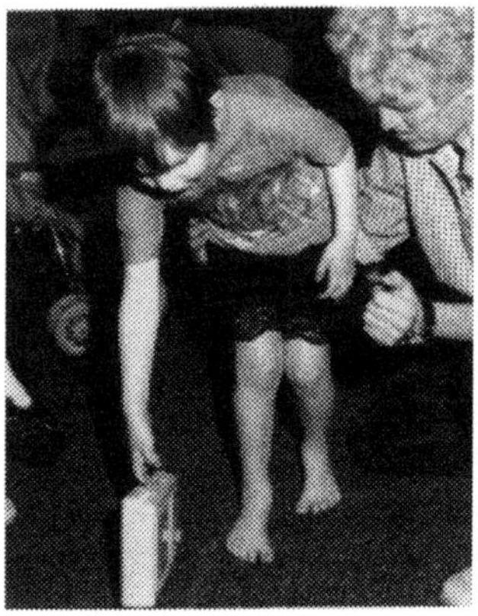

Figure 14. This little girl has poor control over postural adjustments and reaching in standing. She practices in a difficult position with a narrow base of support. The therapist gives a little assistance so that the child gets the idea of the movement.

Early following brain damage patients typically have difficulty eliciting any muscle activity. Initially, the therapist needs to encourage the patient to activate muscles, such as the prime movers responsible for reaching forward. In order to do this it is often necessary to reduce the demands on the system

A

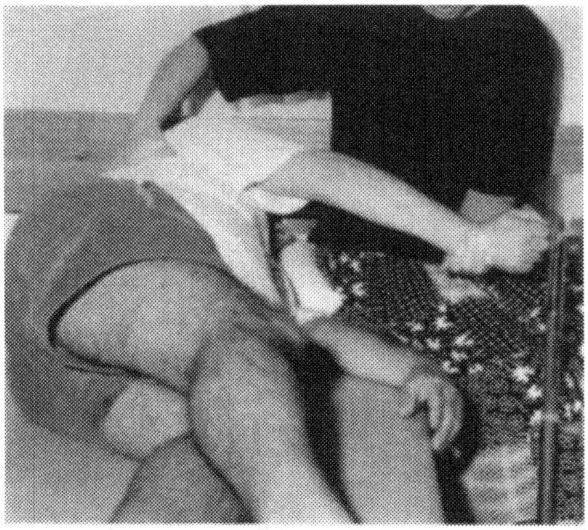

Figure 15. ***A***. *This man following CVA is practicing reaching forward in side lying, a position which not only enables protraction of the shoulder girdle to take place with reduced force requirements but also decreases the demand for complex postural adjustments.* ***B***. *When he gains some control, he practices reaching forward in sitting with his arm on the table.* ***C***. *In order to ensure that he is able to reach and make appropriate postural adjustments, he also practices reaching for objects in sitting.*

B

C

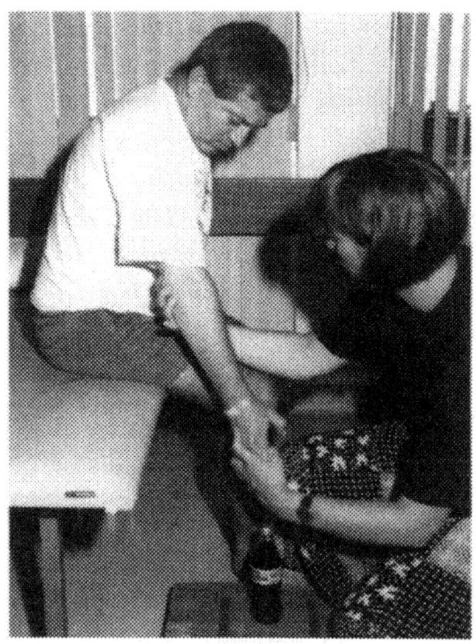

(Fig. 15). However, once the patient has some ability to move the affected arm and hand, it is also necessary to train bimanual activities, since for most everyday functions the two limbs work together on the task. Experiments to date shed some light on how the two hands interact during bimanual tasks (e.g., [2, 3, 80, 81]). In a recent kinematic study of bimanual tasks [81], subjects were asked to pick up a can and pull the small tab on top of the can. The movement duration for both hands was found to be the same, although the hand using a precision grip to pull the tab showed an earlier peak velocity and a longer deceleration phase than the hand grasping the can. Given that the temporal linking between the two hands during bimanual tasks is task-specific, it is logical to train the two limbs in a variety of bimanual tasks (Fig. 16).

Figure 16. ***A****. When this woman following CVA reaches for the box on a high shelf her affected left hand lags behind her intact right. She was not coordinating the speed of her two hands in synchrony and, in addition, she was unable to control the muscles which flex her shoulder in this very shortened range.* ***B****. First she practices lifting her hand off the cupboard one inch and putting it back again, avoiding the adaptive behavior of extending her upper body.* ***C****. Then she practices reaching for the box with both hands, concentrating on contacting the box with both hands simultaneously and increasing her speed.*

Figure 17. As this woman pours water from one container to another, she needs to make preparatory and ongoing adjustments to the grip forces of both hands as the weight of the objects change. Her attempt demonstrates a typical adaptive behavior to the dyscontrol characteristics of cerebellar dysfunction.
***A**. She compensates by resting the container on the table and holding the right arm close to her body.*
***B**. and **C**. She practices pouring water ensuring that the cups don't touch and are held above the table. Practice is progressed by increasing the speed and by varying the amount of water in the cup to increase the range of grip forces required. The patient can be further challenged by changing the frictional qualities of the containers, for example, by using a wet glass.*

An individual's ability to produce appropriate grip forces as the two hands interact also needs to be considered in rehabilitation. The two hands have to interact with objects in a cooperative manner. For example, unscrewing the lid of a jar involves opposing forces being applied to the lid and to the jar

[16]. Johansson and Westling [82] have demonstrated that, when a subject holds a ball in one hand and drops it into a cup held with the other hand, the grip force of the receiving hand increases in anticipation of the impact. Furthermore, the size of the grip force depends on the frictional properties of the object so that if the cup was wet and slippery, the grip force would be higher [83]. This is an example of a task which relies on the accurate processing of visual and tactile information. Practice of this type of task, as illustrated in Fig. 17, aims to improve not only visual and tactile processing but also preparatory and ongoing adjustments to grip force.

In conclusion, the aim of rehabilitation/habilitation of the person who has upper limb dysfunction following damage to the CNS is to enable that individual to take part in everyday life and to function as effectively as possible. If, as seems likely, motor behavior emerges as a result of both internal and external factors, what occurs in rehabilitation, that is, the process of rehabilitation, must be important. Individuals need training which is designed to elicit muscle activity, to eliminate unnecessary muscle activity and to enable the frequent practice of a variety of tasks with a variety of different objects. This chapter has provided clinical examples of intervention derived from the literature about normal reaching, grasping and manipulation. It is important to recognise, however, that only by measuring the outcome of intervention will we know whether such clinical applications are effective.

REFERENCES

1 M. Jeannerod, in: J. Long and A. Baddeley (eds.), Attention and Performance, Vol. IX, Erlbaum, Hillsdale, 1981, p. 153.
2 M. Jeannerod, J. Mot. Behav., 16 (1984) 235.
3 R.G. Marteniuk, C.L. Mackenzie and D.M. Baba, Q. J. Exp., Psych., 36a (1984) 335.
4 A. Wing, A. Turton and C. Fraser, J. Mot. Behav., 18 (1986) 245.
5 C. von Hofsten, Devel. Psychol., 20 (1984) 378.
6 C. von Hofsten and L. Ronnquist, J. Exp. Psychol., Hum. Percep., Perf., 14 (1988) 610.
7 M.A. Arbib, T. Iberall and D. Lyons, in: A.W. Goodwin and I. Darian-Smith (eds.) Hand Function and the Neocortex, Springer-Verlag, Berlin, 1985, p. 111.
8 T. Iberall, G. Bingham and M.A. Arbib, in: H. Heuer and C. Fromm (eds.), Generation and Modulation of Action Pattern, Exp. Brain Res. Suppl., 15 (1986) 153.

9 D.A. Rosenbaum, J. Vaughan, H.J. Barnes, F. Marchak and J. Slotta, in: M. Jeannerod (ed.), Attention and Performance, Vol. XIII, Erlbaum, Hillsdale, 1990, p. 321.
10 J.H. Carr and R.B. Shepherd, A Motor Relearning Program for Stroke, Butterworth-Heinemann, Oxford, 1982.
11 J.H. Carr and R.B. Shepherd (eds.), Movement Science Foundations for Physical Therapy in Rehabilitation, Aspen Publications, Rockville, 1987.
12 J.H. Carr and R.B. Shepherd, A Motor Relearning Program for Stroke, Butterworth-Heinemann, Oxford, 1987.
13 L. Ada and C. Canning (eds.), Key Issues in Neurological Physiotherapy, Butterworth-Heinemann, Oxford, 1990.
14 G. Bard and G.G. Hirschberg, Arch. Phys. Med. Rehab., 46 (1965) 567.
15 C. Gowland, Physio. Can., 34 (1982) 77.
16 V.M. Parker, D.T. Wade and R. Langton Hewer, Int. Rehab. Med., 8 (1986) 69.
17 A. Heller, D. Wade, V. Wood, A. Sunderland, R. Langton Hewer, E. Ward, J. Neurol. Neurosurg. Psychiat., 25 (1987) 714.
18 C. Dean and F. Mackey, Aust. J. Physio., 38 (1992) 31.
19 A. Sunderland, D.J. Tinson, E.L. Bradley, D. Fletcher, R. Langton Hewer and D.T. Wade, J. Neurol. Neurosurg. Psychiat., 55 (1992) 530.
20 L.F. Marshall, Curr. Opin. Neurol. Neurosurg., 3 (1990) 4.
21 D. Burke, Adv. Neurol., 47 (1988) 401.
22 W.R. Gowers, A Manual of Diseases of the Nervous System, J. and J. Churchill, London, 1893. Reprinted 1970. Darien, Cn: Hafner.
23 C. Wernicke, Berliner Klinische Wochenschrift, 26 (1889) 969.
24 T.E. Twitchell, Brain, 74 (1951) 443.
25 J.G. Colebatch, S.C. Gandevia and P.J. Spira, J. Neurol. Neurosurg. Psychiat., 49 (1986) 1019.
26 R.C. Bohannon and M.B. Smith, Phys. Ther., 67 (1987) 522.
27 J.G. Colebatch and S.C. Gandevia, Brain, 112 (1989) 749.
28 J.F.M. Clough, D. Kernell and C.G. Phillips, J. Physiol., 198 (1968) 145.
29 E. Palmer and P. Ashby, J. Physiol., 418 (1992) 397.
30 R.N. Lemon and R.B. Muir, Exp. Brain Res. Suppl., 7 (1983) 230.
31 T. Warabi, K. Inoue, H. Noda and S. Murakami, Brain, 113 (1990) 177.
32 R.D. Jones, I.M. Donaldson and P.I. Parkin, Brain, 112 (1989) 113.

33 A. Brodal, Neurological Anatomy in Relation to Clinical Medicine, Oxford University Press, New York, 1981.
34 J.W. Lance, in: R.G. Feldman, R.R. Young and W.P. Koella (eds.), Spasticity: Disordered Motor Control, Symposia Specialists, Miami, 1980, p. 485.
35 S.A. Sahrmann and B.J. Norton, Ann. Neurol., 2 (1977) 460.
36 A. Tang and W. Rymer, J. Neurol. Neurosurg. Psychiat., 45 (1981) 320.
37 P.D. Neilson, N.J. O'Dwyer and J. Nash, Devel. Med. Child Neurol., 32 (1990) 778.
38 P.W. Nathan, Devel. Med. Child Neurol., 11 (1969) 384.
39 D.L. McLellan, J. Neurol. Neurosurg. Psychiat., 40 (1977) 30.
40 C.E. Chapman and M. Wiesendanger, Physio. Can., 34 (1982) 125.
41 V. Dietz, J. Quintern and W. Berger, Brain, 104 (1981) 431.
42 V. Dietz and W. Berger, Exp. Neurol., 79 (1983) 680.
43 J.C.M. Brust, in: E.R. Kandel, J.H. Schwartz and T.M. Jessell (eds.), Principles of Neural Sciences, Elsevier, New York, 1991, p. 1041.
44 S.C. Gandevia, G. Macefield, D. Burke and D.K. McKenzie, Brain, 113 (1990) 1563.
45 R.S. Johansson and G. Westling, Exp. Brain Res., 71 (1988) 59.
46 D. Marr, J. Physiol., 202 (1969) 437.
47 J.S. Albus, Math. Biosci., 10 (1971) 25.
48 M. Ito, The Cerebellum and Neural Control, Raven Press, New York, 1984.
49 R.F. Thompson, Science, 223 (1986) 941.
50 C.H. Yeo, Ann. N.Y. Acad. Sci., 627 (1991) 292.
51 J.R. Bloedel, Behav. Brain Sci., 15 (1992) 666.
52 E. Taub, in: L.P. Ince (ed.), Behavioral Psychology in Rehabilitation Medicine: Clinical Implications, Williams and Wilkins, Baltimore, 1980, p. 371.
53 C.E. Chapman and M. Wiesendanger, Electroenceph. Clin. Neurophysiol., 53 (1982) 374.
54 R.J. Schwartzman, Am. J. Phys. Med., 53 (1974) 75.
55 S.L. Wolf, D.E. Lecraw, L.A. Barton and B.B. Jann, Exp. Neurol., 104 (1989) 125.
56 T.E. Le Vere, Physiol. Psychol., 8 (1980) 297.
57 P. Bach-Y-Rita, in: P. Bach-Y-Rita (ed.), Recovery of Function: Theoretical Considerations for Brain Injury Rehabilitation, Hans Huber, Bern, 1980, p. 225.
58 F. Chollet, V. Dipiero, R.J.S. Wise, D.J. Brooks, R.J. Dolan and R.J.S. Frackowiak, Ann. Neurol., 29 (1991) 63.

59 C. Weiller, F. Chollet, K.J. Friston, R.J.S. Wise and R.J.S. Frackowiak, Ann. Neurol., 31 (1992) 463.
60 M. Yekutiel and E. Guttman, J. Neurol. Neurosurg. Psychiat., 56 (1993) 241.
61 M. Gentile, in: J.H. Carr and R.B. Shepherd (eds.), Foundations for Physical Therapy in Rehabilitation, Aspen Publications, Rockville, 1987, p. 93.
62 A. Bernstein, The Coordination and Regulation of Movement, Pergamon Press, New York, 1967.
63 J. Blasky, in: S.K. Campbell (ed.), Pediatric Neurologic Physical Therapy, Churchill Livingstone, New York, 1991, p. 213.
64 J.M. Wilson, in: S.K. Campbell (ed.), Pediatric Neurologic Physical Therapy, Churchill Livingstone, New York, 1991, p. 301.
65 Y.R. Burns, R.M. Ensby and M.A. Norrie, Aust. J. Physio., 35 (1989) 141.
66 C. von Hofsten, Devel. Psychol., 18 (1982) 1450.
67 C. Amiel-Tyson, in: J. Mehler and R. Fox (eds.), Neonate Cognition: Beyond the Blooming Budding Confusion, Erlbaum, Hillsdale, 1985, p. 365.
68 P.J. Rochat, J. Mot. Behav., 24 (1992) 210.
69 E.J. Gibson, Perceptual Learning and Development, Appleton-Century Crofts, New York, 1969.
70 C. von Hofsten, J. Hum. Mvt. Stud., 5 (1979) 150.
71 C. von Hofsten, J. Mot. Behav., 23 (1991) 280.
72 J. Napier, J. Bone Jt. Surg., 38B (1956) 902.
73 R.G. Marteniuk, C.L. MacKenzie, M. Jeannerod, S. Athenes and C. Dugas, Can. J. Psychol., 41 (1987) 365.
74 A. Wing and C. Fraser, Q. J. Exp. Psychol., 35 (1983) 207.
75 G. Goodman and S. Bazyk, Am. J. Occup. Ther., 45 (1991) 726.
76 V.Ye. Belenkii, V.S. Gurfindel and Ye.I. Paltsev, Biophysics, 12 (1967) 135.
77 W. Lee, J. Mot. Behav., 12 (1980) 185.
78 S. Moore, D. Brunt, M. Nesbitt and T. Juarez, Phys. Ther., 72 (1992) 335.
79 R.B. Shepherd, J. Crosbie and T. Squires, Proc. Biomech., XIV ISB Congress, in press.
80 J.A.S. Kelso, D.I. Southard and D. Goodman, Hum. Percep., 5 (1979) 229.
81 U. Castiello, K.M.B. Bennett and G.E. Stelmach, Behav. Brain Res., 56 (1993) 43.
82 R.S. Johansson and G. Westling, Exp. Brain Res., 71 (1988) 72.

83 R.S. Johansson, in: M. Jeannerod (ed.), Attention and Performance, Vol. XIII, Erlbaum, Hillsdale, 1990, p. 677.
84 R.B. Shepherd, in: S.K. Campbell (ed.), Pediatric Neurologic Physical Therapy, Churchill Livingstone, New York, 1991, p. 141.

Section Five: Biomechanics and Robotics

Insights into the Reach to Grasp Movement
K.M.B. Bennett and U. Castiello (Editors)

CHAPTER 13

BIOMECHANICAL ANALYSIS OF REACH AND GRASP

M.E. JOHNSON AND K.-N. AN

Biomechanics Laboratory, Department of Orthopedics,
Mayo Clinic/Mayo Foundation, Rochester, Minnesota 55905, U.S.A.

SUMMARY

Reach to grasp function requires coordinated motion and strength of the upper extremities. Interaction between the three-dimensional motion of the shoulder complex and the elbow joint involved in reach function is analyzed based on kinematic principles. The biomechanical factors influencing grasp strength are examined.

INTRODUCTION

The complexity of the function of the upper extremities in performing reaching and grasping tasks can be described and analyzed based on biomechanical principles. Motion and force are the two basic concepts or parameters commonly encountered in a mechanics perspective. Rigorous coordinate system definition is required in order to define the three-dimensional movement of the upper extremities. In addition, grasp function requires the control of force generation of multiple muscles. In this chapter, studies that have analyzed how shoulder and elbow joint motion are coordinated will be used to illustrate the application of kinematic or motion analysis methods. A review of investigations that have looked at grasp strength as a function of wrist joint configuration are presented to illustrate the important physiological and biomechanical factors determining the functional strength of the hand.

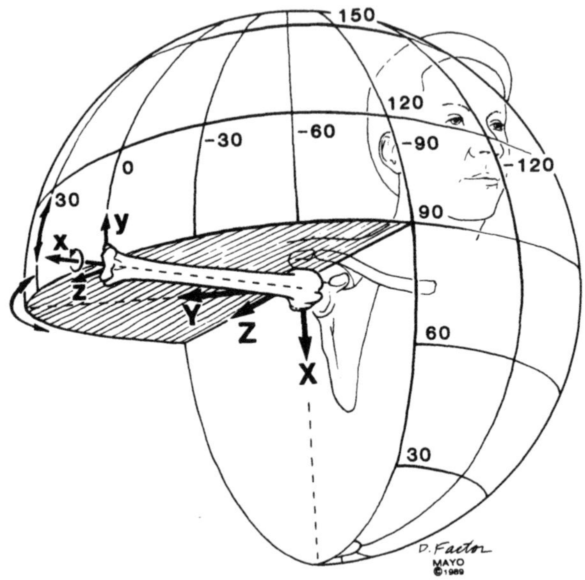

Figure 1. Definition of reference coordinate system. The ***XYZ*** *represents the reference fixed coordinate system affixed on the thoracic body. The* ***xyz*** *represents the moving coordinate system affixed on the humerus. The latitude on the globe represents the amount of humeral elevation, and the longitude on the globe represents the plane of elevation. The humeral axial rotation takes place about the* ***x****-axis along the humeral shaft. From [1]*

MOTION

Description of shoulder motion

The shoulder complex exhibits more motion than any other joint in the human body. Movement of the shoulder complex is accomplished by movement at the glenohumeral and scapulothoracic articulations. However, due to the capability for large range of motion in three-dimensional space, the terminology or system used for measuring and defining shoulder motion has been controversial. From a kinematics perspective, shoulder joint motion cannot be adequately described solely as being in sagittal, coronal, and transverse planes because of the sequence dependent nature of rotation about orthogonal axes. Definition of shoulder motion based on rigorous mechanical principles is therefore necessary. The description of motion is made more clear by the use of reference systems. To facilitate the description of shoulder complex motion, two sets of coordinate systems have been defined, fixed and moving (Fig. 1). The fixed reference system (**XYZ**) is fixed on the thoracic body with the **X**-axis directed inferiorly, the **Y**-axis directed laterally, and the **Z**-axis pointed posteriorly for the right shoulder and anteriorly for the left shoulder. The moving coordinate system (**xyz**) is attached to the humerus. The **x**-axis is defined to be along the humeral shaft and directed distally, the **y**-axis is defined based on the medial and lateral epicondyle and pointed laterally, the **z**-axis follows the right-hand rule. The moving coordinate system is parallel to the fixed coordinate system when in a neutral position (Fig. 2**A**).

Due to sequence dependence of the three-dimensional finite rotation, motion of the shoulder complex is described by the concept of the Eulerian angle system based on a 1-3'-1" rotational sequence [1]. This implies that for a given shoulder configuration, the orientation could be described by the first rotation (ϕ) about the **X**-axis (Fig. 2**B**) fixed on the reference thoracic body. This rotation defines the plane of arm elevation or sets the **Z**-axis for humeral elevation. The second rotation (θ) about this **Z'**-axis corresponds to arm elevation (Fig. 2**C**). As the arm moves into an abduction or elevated position the **Z'**-axis defines the degree or amount of arm elevation that is occurring. The third rotation (ψ) about the **X"**-axis corresponds to humeral axial rotation (Fig. 2**D**). This method of expressing shoulder complex motion can then be interpreted as follows: the angle ϕ, which defines the plane of arm elevation, and the angle θ, which defines the amount of arm elevation, can be thought of as longitude and latitude defining the position of the distal humerus on the surface of a globe.

Based on the definition described above, the three angles can be obtained if the transformation matrix between the fixed reference coordinate system (**XYZ**) and moving coordinate system (**xyz**) is available [1]. Experimentally, such a transformation matrix can be measured by numerous methods. It should be noted that the above description is for the shoulder complex, if only glenohumeral motion were to be described then it would be described in reference to the scapula.

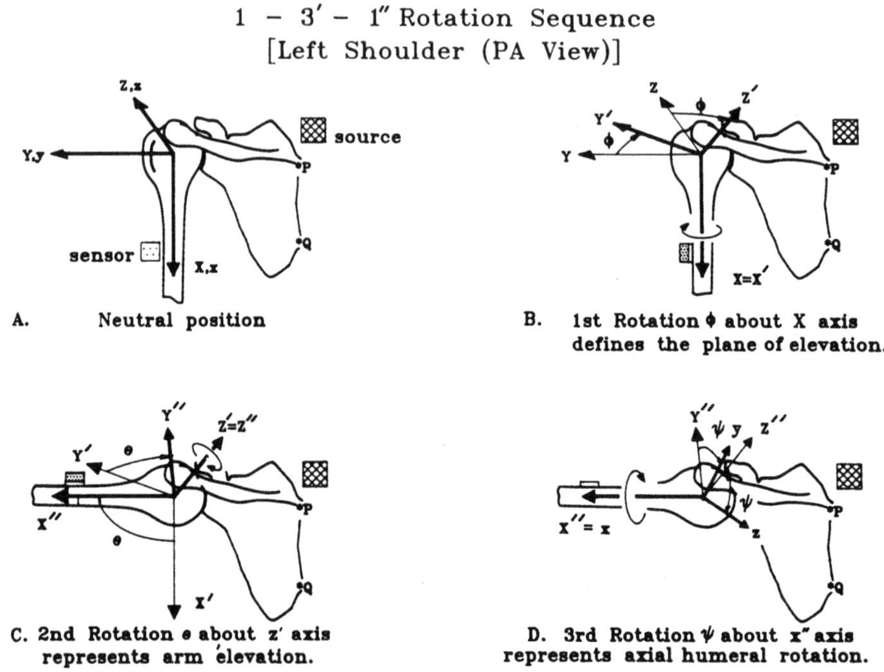

*Figure 2. The shoulder complex motion is described based on the coordinate systems **XYZ** and **xyz**. The three-dimensional rotation is defined based on the 1-3'-1" rotational sequence. **A**. In the initial neutral position, the humeral and scapular coordinate systems coincide. **B**. The first rotation about the **X**-axis defines the plane in which elevation will occur. **C**. As the arm moves into an abducted or elevated position the second rotation about the **Z'**-axis represents the degree or amount of arm elevation. **D**. The third rotation about the **x**-axis represents humeral axial rotation. From [1]*

Measurement of shoulder complex motion

Numerous experimental methods are available for the measurement of shoulder complex motion which include goniometer and stereophotographic methods. More recently, a device based on the electromagnetic tracking concept, 3 Space Tracker (Polhemus Navigation Sciences Division, McDonnell Douglas Electronics, Colchester, Vermont), has been available for real-time monitoring of shoulder complex motion [1]. The system consists of a source which generates a three-orthogonal magnetic field in space and a set of sensors which are able to sense the magnetic field. Based on this technology, the distance and orientation between the source and sensors can be obtained in real time. The accuracy of the Tracker has been reported to be 0.5 mm or 0.5 degrees.

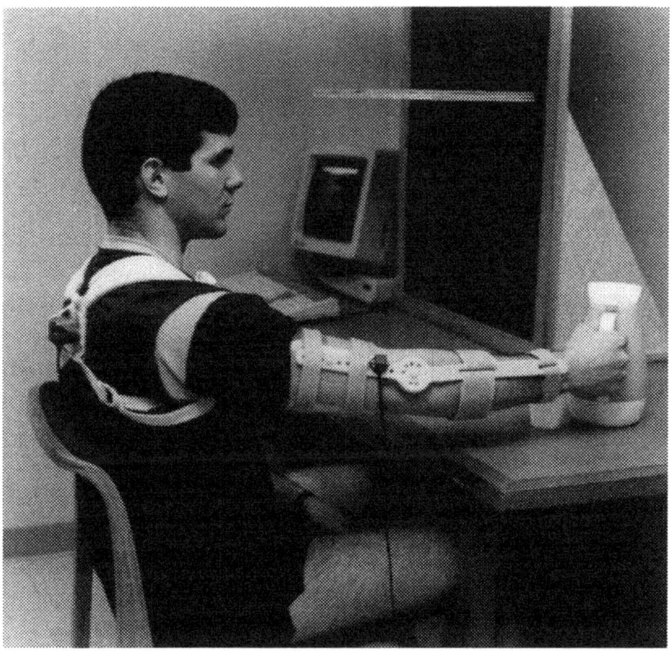

Figure 3. The electromagnetic tracking system. With source on the elbow brace and source attached on the thoracic spine, the motion of the shoulder complex was measured during reach activities. From [2]

In a study by O'Neill et al. [2] which investigated shoulder complex motion, the source was placed on the thoracic spine with a modified figure-eight clavicular splint, and the sensor was mounted on the distal humerus by a custom brace (Fig. 3). The relationship between the fixed and moving coordinate systems (anatomic coordinate system) to those of the source and sensors (global coordinate system) was established. Each subject was requested to sit in a chair with the sacrum, lumbar spine, and scapulae in the plane perpendicular to the floor, and the shoulders were positioned parallel to the floor. In this position, anatomic landmarks defining the anatomic coordinate system were digitized and then used to establish the transformation matrix between the measurement devices and the body coordinate systems.

Ten male subjects with right-hand dominance and no previous injury to the upper extremities were studied. A spectrum of tasks involving reaching movements in activities of daily living were examined. These activities included reaching a pitcher on a table; moving an object between three levels of shelving; turning a steering wheel; reaching and tying the shoes; and actions of personal hygiene, such as moving the hand to the occiput, opposite shoulder, mouth, and sacrum. For activities performed in a standing position, subjects were required to keep their feet stationary; and for sedentary activities, the subjects were required to remain seated. For extrapersonal activities, the distance between the body and target was standardized. The target was adjusted to be the same relative distance from each subject to minimize arm length and height differences. The shelf heights were based on architectural standards with the bottom shelf being 56 inches from the floor, the middle shelf 62 inches, and top shelf 68 inches.

To further examine the coordination and interaction between motions of the shoulder complex and elbow joint, a custom, adjustable brace was made which limited elbow joint motion or simulated elbow arthrodeses at 50, 70, 90, and 110 degrees of elbow flexion (full extension is equivalent to 0 degree). Each activity was then repeated and the results compared with those without elbow constraint.

Interaction of elbow and shoulder motion in reach movement

For the spectrum of tasks studied, most of the activities required reaching movement in front of the body. This movement is well represented by the rotation describing the plane of arm elevation or circumduction (Table 1). Except for reach to the sacrum, which requires placing the humerus in the plane 77 degrees behind the body, the rest of the activities required placement of the humerus in a plane in front of the body within the range of 40 to 125 degrees for pouring the pitcher and reaching the opposite shoulder, respectively. The amount of arm elevation required depended on the relative location of the target to the shoulder joint. For the spectrum of tasks studied, the amount of arm elevation required ranged from 25 to 105 degrees to reach the bottom and top shelves, respectively.

Table 1.

Average shoulder position in space (in degrees) and standard deviation (SD) for all subjects tested to reach and maintain specified activities.

Activity	**Circumduction**		**Arm Elevation**	
	Degrees	SD	Degrees	SD
Grasp Pitcher	- 65.5	7.4	66.0	6.9
Pour Pitcher	- 41.7	7.7	73.7	7.9
Reach to Top Shelf	- 61.3	7.1	104.9	6.8
Reach to Mid Shelf	- 66.3	6.3	62.2	7.1
Reach to Bottom Shelf	- 41.1	14.5	25.6	11.9
Steering Wheel Top	- 89.5	6.5	70.2	7.7
Steering Wheel 1/4	-100.9	8.0	60.4	7.7
Steering Wheel 2/4	- 79.6	9.5	40.3	10.0
Wheel 3/4	- 74.6	7.3	49.0	8.2
Tying Right Shoe	- 87.8	16.7	72.1	13.9
Tying Left Shoe	- 72.5	7.0	63.2	12.3
Hand to Occiput	- 57.3	16.1	127.2	11.4
Hand Opposite Shoulder	-124.3	9.1	68.7	11.1
Hand to Mouth	- 77.0	10.6	86.6	14.9
Hand to Sacrum	76.9	10.9	31.3	3.4

\- *Plane in front of body*
\+ *Plane behind the body*

The reaching movements with the elbow joint motion limited were interesting. Under simulated elbow fusion, not all reaching tasks could be performed. For certain activities that could be performed, significant changes in arm elevation and circumduction were identified from those of normal elbow conditions. With the elbow fused at lesser flexion angles, less than 90 degrees, the arm elevation of shoulder movements showed little statistical change from normal. However, with the elbow fused at greater flexion angles, more than 110 degrees, significant changes from normal in arm elevation were demonstrated in half of the activities.

The circumduction motion of shoulder movement showed greater change with limitation of elbow joint motion in reaching activities. Again, little change existed for the elbow joint fused at 50 degrees. However, with the elbow joint fused in more than 90 degrees of flexion, all activities showed a significant change in circumduction motion. A trend has been observed indicating that with the increase of elbow fusion angle, the plane of arm elevation for the shoulder complex decreased, that is, humeral functions were closer to the side of the body.

The data in this study suggested that with elbow joint function limited, the shoulder complex is not a major contributor to compensatory motion. The function of the shoulder is different from that of the elbow joint. By examining the linkage of the upper extremities, we can identify that the hand is the effector; and the wrist, elbow, and shoulder act to place the hand into a functional position. Flexion and extension of the elbow joint change the regional position of the hand relative to the body or the external environment. On the other hand, the shoulder complex moves the arm within the range of a sphere. The primary functions of elbow and shoulder are mutually exclusive. Therefore, minimum compensatory motion occurs at the shoulder complex when the elbow motion is limited.

STRENGTH

Grasp strength and wrist position

There are many biomechanical questions still to be answered, also regarding the distal upper extremity linkage of the wrist and hand. The function of grasp is dependent on many factors, two of which especially lend themselves to biomechanical analysis: the effect of the muscle length/tension relationship and the moment arm contribution to grasp strength. The effect of wrist joint position on grasp strength has been observed [3]. However, the precise relationship between wrist position and grasp strength has not been well established. In a study performed in our laboratory, we attempted to ask two questions [4]. First, is a consistent position assumed by the wrist joint during maximal unconstrained grip function? Second, is that position optimum for maximal grasp strength?

Twenty healthy, normal right-handed adults were recruited for this study. Grip strength for all subjects was measured with a grip dynamometer. The subjects were instructed to perform maximal grip efforts. Unconstrained wrist motion in flexion-extension and radial-ulnar deviation was allowed. Wrist joint motions were measured by using biaxial electrogoniometers which have been developed and used in standardized testing in our laboratory. The electrogoniometer was positioned so that it was carefully aligned with the radius in both sagittal and coronal planes, as well as the centers of rotation with the center of the head of the capitate for radial and ulnar deviation and at the midpoint between the ulnar styloid and the triquetrum for flexion and extension. The electrogoniometers were connected to a four-channel strip chart recorder (Model 3314, Soltec Corporation, Sun Valley, California).

The subjects were instructed to exert maximal grip effort in any position in which the wrist felt most comfortable. This comfortable position was then referred to as the self-selected position. Maximal grip strength measurements were then performed at wrist positions which deviated 10 to 15 degrees in either flexion, extension, radial deviation, or ulnar deviation away from the self-selected wrist position. The order of testing was selected from a random table. Each grasp effort was held for two to three seconds with a 15 second rest interval after each trial to minimize the risk of muscle fatigue.

The results of this study demonstrated that the self-selected position chosen by the subject when performing a maximum grip effort was consistent. For the size of grasp dynamometer tested, the wrist joint position was at 35 $\pm$ 2 degrees of extension and 7 $\pm$ 2 degrees of ulnar deviation. Grasp strength was 41 $\pm$ 3 kg in that position. Furthermore, the corresponding grasp strengths were significantly reduced in each of the four deviated wrist positions as compared to that in the self-selected position, with average decreases ranging from 9 to 13 kg. The self-selected position was highly reproducible irrespective of gender or handedness. Overall, grasp strength is significantly decreased when wrist position deviates from a self-selected optimal position.

Factors affecting grasp strength

The results of the above study suggest that there are several important biomechanical factors which determine the potential power and strength of grasp function. The size and shape of the object to be grasped determines the joint configurations of the thumb and fingers involved in grasp function. The corresponding moment arms of both intrinsic and extrinsic muscles at a particular joint configuration determine the mechanical advantage, tendon excursion, and the corresponding muscle length. Therefore, the object size and shape are important considerations in determining grasp power and strength. The extrinsic muscles of the fingers and thumb originate from the forearm. Wrist joint motion will, therefore, create excursion of these tendons and modify the muscle contraction characteristics due to the length-tension relationship. Grasp power and strength are thus regulated by the wrist joint configuration. For a given size of object, there is one optimum wrist joint position which provides comfortable and maximum grasp strength as observed in the above study.

Three anatomic parameters of muscle morphology have been recognized to be important to define the biomechanical potential of the muscle [5, 6]. Muscle fiber length is related to the potential for physiological excursion of the tendon and muscle. The physiological cross-sectional area of a muscle is proportional to the maximum tension of the muscle. Physically, the product of the force and distance is the work; thus, the muscle mass or volume has been considered to be proportional to the work capacity of the muscle.

It is well known that potential muscle force generation depends on the muscle length at contraction. Usually, there is an optimum length where maximum contractile force is generated. The force potential of muscle with contraction at either a shorter or longer muscle length will be less. The arrangement of the muscle fiber architecture will further influence the characteristics of the muscle contraction [7, 8]. It has been demonstrated that parallel muscle fibers produce a length-tension curve with maintained force throughout a wider range of excursion than sharply peaked curves for muscles with shorter fiber pennate structure. The index of muscle architecture has been used to define such characteristics.

The orientation or constraint of muscles or tendons crossing a joint determines the characteristics of excursion and moment arm. In general, the slope of the curve between the tendon and muscle excursion as a function of the joint angle represents the magnitude of the moment arm or mechanical advantage of the muscle and tendon responsible for that particular function or movement of the joint. The larger the moment arm, the better the mechanical advantage for the same amount of tendon or muscle force. On the other hand, the larger the moment arm, the more tendon excursion is expected for the same amount of joint rotation. The excursion of the tendon eventually affects the muscle length of contraction and ultimately determines the potential force generation according to the muscle length-tension characteristics.

CONCLUSION

Overall, it is felt that biomechanical factors of the musculoskeletal system play an important role in determining limb position and force control in reach and grasp function. The use of referent systems to describe complex motion of the upper extremity is advocated to be able to characterize the three-dimensional nature of movement in space. The kinematic measurement of three-dimensional motion has been aided recently by the use of electromagnetic tracking measurement methods. Biomechanical studies using this technology have shown that the shoulder complex moves the arm in a sphere of motion placing the humerus at a fixed angle while flexion and extension of the elbow changes the regional position of the hand relative to the body or external environment. It has been concluded that the motion of the shoulder is not able to compensate for the loss of elbow motion and elbow fusion at any angle results in significant limitations in reach activities.

The main effector of the upper extremity is the hand. The positioning of the wrist is a key factor in control and optimization of hand grasp strength capabilities. When attempting unconstrained grip individuals naturally select the position of the wrist that is consistent, reproducible, and optimum for strength production. Since the extrinsic muscles of the fingers and thumb cross the wrist joint it is believed that wrist joint motion effects the excursion of the tendons and modifies the muscle-contraction characteristics due to the length-tension relationship. Further study is necessary to refine the biomechanical analysis of reach to grasp function as well as collaboration with neuroscientists to more critically look at control factors.

REFERENCES

1 K.N. An, A.O. Browne, S. Korinek, S. Tanaka and B.F. Morrey, J. Orthop. Res., 9 (1991) 143.

2 O.R. O'Neill, B.F. Morrey, S. Tanaka and K.N. An, Clin. Orthop. Rel. Res., 281 (1992) 89.

3 J. Pryce, J. Biomech., 13 (1980) 505.

4 S.W. O'Driscoll, E. Horii, R. Ness, T.D. Cahalan, R.R. Richard and K.N. An, J. Hand Surg., 17 (1992) 169.

5 P.W. Brand, R.B. Beach and D.E. Thompson, J. Hand Surg., 6 (1981) 209.

6 K.N. An, F.C. Hui, B.F. Morrey, R.L. Linscheid and E.Y.S. Chao, J. Biomech., 14 (1981) 659.

7 R.D. Woittiez, P.A. Huijing, H.B.K. Boom and R.H. Rozendal, J. Morph., 182 (1984) 95.

8 K.R. Kaufman, K.N. An and E.Y.S. Chao, J. Biomech., 22 (1989) 943.

Insights into the Reach to Grasp Movement
K.M.B. Bennett and U. Castiello (Editors)

CHAPTER 14

REINFORCEMENT LEARNING FOR ROBOTIC REACHING AND GRASPING

A.H. FAGG

Center for Neural Engineering, Computer Science Department, University of Southern California, Los Angeles, California, 90089-0781, U.S.A.

SUMMARY

A reinforcement learning approach is used to train a neural controller to perform a robotic reaching task. Unlike supervised learning techniques, where the teacher must provide the correct sequence of motor actions, only an evaluation of the robot's performance is provided. From this limited information, the robot must discover the appropriate motor programs that best satisfy the teacher's evaluation criterion. This type of learning approach is important because in a real-world environment, the teacher is generally not able to describe the motor program that performs the desired motor skill. This chapter utilizes the language of schema theory [1] as a mechanism for describing functional decompositions of motor programs. A connection is made from schema descriptions to a neural-level implementation of the schemas. It is at this low level of processing that we define a reinforcement learning algorithm that acquires motor programs that satisfy the reinforcement policy defined by the teacher.

INTRODUCTION

In a laboratory situation, a primate learns to perform the task designated by the experimenter through a reward/penalty or reinforcement-based paradigm. This reinforcement information, however, is extremely sparse relative to all of the things the monkey must do in order to obtain a reward. Even with the simplest tasks (e.g., reaching to grasp a handle), a monkey has many different motor acts that are available, from which it must select some sequence. When a reinforcement signal is provided, it must somehow infer the critical elements of its actions that caused it to receive the reward, so that these elements may be repeated the next time that the same situation arises. Despite this very limited amount of information, the monkey is often able to learn the desired task.

Within the robotics domain, we find a somewhat similar problem, in that it is typically difficult to specify a robust motor program. A very common technique is to specify in great detail the trajectory through space that the manipulator is to take in performing a task. This has worked fairly well in structured environments, but as the environment becomes more uncertain, it becomes more difficult for the programmer to anticipate all possible situations, let alone the appropriate actions that must be taken. We would therefore prefer to specify programs at a higher level: one in which it is more natural for a programmer or teacher to communicate. Our approach draws inspiration from learning in monkeys, using reinforcement (or reward-based) information to specify the desired behavior of the robot, as opposed to specifying the motor program that produces that behavior (note that *supervised learning* techniques are one way of implementing this latter case).

Learning within a reinforcement-based paradigm, however, presents several key difficulties, which have been explored by a number of authors, including Barto, Bradtke, Dayan, Sutton, Watkins, Werbos, and Williams [2-6]. These are:

A) The reinforcement signal is only a scalar measure of the performance and does not provide explicit corrective information.
B) The reinforcement signal is not necessarily continuous in time (i.e., it may only be available at very discrete events).
C) The reinforcement signal can be temporally delayed relative to the *critical actions* taken by the neural controller. *Critical actions* are actions in the sequence generated by the controller that determine the final result in the environment (and indirectly determine the reinforcement).

Some elements of this reinforcement-based approach have also been explored in Fagg and Arbib [7], which presents a model of the work of Mitz et al. [8]. Their work examined the changes in behavior and in neural responses in the premotor cortex as a monkey learned an association task. In these experiments, monkeys were first taught to associate a set of four distinct visual patterns with a particular movement of a joystick. For example, when the monkey is shown the character *A*, then it is expected to move the joystick to the right. If the monkey responds correctly to the stimulus, then it is rewarded with a squirt of juice. Once the monkey has learned the overall task, sets of novel stimuli are presented. The monkey is to infer the appropriate motor response to each stimulus based upon the reward information that it is given. The key results of the model [7] were:

- The model produced a similar pattern of behavior as was observed in the experiments with the monkey.
- The modeled neural units behaved similarly to the premotor cortex cells observed by Mitz et al. [8].
- The model represented the visual-motor transformation in a distributed manner and updated this transformation based only upon the reinforcement signal received from the teacher. By *distributed representation,* it is meant that a particular transformation did not depend exclusively upon a single computational unit, but rather on the co-activation of a set of units.

In this work, these ideas are extended in several dimensions:

- Sensing and generating actions now become continuous processes, rather than a one-step sensor-to-motor transformation.
- The teaching signals are no longer in one-to-one correspondence with the actions taken by the learning system. In general, a whole sequence of actions is taken before reinforcement information is available. In addition, it is possible that this signal is delayed relative to the critical actions taken by the system. These problems are approached by modifying the learning algorithm such that the reinforcement signal is propagated backwards through time in an efficient and biologically plausible manner.
- We begin to approach the issue of different neural regions being involved in a computation and how their relative functions might work together to perform a task. It is of special interest to understand how learning may occur at different levels within a control hierarchy. For example, when a neural system is learning a new task, not only must it decide what must be learned, but also at what level the new information needs to be encoded. In some cases, the low-level components of the controller for the new task are already in place, and it is only necessary for the higher-level components to make adjustments to bind them together in a unique way.

Schema theory to neural networks

Schema theory [1] provides a language for describing functional decompositions of sensory and motor processes. An individual *schema* is a parameterizable description of a computational element that may actually be implemented as a network of *sub-schemas*. Traditionally, the lowest-level schemas are implemented as either C processes or as encapsulated neural networks. A *schema instance* is a parameterized copy of a *schema* that performs the specified computation based upon the schema description and the provided parameters. The theory allows for the simultaneous existence of multiple *schema instances*, each with their own set of parameters.

From a biological stand-point, however, schema theory does not provide a sufficient language for mapping between schemas and neurons. Although we allow a schema to be implemented as a neural network and then connect it into a network of other schemas, this is only done at a functional level. What is missing is a bridge from the functional level of analysis to an implementational one. At such a level, we would like to explicitly address the issues of:

- The distributed representation of schemas across sets of neurons, and potentially over multiple layers.
- The participation of a single neuron within one or more schemas. These schemas may, in fact, be functionally distinct from one another, and the task that the neuron performs for each case may also significantly differ.
- Neural representation of information and the operators that act on these representations. Schemas tend to exchange state information as sets of real-valued numbers and/or symbols. However, on the neural side, we have (somewhat ill-defined) notions of firing rates, spatial codes, and cosine tuning functions.

In order to bridge the gap from *schema descriptions* to the *implementation of schemas* using neural hardware, we introduce the concept of a *μ-schema*. A *μ-schema* is a simple processing unit that is still at a level higher than that of a neuron. A *schema* is implemented by recruiting a collection of *μ-schemas*. Even though different *schemas* can take on radically different computational structures, all *μ-schemas* utilize a fixed computational structure. Thus, the different computational structures of two *schemas* are achieved by recruiting different (but potentially overlapping) sets of *μ-schemas*.

In the remainder of the chapter, we first present the task to be learned: reaching towards a specified target from different points in the workspace. A global (schema-level) view of the neural controller is then presented, followed by a description of the neural implementation of the model. We then present the learning algorithm that is used to acquire schemas that perform the desired task. Finally, through a set of simulation results, the behavior of the model is illustrated.

TASK TO BE LEARNED

In this chapter, we will illustrate the design and behavior of the neural system described in the next section with a simple reaching task. The learning system controls an X-Y robot (a robot with two prismatic joints) and is to learn how to reach towards a specified target location. The inputs to the system are a teacher-provided command signal and goal location, as well as a feedback signal that informs the system as to the current location of the arm. The neural controller specifies outputs in the form of incremental changes in position of the arm. Reinforcement learning techniques have been applied to a similar problem by Barto et al. [9, 10], except in their case the target position was always fixed.

The robot arm is located in a closed workspace. The teacher provides the system with several different types of reinforcement information, which are summed to create a global reinforcement signal (this measure is scalar and continuous). First of all, the system is positively rewarded when the endpoint of the arm reaches the target location. Secondly, if the endpoint of the arm reaches the edge of the workspace, it is prevented from further movement and it is given negative reinforcement. Finally, the system also receives a small amount of positive (negative) reinforcement if the movement in the last time-step was towards (away from) the target location. One important question to be examined is the degree to which this third type of reinforcement is necessary for the system to learn the task within a reasonable amount of time.

This two degree-of-freedom manipulator provides a simple example through which to illustrate the neural architecture, but still presents interesting challenges. One primary difficulty in learning is that the controller must output both the correct x-dimension increment and y-dimension increment in order to move towards the goal and receive a reasonable amount of positive reinforcement for doing so (Fig. 1, action ***a***). If the increment for only one dimension is correct (action ***b***), but the other is incorrect (e.g., opposite in sign), then the system could receive either negative reinforcement or none at all. In a case such as this, it is impossible to determine which of the two control outputs was correct. Therefore, the system must rely upon multiple samples with different output actions to infer what the correct action is given a particular situation. Note that in this reinforcement scheme, the feedback resulting from the production of an action that takes the arm directly away from the target (action ***c***) provides just as much information as does a movement towards the target.

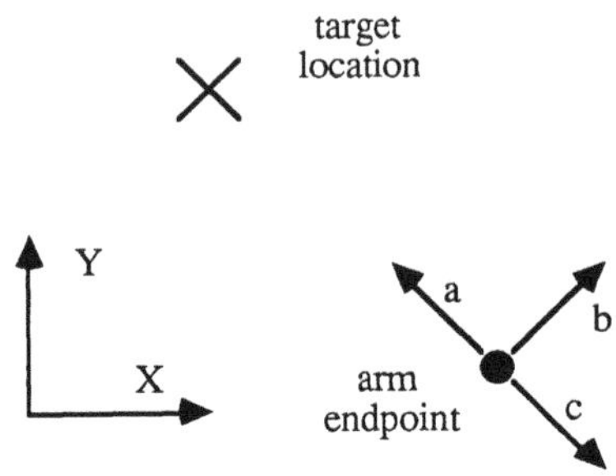

Figure 1. Several possible actions given a particular situation. ***a****. Move towards the target, receiving a significant amount of positive reinforcement.* ***b****. Move perpendicular to the correct direction, receiving no reinforcement.* ***c****. Move directly away from the target, and therefore receive negative reinforcement (of the same magnitude as* ***a****).*

MODEL DESIGN

The neural model described in this section can be viewed at several levels of abstraction. We will first look at the overall organization of the model, and then look closer at the details of the implementation.

Global network architecture

The global view of the network design is depicted in Fig. 2. The network consists of two main processing layers and several input/output layers. The *Command Vector* is the input into the network that calls up specific motor programs (high-level schemas). In other words, this defines the current task that the system is to perform. One such task (the one upon which we are concentrating in this chapter) is reaching towards a target. Another task might be to reach towards a specific location (regardless of the target input), or to reach towards a location opposite the target. For this example, however, it will be assumed that this input is fixed.

The *Planning Layer* is the neural structure that implements the *reach-toward-target* schema (which has been selected by the *Command Vector*). The *Target Location* vectors are a neural representation of the X-Y position of the target object. The *reach-toward-target* schema in the *Planning Layer* makes use of this position information to select the schema within the *Execution Layer* that is responsible for moving the arm to the specified target location. This *reach-toward-a-specific-target* schema utilizes the current state of the arm (input from the *Arm Location* vectors) to generate movement commands for each of the two degrees of freedom. These output commands are sent to the *dX* and *dY* layers, where they are executed by the manipulator.

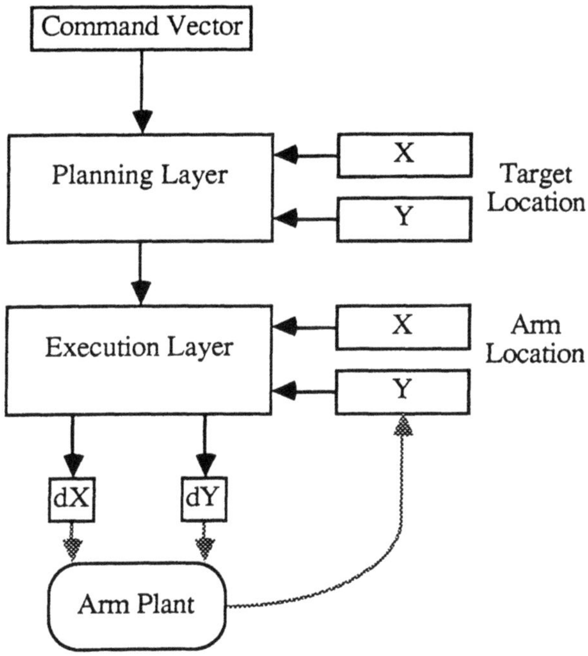

Figure 2. Overall view of the reach control network.

It is important to keep in mind that the global architecture presented in this section is specific to the particular reaching task. It is our intention that the neural-level implementation of the layers (described in the following sections) be generic in the sense that given other control problems, the same implementation would be useful, even if the global architecture has changed in some way (e.g., the addition of more processing layers).

Input/output coding

Each *Target Location* and *Arm Location* vector is a linear array of neurons that code a continuous value using a spatial code. In this case, a *Gaussian distribution* is used, where the location of the mean of the distribution is determined by the value being coded. The variance of the distribution is adjusted such that two values must be relatively close to one another to have significantly overlapping representations. The coding scheme is such that we are also able to represent a continuous range of values while utilizing only a finite number of neural units. Fig. 3 shows an example of Gaussian coding for three different values.

The *dX* and *dY* layers represent the output of the control network. Each layer consists of a linear array of units that spatially code one output variable (the increment of robot position along the X- or Y-dimension). A value is read out from a linear array of units by computing the *center of mass* of the activity levels:

$$x = \frac{\sum_{i=0}^{N-1} i * a_i}{\sum_{i=0}^{N-1} a_i}$$

where:
$\boldsymbol{x}$ is the decoded value.
$\boldsymbol{i}$ is the linear position along the array.
$\boldsymbol{N}$ is the length of the array.
$\boldsymbol{a_i}$ is the activity level of the **i**th unit.

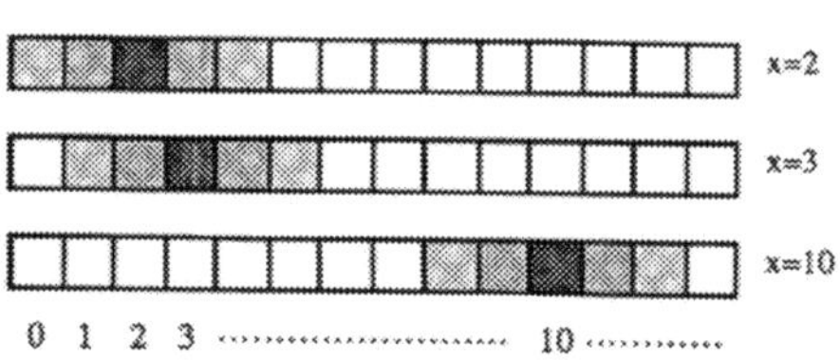

*Figure 3. Example of Gaussian spatial coding in a linear array of neurons. Neuron firing rate is indicated by shading (white = not firing, dark = maximum firing rate). The values coded by the three linear arrays are **2**, **3**, and **10**, respectively.*

Processing layer implementation

The *Planning* and *Execution* layers of the global architecture (Fig. 2) consist of collections of *μ-schemas*. Physically, the *μ-schemas* are arranged into a two-dimensional grid. We first define the primary computational concepts that make up a *μ-schema* and then show how these concepts are implemented within neural hardware.

A *μ-schema* produces an output when it detects an incoming sensory feature. However, the activation (selection) of this mapping is constrained by both global and local inputs. In order for a μ-schema to become active, it must first be *primed* by some external input. When a higher-level process primes a μ-schema, it effectively grants its permission for the μ-schema to participate in a computation. For example, a high-level *grab object* process will need to recruit sub-schemas to execute the reach and grasp elements of the task. This is implemented by priming the μ-schemas that make up the two sub-schemas. In turn, these sub-schemas may be further broken down into more specific sub-schemas. Also, these two schemas may establish lines of communication for the purposes of coordinating their execution.

Due to the inherent simplicity of a μ-schema, implementing a single schema requires the activation of an entire set of μ-schemas. It is therefore necessary on an implementational level to provide a mechanism that ensures the co-activation of this set. On the other hand, some μ-schemas conflict with others by producing conflicting commands to a lower level or to an actuator. The co-activation of these μ-schemas is therefore not desirable.

These constraints are implemented through interactions between primed μ-schemas within a single layer. By allowing such an interaction to take place within each layer, the problem of deciding which processes are appropriate for execution is distributed throughout the network. As a result, the decision as to which sub-schema is appropriate for a given situation is left to the layer that has the contextual and sensory information necessary to make such a decision. In this work, these interactions are implemented as inhibitory and excitatory connections between the primed μ-schemas.

One simple way to implement μ-schema interaction is by connecting μ-schemas through a *Mexican-hat* type operator. This connection scheme provides excitatory connections to physically near neighbor μ-schemas, and inhibitory connections to a ring of neighbors just outside the circle defined by the excitatory connections. Computationally, this scheme implements a contrast-enhancement operation that forces only a few neurons to be active in a layer at any one time.

Given this context, it is possible to more precisely interpret the concepts of *schema* and *schema instance*. A schema instance is a particular pattern of activity within a layer. A schema can be considered as the set of constraints imposed by a particular class of priming input and the interaction between the set of μ-schemas.

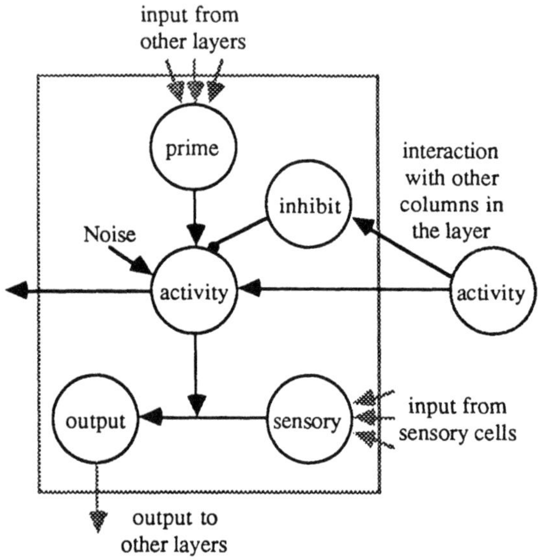

Figure 4. The structure of the column. Arrowheads indicate excitatory connections. Inhibitory connections are indicated by small circles. Gray arrowheads represent modifiable synapses.

Neural implementation of the μ-schema

The *μ-schema* is implemented using a tightly-coupled collection of neurons. These collections are referred to in this work as columns of neurons. Fig. 4 depicts the neural implementation of the column. We first define the behavior of the generic neuron model, and then discuss the detailed implementation of the column.

$$\tau \frac{d\, mem}{dt} = -mem - threshold + \sum inputs$$

$$firing = f(mem)$$

Leaky integrator model of the neuron. The neurons in the model are implemented using the leaky-integrator model [1] as a basis. Each neuron is represented by a membrane potential and a firing frequency (Fig. 5). The dynamics of the generalized neuron are defined by the two above equations.

where :
τ is the time constant of integration.
mem is the membrane potential of the neuron.
threshold is the neuron's internal threshold.
inputs is the set of external inputs into the neuron.
firing is the firing rate of the neuron.
$f(\,)$ is a function (typically non-linear).

The priming unit. Inputs from other layers (primarily higher-level layers) prime the column. The inputs are summed to determine changes in the *priming unit's* membrane potential:

$$\tau_P \frac{d\, pr_i}{dt} = -pr_i - thresh_P + \sum_j w_{ji}^P * out_j$$

where:
pr_i is the membrane potential of the priming unit of column **i**.
$thresh_p$ is the threshold parameter for the priming units.
w^P_{ji} is the strength of the connection from column **j** (another layer) to priming unit **i**.
out_j is the output activity of column **j** of a preceding layer.

The firing rate of the priming unit is then computed by:

$$prime_i = NSLsat(pr_i)$$

where:
$prime_i$ is the firing rate of priming unit **i**.

$$NSLsat(x) = \begin{cases} 0 & x < 0 \\ x & 0 \le x \le 1 \\ 1 & 1 < x \end{cases}$$

The activity unit. The activity of a column is determined by the firing rate of the *activity unit.* Excitatory interactions between two columns are implemented as a positive connection between the corresponding *activity units.* Inhibitory interactions are implemented through the use of an *inhibitory unit*. When one column inhibits the activity of another column, it excites the *inhibitory unit* of the target column, which in turn inhibits the *activity unit*. The dynamics of these two units are as follows :

$$\tau_A \frac{d\, ac_i}{dt} = -ac_i - thresh_A + pr_i - inhibit_i + noise_i + \sum_{j \neq i} w_{ji}^A * act_j$$

$$\tau_I \frac{d\, inh_i}{dt} = -inh_i - thresh_I + \sum_{j \neq i} w_{ji}^I * act_j$$

$$act_i = NSLsat(ac_i)$$

$$inhibit_i = NSLsat(inh_i)$$

where:
w^A_{ji} and w^I_{ji} are connections from other columns (specifically their *activity* and *inhibitory units*) within the same layer. In this implementation, connections are made only to adjacent columns.
$noise_i$ is a noise signal that is injected into the membrane potential of the activity unit. This signal changes slowly relative to the time constant of the processing units; the distribution is uniform over a small range: $[-\alpha, \alpha]$.

The noise term plays an important role in the behavior of the system. In the early stages of learning, this noise helps to drive the search process when the controller is unsure of the correct action to be taken. As a result, the control space is more efficiently explored. When an *activity unit* is firing, the column (μ-schema) is participating in the current computation.

The sensory unit. The *sensory unit* detects sensory events from the environment or from the state of the robot. For example, the input into the *Execution layer* of the network consists of two arrays of units that specify the current end-point position of the arm. Thus, a *sensory unit* at this level can detect events such as the arm moving into:

A) A particular range of the Y dimension.
B) A specific region of the workspace (requires inputs from both the X and Y dimensions).
C) Multiple regions of the workspace.

A similar type of incidence coding scheme has been used by Mel [11], except that each sensory unit receives exactly one input from each dimension (e.g., X and Y). This results in only units of class B, which requires many more units to cover the entire space. The sensory units in the *Planning* layer receive input from the *Target Object* vectors.

Figure 5. The leaky integrator model of the neuron. The membrane potential of the neuron is affected by the current set of inputs, the neuron's threshold, and the current state of the neuron. The firing rate is a non-linear function of the membrane potential.

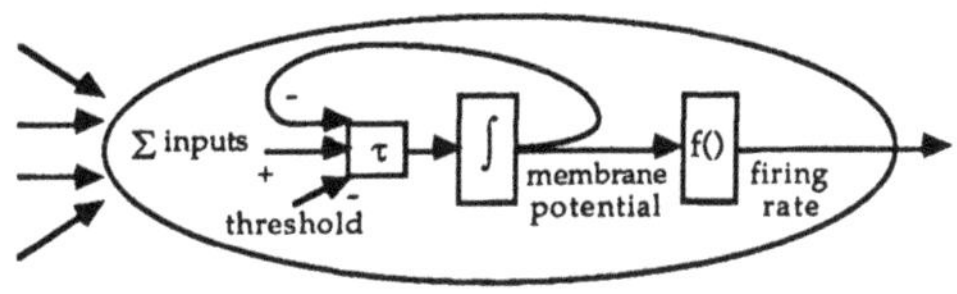

The dynamics of the *sensory unit* are similar to those of the *priming unit:*

$$\tau_S \frac{d\, sen_i}{dt} = -sen_i - thresh_S + \sum_{j \neq i} w_{ji}^S * input_j$$

$$sensory_i = NSLsat(sen_i)$$

where:
$\boldsymbol{input_j}$ is an element of the sensory input vector for this particular layer.

The output unit. The firing rate of the *output unit* is the *sensory unit* firing rate gated by the *activity unit*:

$$output_i = activity_i * sensory_i$$

The activity level of the *output unit* represents the output of the column, which then connects to other layers in the network. The output from one layer provides input to *priming units* of the destination layer. Thus, schemas at one level prime sub-schemas implemented at lower layers of the network.

LEARNING SYSTEM

Given the overall structure of the network, the task of the learning system is to tune the connection strengths between the various sets of units to develop a set of μ-schemas that accomplish the task specified indirectly by the reinforcement signal. This tuning must be done based upon the experience of the system interacting with its environment and the reinforcement signal that it receives from the teacher. As discussed in the introduction, this problem is difficult because:

A) The reinforcement signal is a scalar measure of the performance and does not provide explicit corrective information.
B) The reinforcement signal is not necessarily a continual signal.
C) The reinforcement signal can be delayed temporally relative to the critical actions that were taken by the neural controller.

Determining which connections should be updated is referred to as the *credit assignment problem*. Given that the teacher provides some instantaneous reinforcement signal, the learning system must identify which computational elements (columns) were responsible for generating the actions that ultimately led to the reinforcement (*structural credit assignment*), and at what time did these elements make the critical decisions (*temporal credit assignment*).

In the columnar structure, two sites are subject to updates in connection strength: connections from the *output units* of one layer to the *priming units* of another layer, and the connections from the sensors to the *sensory units* (gray arrowheads of Fig. 4). By tuning the set of connections from the output units of one layer to the priming units of another, the learning system adjusts the set of μ-schemas that are to be primed at the lower level, and thus determines the set of μ-schemas that make up the sub-schemas. At this level, the learning scheme effectively implements the following rule:

A) If the lower-level μ-schema is active (and thus is participating in the computation) during periods of time when the control system *tends* to receive positive reinforcement, then increase the connection strength from the higher-level column (which is active) to this column.
B) If this μ-schema tends to participate during times when the system is receiving negative reinforcement, then reduce the connection strength.

The connections from the sensors to the sensory units are adjusted in a similar manner. This adjustment implements the rule:

A) If a μ-schema is active and it is producing a non-zero output during a period of time in which the system *tends* to receive positive reinforcement, then increase the connection strength from those sensor inputs that are currently firing.
B) If the system *tends* to receive negative reinforcement, then it is possible that the μ-schema is attending to the incorrect sensory feature; therefore, reduce the connection strength to those sensor inputs that are currently firing.

In both of these loosely-defined learning rules, the terms ***tends to receive***

positive reinforcement and ***tends to receive negative reinforcement*** are very important. Even though the system finds itself in several very similar situations, the control system may produce different control actions (due to the noise injected into the *activity units*), or the teacher may provide apparently inconsistent reinforcement information (due to the inexactness or more qualitative nature of the reinforcement signal). As a result, the learning system must not make large adjustments based on the instantaneous reinforcement signal, but rather must take into account many experiences in constructing an effective control program.

The challenge, then, is to consolidate these experiences in an efficient manner - both in time and storage space. The algorithm below presents one approach to solving this problem.

Eligibility as a temporal measure of credit assignment

The eligibility of a weight (connection between two units) measures the participation of the connection within the computation that is currently taking place. The *instantaneous eligibility* is defined as the coincidence between the pre- and post-synaptic cell activities. In our case, this product is also modulated by the strength of the connection between the two cells (this definition of eligibility was inspired by the work of Klopf [12], and Barto et al. [2]). Thus, the *instantaneous eligibility* is:

$$\hat{e}_{ij} = a_i * a_j * w_{ij}$$

where:
$\hat{e}_{ij}$ is the *instantaneous eligibility* between unit **i** and unit **j**.
a and a_j are the activity levels of the pre- and post-synaptic columns, respectively.
w_{ij} is the weight from unit **i** to **j**.

An exponentially-decaying memory of the *eligibility* can be implemented by applying a low-pass filter to the time series of instantaneous eligibilities:

$$\tau_e \frac{de_{ij}}{dt} = -e_{ij} + \hat{e}_{ij}$$

where:
τ_e is the time constant of integration, or the decay of the memory.

e_{ij} is the eligibility of the connection.

When a reinforcement signal ($\boldsymbol{R}$) is provided by a teacher, the eligibility of a connection is used to update the weight. More specifically:

$$\Delta w_{ij}(t) = \alpha * R(t) * e_{ij}(t)$$

where:
$\Delta w_{ij}(t)$ is the change in weight.
α is the learning rate.

This update equation says if a connection has recently been participating in a computation, then make a small incremental change to the connection strength. The sign of this incremental change is determined by the sign of the instantaneous reinforcement signal, $\boldsymbol{R(t)}$. The magnitude of this increment is determined by the magnitude of the reinforcement signal, and by the degree of participation of the connection, $e_{ij}(t)$.

It is important that the rate of learning, α, is adjusted appropriately. When set at a value that is too small, the learning time can be longer than practical. If set too large, the noise component of the weight increments (due to the noise injected into the controller or to noise in the reinforcement signal) can be amplified above the level of the meaningful information.

Weight normalization

Once the change in weights is computed, the actual connection strength is updated according to:

$$w_{ij}(t+1) = \boldsymbol{Normalize}(w_{ij}(t) + \Delta w_{ij}(t))$$

Biologically, normalization comes out of the limited resources that a neuron has to establish connections to other neurons. Computationally, normalization performs two important functions:

A) Individual weights are bounded within a finite range, thus alleviating some computational difficulties.
B) Normalization implements a form of competition between the individual weights. This weight competition can take one of two forms: presynaptic or postsynaptic.

For presynaptic normalization, the function *Normalize* ( ) maintains the conditions

$$\sum_i w_{ij} = 1 \text{ and } 0 \leq w_{ij} \leq 1$$

(note that w_{ij} is defined as the connection strength from unit **i** to unit **j**). In other words, the total output from the presynaptic unit is a constant value; as the weights change, it is only the distribution of the output that changes. This type of normalization is used for the connections from the output unit of one layer to the priming unit of another layer. Within this context, normalization can be interpreted as an active column (at the presynaptic side) searching for the appropriate set of lower-level columns to which to distribute its priming support.

Fig. 6**A** shows the effect of positive reinforcement on the connections from one column to a set of columns. Initially, the connections to columns ***a*** and ***b*** have significant strengths. However, columns ***b*** and ***c*** are the ones that are currently active. Thus, when positive reinforcement is received, the strength of these two connections increases. Due to normalization, the connection strengths to columns ***a*** and ***d*** are reduced. One way to interpret this behavior is that the higher-level column is becoming more *sure* of the correct set of sub-columns that it should prime so as to receive positive reinforcement in the future. Thus, it becomes more committed towards these columns through the increase of the weights.

When negative reinforcement is received, the opposite situation occurs (Fig. 6**B**). The weights leading to the active columns are decreased slightly. The remaining weights are then increased as a result of normalization. In this case, the higher-level column is not so sure about whether it should be priming the currently active columns, and hence the decrease in connection strength. By reallocating the connection strength removed from these active columns, the higher-level column gives other columns more of a chance to become active the next time the same situation occurs, thus driving the search process for more appropriate sub-columns.

When postsynaptic normalization is used, the weights are normalized across the opposite dimension of what is done in presynaptic normalization. So the function *Normalize* ( ) maintains the condition:

$$\sum_j w_{ij} = 1 .$$

It is this type of normalization that is used for the projection from sensor inputs to *sensory units*. From the *sensory unit* point of view, the unit is attempting to select the sensor elements that tend to yield positive reinforcement when used as triggering signals.

MODEL BEHAVIOR

The network used for the experiments described here consisted of a 7 x 7 grid of columns for the *Planning* layer, and a 16 x 16 grid of columns for the *Execution* layer. The *dX* and *dY* layers each consisted of 3 columns. The input layers representing positional information consisted of arrays of 15 units each. On network creation, the connections and their strengths are randomly generated, yielding a network that is not committed to any particular set of schemas.

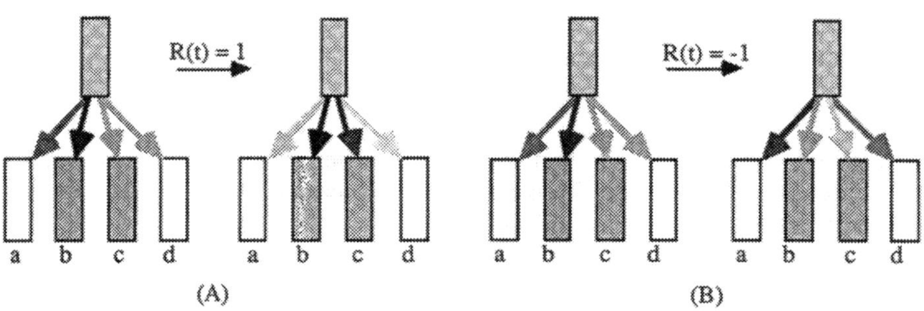

*Figure 6. Demonstration of the effect of positive (**A**) and negative (**B**) reinforcement with presynaptic normalization. The upper box represents the higher-level column, whose output unit connects to the priming units of the lower-level systems (lower boxes). Column activity is represented by degree of shading. Connection strengths are also represented with different shading levels (light = small weight, dark = large weight).*

A learning trial begins by selecting one of two opposite corners as a starting position for the arm endpoint, with the target located roughly in the center of the workspace. The system is then allowed to drive the arm until one of

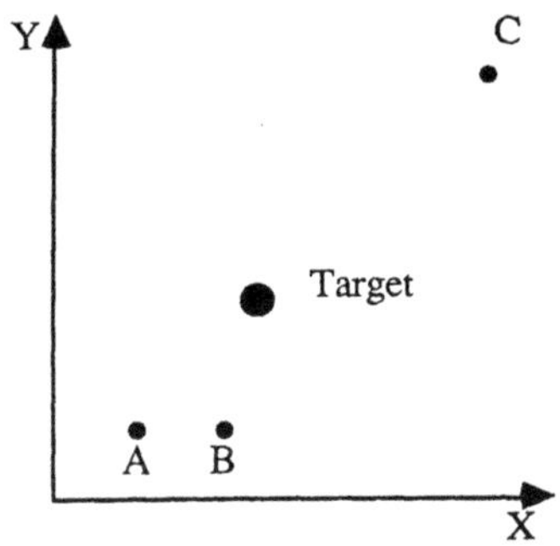

Figure 7. The workspace layout for three different arm positions.

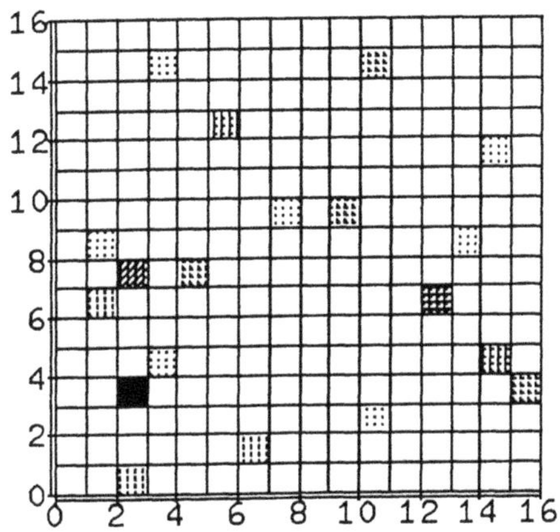

Figure 8. Pre-learning state of the output units of the Execution layer when the arm is located at position A. Each small box represents the state of one unit, with darker boxes indicating high firing rate. The high firing rate in the activity units of the execution layer indicates those columns that are currently participating in the motor program for this particular target location. It is these active units that prime the dX and dY layers.

two events occurs: the arm arrives at the target position, or the arm reaches the side of the workspace. At this point, the final reinforcement is given and the arm position is again reset to a starting location. By using more than one starting location, the system is forced to explore a large region of the state space. However, using only a small number of starting locations allows for efficient experimentation and a more controlled analysis.

When a target position is specified by the teacher, a small subset of *Execution* layer columns becomes active (about 25%). For a given arm position, the *sensory units* of some of these active columns fire in response to the arm position input. These columns then prime the *dX* and *dY* layers. Because the connections are randomly generated, the priming of each of the six columns in the *dX* and *dY* layers tends to be at about the same level as the others. Moving the arm position to different locations in the workspace yields very little change in the priming levels.

Figs. 7-11 demonstrate the responses of the *Execution*, *dY*, and *dX* layers before learning has occurred. Fig. 8 shows the response of the *output units* in the *Execution* layer when the arm is in position **A** (as defined in Fig. 7), and Fig. 9 shows the response of the *dX* and *dY* layers. The responses to the other two arm positions (**B** and **C**) are depicted in Figs. 10 and 11, respectively.

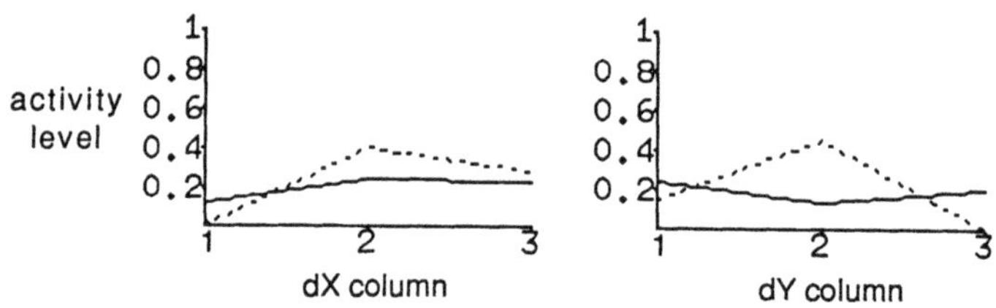

Figure 9. Firing rate plotted against linear position for the priming units (solid) and output units (dotted) for the **dX** *and* **dY** *layers (same conditions as in Fig. 8). Each layer has three units. An activity peak centered near the second unit implies a command of zero magnitude; biased towards the first unit implies movement in the negative direction. Note the relatively uniform priming input in both cases. However, once this signal is contrast enhanced (output unit activity), it is possible to see slight bias in one direction. In this case, the system is commanding a slight positive movement for the X direction, and a slight negative movement in the Y direction.*

Note that Figs. 8 and 10 show only some difference in the output activity of the Execution layer. This is due to the fact that the two arm positions are very near each other. There are, however, several *output units* that change significantly in their activity levels. After learning takes place, it is these units that will encode the essential differences in motor output between these two positions. Those units that are active in both cases will encode the commonalities of the two motor commands.

The differences between Figs. 8 and 11 are much more significant. This is due to the physical separation of the two locations, which implies that the arm position input patterns for the two cases (**B** and **C**) do not overlap and therefore do not activate many common *sensory units*. This will make it much easier for the system to learn radically different motor outputs for these two conditions.

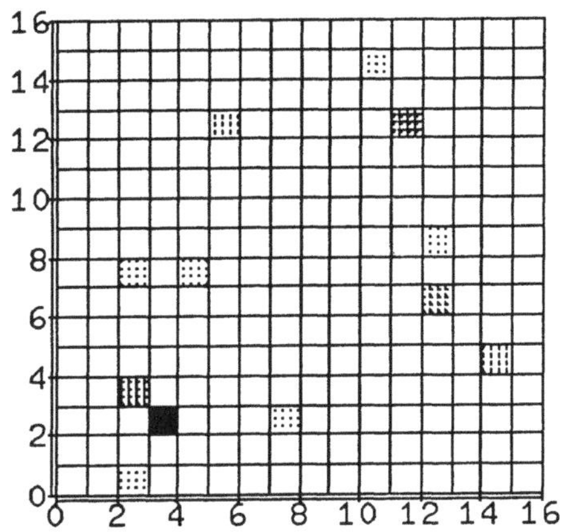

*Figure 10. Pre-learning state for the output units of the Execution layer for the arm position **B**. Note similar activity pattern as that in Fig. 8.*

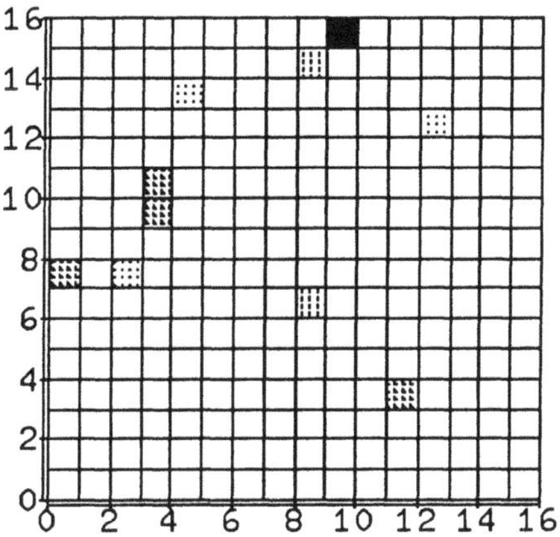

*Figure 11. Pre-learning state for the output units of the Execution layer for arm position **C**. Note significant differences in the activity pattern as compared with Figs. 8 and 10. These differences are due to the radically different arm positions.*

In all three cases, however, the priming signals for the *dX* and *dY* layers are all at approximately the same level (the response to arm position **A** is depicted in Fig. 9), with only very small biases in one direction or another. Despite this relative non-commitment on the part of the *Execution* layer, the contrast enhancement that occurs between the activity units forces a choice of movement in one of the directions (this is especially evident in the firing rates of the output units). As learning proceeds, a particular schema instance will begin to prime very specific columns in these layers. As a result, the layer will rely less on the contrast enhancement as a way of selecting a specific output.

The small amount of variation between the *dX* and *dY priming units* confirms that the network is not significantly biased to produce any particular action before learning has occurred. Therefore, the choice of action to output is primarily driven by the noise that is injected into the *activity units* of the *dX* and *dY* layers. The result of this random set of actions is that the endpoint of the arm tends to wander around the workspace through a random trajectory (e.g., Fig. 12**a**).

As learning progresses, the controller begins to bias the noise signal in regions of the workspace that have been visited several times. This biasing must be done slowly so as to allow the system to explore several different possibilities before committing to one particular movement direction. It is at the point of full commitment that the bias provided by the controller reaches a level above that of the noise.

Examining the behavior of the system relative to a single starting position over many trials, one can observe the general strategy taken by the system. The system first begins by exploring a small local area around the starting point before wandering off to another region of the workspace (Fig. 12**a**). Over several trials, however, the same local area is explored every time (Fig. 12**b**). This common experience allows the system to decide upon the best action for this region of the workspace (Fig. 12**c**). In subsequent trials, the system executes this action, taking the arm to a different location of the workspace (closer to the target), where the controller is now relatively inexperienced. It then proceeds to repeat this process, overall taking small steps towards the target location (Fig. 12**d**). Once the entire path is discovered, repeating it several times solidifies the set of actions in memory (Fig. 12**e** and **f**).

For starting location **12,12** the control system caused the robot to collide with the edge of the workspace only in the first three trials. By the 12th trial it had learned how to perform the task perfectly. For starting location **2,2**, the robot collided with the side of the workspace a total of seven times, and learned to navigate to the target by the 16th trial.

After learning has completed, the *Execution* layer is much more committed towards particular output commands. This is evident in the distribution of the firing rate of the priming units in the *dX* and *dY* layers (Fig. 13). Instead of the roughly uniform distribution of activity that was seen before learning, the *Execution* layer now forces the system to execute specific motor commands.

KEY NETWORK DESIGN ISSUES

Through the design and implementation of this neural network model, we have touched on a number of important network design and neural computation issues. This section explores a number of these issues further.

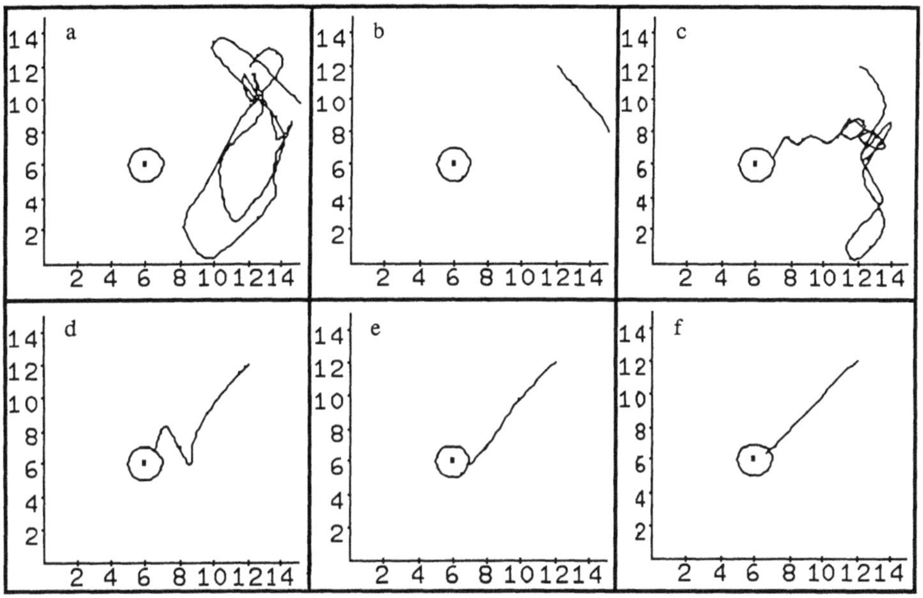

*Figure 12. Trajectories taken during the learning process. Panels **a-f** (top left to right, then bottom left to right) are several trajectories taken by the control system during learning. The starting position for each case is at coordinate **12,12**, with the target located at coordinate **6,6**. The circle around the target position is the area within which the system receives large positive reinforcement signals. Panels **a** and **b** are cases where the system ran into the side of the workspace, where the trial was ended.*

Overlapping state representations

The representation of state is a key problem that plagues neural learning systems in general. Some systems, such as the work of Barto, Sutton and others [2, 3] rely on orthogonal or linearly independent state representations. This leads to two problems:

A) A large number of discrete states must be learned and represented.
B) No sharing of information is done between distinct states that might actually yield similar control decisions.

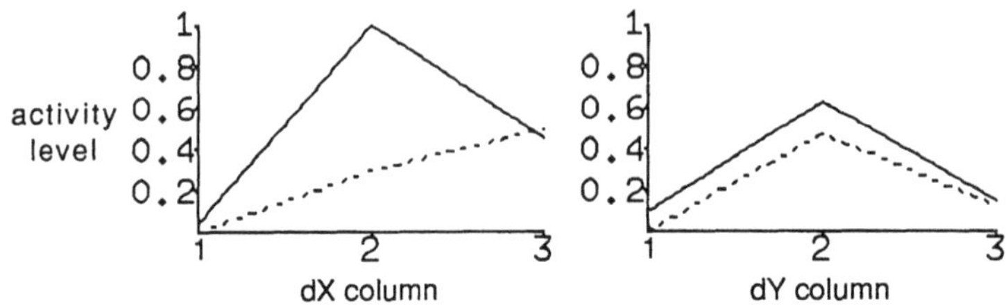

Figure 13. Firing rate plotted against linear position of the priming units (solid) and outputs units (dotted) of the ***dX*** *and* ***dY*** *layers, after learning has taken place (arm position* ***A****). Note the difference with Fig. 9: the Execution layer now has a much clearer idea as to which units should be primed. In this case, we see a significant commitment to move along the positive* ***X*** *direction, and somewhat along the positive* ***Y*** *direction.*

On the other hand, many error propagation techniques, such as backpropagation [13] often require that all computational units participate in every mapping that is learned (this is especially true early in the learning of a training set). Because every hidden unit participates in every mapping, it and the connections that synapse onto it are subject to increments for each input/output mapping. Typically, the modification to a weight for a single mapping conflicts with that of another mapping. These types of conflicts increase the amount of time that is necessary to learn the set of mappings, especially as the number increases.

We would therefore like to find some sort of middle ground where a mapping need not require its own representation, nor should all mappings be encoded by all computing units. Some of these issues have been approached by Jacobs et al. [14]. The network architecture presented in this chapter is designed to address these issues explicitly. The key features that accomplish this include:

- **Sensory coding.** Input variables (from sensors) represent values spatially, using a Gaussian distribution of activity. Thus, two similar values are represented by overlapping patterns of activity, but two very different values have representations that do not overlap.
- **Schema coding.** Generally, different schemas that share common functions will utilize overlapping sets of columns. This is important in terms of efficiently representing the set of schemas, and by the fact that learning in one schema can provide important information for

another.

- **Contrast enhancement.**

Importance of contrast enhancement for learning

The contrast enhancement operation provided by the *Mexican-hat* operator possesses three key computational properties.

- The number of active columns within a layer is limited to be some subset of the entire layer. This implies that only a subset of the columns are allowed to be involved in the representation of a schema, leaving other columns for the representation of other schemas. Consider the case where a large number of columns are allowed to become active and thus participate in a computation. When a reinforcement signal is received, all active columns are considered to be responsible, and are therefore updated in an attempt to improve the system's performance. When learning several different schemas at the same time, this can result in a high degree of overlap, causing the learning due to one schema to interfere significantly with another. Because the contrast enhancement operation reduces this overlap, the interference can be greatly reduced.
- During the learning process, the *Mexican-hat* operator tends to induce a topological representation of the schemas that it learns to represent [15, 16]. Because a μ-schema tends to be active in correlation with its close neighbors, they will tend to acquire similar (but not exactly the same) functions. Thus, a slight shift in activity within a layer (due to some small change in priming) will tend to produce a small difference in the function of the active schema.

Importance of noise

Noise plays a key role in the search mechanism employed by the system as it attempts to identify the correct outputs given a particular situation. Before learning has occurred, the noise injected into the *activity units* (of the *dX* and *dY* layers) drives the outputs of the system, forcing the exploration of the local workspace. As learning proceeds, the noise is biased by the outputs of the *Execution* layer, until the bias overcomes the noise altogether. This form of stochastic search and biasing of noise is reminiscent of the SRV (stochastic real-valued) units of Gullapalli et al. [17, 18].

Noise is also injected into the *activity units* of the *Execution* layer. This noise affects the set of columns in the execution layer that participate in a control computation. This random switching on and off of columns allows the controller to experiment with different subsets of columns. Through learning, those columns that tend to participate at times when positive reinforcement is received will begin to participate more often. This is done until a set of columns is selected that is most able to learn the control problem at hand.

Modularity of structure

The column and layer structures have been designed with some degree of modularity. The difference in function between two layers should not be determined by a difference in layer implementation, but by the type of information that flows into and out of a layer. The network described in this chapter utilizes the same structure for the *Planning* and *Execution layers*. However, the inputs and outputs of these two layers differ significantly.

Using this same modular structure, it is possible to build up more interesting network architectures, such as one that controls a combined arm/hand system. This type of network may be *grown* from the one presented in this chapter by adding additional processing layers for control of the hand, and adding cross-connections between the hand and arm layers.

FUTURE DIRECTIONS

One dimension of future exploration will be the implementation of such a reaching and grasping network, which will ultimately interface to a Puma 560 Arm and a Belgrade/USC Hand for experiments in reaching and grasping within a real environment. We are interested in the development of real-time, on-line control and learning systems that can be used to teach robots to perform interesting tasks within a short period of time. One approach to this problem of learning efficiency is the use of teaching by example, where the teacher demonstrates a motor program to the robot [19]. The robot first learns to mimic the teacher, and then through reinforcement-based feedback (either provided by the teacher or generated internally), refines the motor programs to increase their success and generality.

In our primate modeling research, we see this work as providing one possible way to understanding why certain brain regions take on particular functions in sensing and motor control, and why different regions are connected together in specific ways. One question that can be asked is that given a set of constraints from a particular network architecture (set of regions and connections) and a set of behavioral requirements (as specified by the environment or experimenter), what functions do specific regions and even neurons take on through the learning process?

The work in modeling of primate behavior and neural systems, and the work in learning in robots has progressed in parallel. The primate domain provides important hints as to how a learning system is able to efficiently acquire the ability to perform new tasks, both from a behavioral and a functional point of view. Robotics provides an environment in which models from the primate side may be tested, analyzed, and improved in agents that must also behave in a real environment. Ultimately, predictions that arise from these models may be brought back to the primate domain in the form of new experiments to be tried or as a better understanding as to how the primate system functions.

ACKNOWLEDGEMENTS

The author would like to thank Professors Michael Arbib, George Bekey, and Ken Goldberg for their help and support in the development of this work. Many thanks are also due to Peter Dominey, Amanda Bischoff, Mike McHenry, Nicolas Schweighofer and David Lotspeich for their many helpful comments on earlier drafts of this chapter.

The simulation of this network was implemented in NSL (Neural Simulation Language; [20]), which is available by anonymous ftp (from yorick.usc.edu). Contact Alfredo Weitzenfeld (alfredo@rana.usc.edu) for more information.

REFERENCES

1 M.A. Arbib, The Metaphorical Brain 2: Neural Networks and Beyond, Wiley Interscience, New York, 1989.

2 A.G. Barto, R.S. Sutton and C.W. Anderson, IEEE Trans. Sys. Man Cyber., 5 (1983) 834.

3 A.G. Barto and S.H. Bradtke, TR 91-57, Dept. Comp. Sci., University of Massachusetts, Amherst, 1991.

4 R.S. Sutton, in: W.T. Miller, R.S. Sutton and P.J. Werbos (eds.), Neural Networks for Control, MIT Press, Cambridge, 1990, p. 179.

5 C.J.C.H. Watkins and P. Dayan, Machine Learning, 8 (1992) 279.

6 P.J. Werbos, in: W.T. Miller, R.S. Sutton and P.J. Werbos (eds.), MIT Press, Cambridge, Massachusetts, 1990, p. 67.

7 A.H. Fagg and M.A. Arbib, J. Adapt. Behav., 1 (1992) 3.

8 A.R. Mitz, M. Godshalk and S.P. Wise, J. Neurosci., 11 (1991) 1855.

9 A.G. Barto, C.W. Anderson and R.S. Sutton, Biol. Cyber., 43 (1982) 175.

10 A.G. Barto and R.S. Sutton, Biol. Cyber., 42 (1981) 1.

11 B.W. Mel, in: B. Chandrasekaran (ed.), Perspectives in Artificial Intelligence, Academic Press, Boston, Vol. 7, 1990.
12 A.H. Klopf, The Hedonistic Neuron, Hemisphere, Washington, 1982.
13 D.E. Rumelhart, G.E. Hinton and R.J. Williams, in: D. Rumelhart and J. McClelland (eds.), Parallel Distributed Processing: Explorations in the Microstructure of Cognition, MIT Press, Cambridge, 1986, p. 318.
14 R.A. Jacobs, M.I. Jordan and A. Barto, Cogn. Sci., 15 (1991) 219.
15 C. von der Malsburg, Proc. Int. Symp. Synergetics, Springer, Berlin, 1983, p. 265.
16 C. von der Malsburg, in: F.E. Yates, A. Garfinkel, D.O. Walter and G. Yates (eds.), Self-Organizing Systems - The Emergence of Order, Plenum Press, New York, 1987, p. 265.
17 V. Gullapalli, R.A. Grupen and A.G. Barto, Proc. IEEE Int. Conf. Robot. Automat., 1992, p. 1475.
18 V. Gullapalli, Neural Networks, 3 (1990) 671.
19 L.J. Lin, Machine Learning, 8 (1992) 293.
20 A. Weitzenfeld, TR 91-05, Center for Neural Engineering, University of Southern California, Los Angeles, CA.

Section Six: Neural Modeling

Insights into the Reach to Grasp Movement
K.M.B. Bennett and U. Castiello (Editors)

CHAPTER 15

TRENDS IN NEURAL MODELING FOR REACH TO GRASP

M.A. ARBIB AND B. HOFF

Center for Neural Engineering, Computer Science Department, University of Southern California, Los Angeles, California, 90089-2520, U.S.A.

SUMMARY

The large variety of models of limb movement control reflect the diverse fields of researchers. Mechanical engineers and roboticists solve the engineering problems of kinematics, dynamics, and impedance control. Neural network researchers address problems in internal representation, learning, and execution of motor patterns. Researchers in computational neuroscience marry motor control functionality with neural anatomy and physiology. Here, we first review the use of optimality criteria to characterize point-to-point reaching movements in terms of the trajectory of the limb endpoint or joint angles. We then turn to behavioral studies which reveal internal motor representations. We look at neural network models: both artificial networks which generate trajectories using either supervised or reinforcement learning, and models of brain regions which control reach and grasp. Finally, we return to the behavioral level to model several phenomena, including perturbed reach, the speed/accuracy trade-off in reaching movements, and the coordination of reach and grasp.

MODELING TRAJECTORIES OF LIMB MOVEMENT

Characterizing point-to-point reaching movements

Consider the task of reaching to a visually located target. A standard robotics solution is in terms of solving a sequence of kinematic and dynamic problems, and performing coordinate transformations: **i**) The goal of the movement (the target) is first represented in an extra-personal (world) coordinate system. **ii**) This target location is translated into a trajectory for the hand. **iii**) The trajectory is translated into the time-varying configuration of the arm's joints. **iv**) Finally, the configuration trajectory must be translated to the sequence of joint torques (or muscle activations) which drive the arm along this trajectory.

Kinematics involves the specification of a trajectory, while **dynamics** specifies the forces required to achieve a trajectory. **Direct** kinematics takes us from, say, the time course of joint positions to the successive positions of the hand, while inverse kinematics takes us in the inverse direction - from a desired hand trajectory to the time course of joint positions that achieves it. Similarly, direct and inverse dynamics are related to the time course of forces. Thus, the translation in (**iii**) is the solution to what is called the *inverse kinematics problem*, while the translation in (**iv**) solves the *inverse dynamics problem*.

Below, we must address at least three questions posed by the above approach: **A**) Is this sequence of steps necessary? Or, for example, could the brain go from target position and hand position directly to the current state of muscle contraction without passing through such intermediate stages as explicit representation of the trajectory? **B**) Since both hand and object are spatially extended, what is **the** position of the target and **the** position of the hand at any time? and **C**) Since we not only move the hand towards the target using arm muscles, but also use wrist and hand muscles to orient and shape the hand, how are arm control and hand control coordinated? However, in the present section we keep to the simple case in which hand and target are each ***point objects***, and study the ***minimum jerk*** approach to problem (**ii**) above:

Hogan [1] studied elbow rotations in pointing movements of monkeys toward a visually located target. The movements were in the horizontal plane, about 60° in magnitude, and of intermediate speed (about 700 ms in duration). To model limb *kinematics*, he proposed the ***minimum jerk*** hypothesis: He used the optimization criterion that the mean squared jerk (third derivative of position)

$$\int_0^D \left[\frac{d^3x}{dt^3}\right]^2 dt$$

be minimized during the movement, where the duration D is assumed known but the trajectory ***x(t)*** is subject to variation. By applying the calculus of variations, he derived a position function of time given by a fifth order polynomial, uniquely specified by the initial and final values of the elbow's angular position, velocity, and acceleration. If the target has zero velocity and zero acceleration at the start and end of the movement, the velocity profile is symmetric and bell-shaped, much like the low-accuracy pointing movements performed by the subjects.

Flash and Hogan [2] examined subjects performing unconstrained arm movements in the horizontal plane, holding a light-weight manipulandum. The room was darkened, removing visual feedback of arm location. Targets were indicated by light emitting diodes at distances of 20 - 40 cm. Among other experiments, subjects were instructed to move between points in the plane without obstacles. It was found that the hand's path was approximately a straight line (as predicted by the minimum jerk criterion), regardless of the start and end points of the movement. Also, the *trajectory* of the hand was well predicted by the minimum jerk hypothesis, yielding characteristic symmetric bell-shaped speed profiles. Thus the principle that explains elbow rotations also explains whole arm movements. Further, the same cost function was used to predict reaching trajectories constrained to pass through a particular *via* point, and to predict an observed *principle of isochrony*, that the duration of the two sections of the movement, before and after the via point, had approximately the same duration, even when the relative distances of these two sections of the movement varied.

Hollerbach and Atkeson [8] compared the kinematics of the hand and the arm joints to show that simplicity of movement when mapped in a given reference frame is no proof that the brain primarily plans in that reference frame: Under certain conditions the movement of the hand, and the trajectory of joint angular positions (recall [1]) are both (close to) straight lines. Further, under more general conditions they showed that a simple joint-based plan, called *staggered joint interpolation*, produces realistic reaching movements. In this scheme, the onset of one joint's movement is delayed for some amount of time (i.e., the movement onsets of shoulder and elbow are *staggered*), but both finish moving together. The path in joint space is only a little more complex than a single straight line but, with the appropriate choice of onset delay, the hand path comes close to a straight line, with a realistic movement trajectory.

In the models of Hogan [1] and Flash and Hogan [2], where optimization was used to mathematically describe hand trajectories during reaching movement, the duration of the movement was given as part of the boundary conditions. Since the model has duration as an input rather than as an output, the model fails to describe how the duration depends upon the circumstances of the movement (but note that in [2] the relative durations of submovements *are* determined by optimization). Hoff [3] extended the minimum jerk model of reach trajectory planning to include a penalty for duration, extending the cost from

$$\int_0^D \left[\frac{d^3x}{dt^3}\right]^2 dt \qquad \text{to} \qquad D + \int_0^D \left[\frac{d^3x}{dt^3}\right]^2 dt$$

with $\boldsymbol{D}$ as well as the trajectory $\boldsymbol{x(t)}$ subject to variation. This allowed him to predict the duration of unperturbed and perturbed voluntary reaching movements as a function of movement distance and perturbation in several bodies of experimental data.

Controlling limb joints

The minimum-jerk description of movement is purely kinematic, and as such it does not describe the torques or muscle activations needed to drive the limb along this trajectory. Further modeling developments have given us dynamic models of joint torque control [4-7]. Uno et al. [7] proposed that instead of modeling trajectories by optimizing hand kinematics, limb dynamics should be considered. Note that this approach no longer observes the separation of stages (**ii**) to (**iv**) above, since these three stages are combined into a single stage translating the target location into a sequence of joint torques which move the arm to place its endpoint at the target position.

Using a two-link planar arm configuration, they sought the trajectory which minimized the integral of the sum of the squared joint torque derivatives,

$$\int_0^D \left\{ \left[\frac{d\tau_1}{dt}\right]^2 + \left[\frac{d\tau_2}{dt}\right]^2 \right\} dt$$

applying the appropriate limb dynamics model. They found that for movements in the proximal region of reachable space, straight line trajectories were generated similar to those of the kinematic model. However, for reaches into more eccentric regions, human movements showed characteristic curvature, replicated by this ***minimum torque change*** model. The minimum jerk model, however, generates only straight line trajectories, and hence cannot replicate these results. Central to the approach of Uno et al. [7] is that joint torques may be determined directly from the movement goal, instead of through multiple stages of kinematics and dynamics, as previously viewed. Optimization may be done in terms of the joint forces directly.

What the minimum jerk and minimum torque change criteria have in common is that they embody the idea that the whole trajectory is planned (at the kinematic and dynamic levels, respectively) prior to being put into action. We shall see, however, that these criteria may also describe trajectories that are generated ***on-line***. Early examples of ***on-line trajectory generation*** were provided by Berkinblit et al. [9] and Hinton [10] who studied the control of a redundant limb (i.e., one in which the total number of degrees of freedom of the joints exceeds the dimension of the movement of the limb tip). They showed how, at each time step, to rotate the many joints in order to get the tip of the limb to move towards a desired position. A simple iterative algorithm based on the relationship between the movement of each joint and the movement of the joint tip, combined with the error vector between the limb's tip and the target, allows the specifications of movements for an arbitrarily large number of joints in the limb. The search for simple algorithms that do not involve the explicit representation of a trajectory, was in part motivated for Berkinblit et al. [9] by the fact that a spinal frog is able to accurately position its hindlimb to remove an irritant on its forelimb despite changes in limb and body position [11].

Relating limb positioning errors to internal representations

Close to the question of which reference frame is used for planning reaching movements, is the question of which reference frame is used to represent the target location. Helms-Tillery et al. [12] addressed the latter question by studying systematic final position errors in pointing to a remembered target location. In their experimental paradigm, subjects were shown a reachable target. The target was then removed and the lights in the room turned off. The subject then attempted to place his or her finger on the remembered target location. The authors sought to characterize constant error (CE) as a function of target location, under the assumption that CE is due to the use of an approximation of inverse kinematics (in this open loop paradigm), rather than the precise inverse kinematics transformation from finger tip location to joint angles - an approximation which can be readily adjusted by visual feedback in normal reaching, as distinct from this experimental condition. In their model, target position (T) is given by spherical coordinates in a shoulder-centered reference frame ($\boldsymbol{R}$, ϕ, χ). The arm configuration is given by the angles of the upper and lower arm segments, for example, upper arm elevation (UE), forearm elevation (FE), upper arm yaw (UY), and forearm yaw (FY). To allow a concise statement of their constant error formulation, let us here introduce the following abbreviations: Let ***FK*** be the exact forward kinematic transformation, ***IK***, the exact inverse kinematic transformation, and ***IK****, the approximation of inverse kinematics purportedly used by the CNS. ***CE*** is then the difference between the target location and the location of the hand after positioning the arm segments according to the mapping ***IK****,

$$CE = (R, \phi, \chi) - FK[\, IK^*[(R, \phi, \chi)]\,].$$

If ***IK*** = ***IK***, then ***CE*** would be zero everywhere. In explaining the ***CE***, the authors relate the arm segment final position data to the target location data using a particular functional form. They show that the best fit is obtained without using high order terms available in this function, then argue that the function ***fit*** to the data explains the ***CE***. They found that the position reached by the hand was best modeled by a *linear approximation* of the transformation of target location from spatial location to arm segment orientations, that is, the remembered target of reach seemed to be represented in terms of motor coordinates, that is, the joint angles inferred for the given target position by ***IK****.

Note that this is a *behavioral* result. It may *suggest* something of how position is encoded in the brain, but it does not *prove* that, for example, there are two separate neurons, or neuron pools, whose firing rate correlates with the values of the two joint orientations. We shall return to the issue of neural coding later. For now, it is enough to note the warning that the finding that an organism acts as if its behavior is specified in terms of some low-dimensional coordinate system with typical coordinate vector $(x_1, x_2, \ldots x_n)$ does not imply that there are n neurons, or neuronal pools, $N_1, N_2, \ldots N_n$ such that the overall firing rate of N_i correlates with the value of x_i appropriate to the corresponding behavior. On the other hand, it does suggest that such a hypothesis is worth investigation.

EQUILIBRIUM TRAJECTORY: MODELING LIMB IMPEDANCE

Neither the minimum jerk nor minimum torque change models of trajectory formation address biomechanical properties of the limb and muscles. We now turn to studies which seek to connect the nervous system (and its efferent movement commands) with the observed movement kinematics via the biomechanics.

Feldman [13] observed that the CNS does not directly control the muscular force exerted at a joint - there is no one-to-one relation between CNS activity and muscle force. Instead, the relation between force and neural activity varies with muscle length. Equivalently, for each strength of activation by the CNS, there is a length-tension curve relating muscle length (joint position) to the force produced by the muscle when held at that length. Feldman expresses this mathematically as:

$$\mathit{Force} = F(x - \lambda)$$

where x is muscle length, λ is the quantity controlled by the nervous system, and $F(\ )$ is a monotonically increasing function, such as shown in Fig. 1. These curves reflect both the stiffness properties of muscles, and the fact that the higher nervous system commands act not directly on the muscles but through spinal reflex pathways, such as that of the stretch reflex. By activating the muscles in the arm's joints to different degrees, the CNS can both vary the stiffness of each joint (i.e., the change in muscle force with respect to a change in joint position), as well as its equilibrium position, that is, the position at which the joint comes to rest given particular muscle activations.

This led to the equilibrium-point model of trajectory formation [13], which views movement as being due to a step change in the equilibrium positions of the involved joints. Bizzi et al. [15] tested this hypothesis by unexpectedly moving a subject's limb to its final position shortly after movement onset. Instead of remaining at that position, the limb moved backward, toward its unperturbed location within the trajectory. This suggested that the equilibrium configuration of the arm does not shift suddenly to its terminal state, but rather travels gradually along what was dubbed the ***equilibrium trajectory*** by Flash [16], who combined the minimum jerk kinematic model with this idea to argue that it was the hand's equilibrium point which followed a minimum jerk trajectory. Such a model predicts that the arm's kinematic trajectory will slightly deviate from the minimum jerk trajectory of its equilibrium point in a manner dictated by the arm's dynamic characteristics, such deviations being observed in arm movement experiments. Although the modeling results were promising, others [17] argue that the chosen stiffness values were too large and that for realistic stiffnesses, the deviation from the equilibrium trajectory is so great that the model yields unrealistic movement trajectories.

Still, the idea of controlling an adjustable stiffness was valid and accumulating support. Bullock and Grossberg [18] proposed the Factorization of Length and Tension (FLETE) model of the spinal

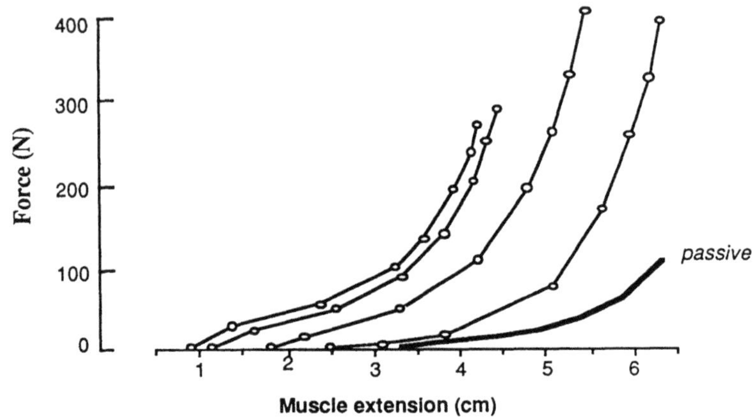

Figure 1. Length-tension curves for elbow flexor muscles at different activation levels. From [14].

cord showing how separate descending commands for target position and movement force may be combined into the appropriate muscle activations for a joint. Spinal cord circuitry also inspired the limb positioning algorithms for the spinal frog discussed below.

The **tunable spring** view of the arm's biomechanics was expanded into multiple dimensions by Mussa-Ivaldi et al. [19]. The apparent stiffness of the hand was found to be direction dependent, yielding an elliptical force field, which changed in orientation and eccentricity with arm configuration. Subjects seemed only able to adjust the field's overall magnitude when ***stiffening up*** their arm, while these other properties remained constant.

Detailed studies of cumulative limb stiffness in frog leg resulted in an equilibrium trajectory control model [20]. Stimulation of certain points in the spinal cord generated measurable stiffness fields, with different equilibrium positions for different stimulation points. This supports the hypothesis that the frog's CNS controls its limbs' movements by shifting their equilibria.

Shadmehr [14] used stiffness fields to generalize the Berkinbit model [9] of limb movement (discussed above), proposing a ***force driving algorithm*** which controls not only the terminal limb position, but also the endpoint compliance. Building on the findings of Giszter et al. [20], he also introduced the concept of ***postural modules*** as building blocks of stiffness and equilibrium position, showing how to superimpose the activity of a small number to produce desired movement trajectories. In this view, instead of the central nervous system controlling muscles individually, it acts through these postural modules, which activate muscle synergies, allowing the central nervous system to exercise control in a lower dimensional space (defined by the number of modules rather than the number of muscles) while still being able to reach any desired target with appropriate levels of force and stiffness in the limb.

Extending this biomechanical modeling paradigm to human perception, Shadmehr [14] showed how control through non-isotropic stiffness fields can produce perceptual illusions regarding the size of objects: If we perceive arm orientation in terms of equilibrium lengths of muscles, then perceived hand movement distance depends on the length of the hand's equilibrium trajectory. Differently oriented object boundaries seem to have different lengths, because of the eccentric stiffness field.

The problem of control of muscular biomechanics also inspired Dornay et al. [17] who took the minimum-torque-change model of Uno et al. [7] further, using a sophisticated seventeen muscle biomechanical model of the macaque arm, to control a two-link planar manipulator. Using *minimum muscle tension change* as the penalty function, they found realistic trajectories, similar to the *minimum torque change* trajectories. Currently, their work is progressing one level inward from the biomechanical periphery, to address the muscle ***command***, rather than muscle tension.

Hogan [21] discussed in depth the implications of arm biomechanics for control. Not only is there a generalized, multi-dimensional stiffness property, but also multi-dimensional viscosity and mobility properties. He discusses the utility of the extra degrees of freedom in a kinematically redundant limb in advantageously modifying the limb's dynamic properties.

NEURAL NETWORK APPROACHES TO MODELING LIMB MOVEMENT

Artificial neural network researchers have taken up limb control modeling to study self-organization of automatic control in a robotic paradigm. Ritter and Schulten [22] employed a neural model called the Kohonen map to learn and store motor information. A Kohonen map is a particular type of topographical neural network [23]. An input vector activates one ***neuron*** in the map, and the map's output is the information stored at that point. The training algorithm allows an input space of some dimensionality to be mapped into the network, which may be of a different dimensionality, while preserving neighborhood relations as much as possible. Another property of the Kohonen map is generalization, where the training of one part of the map spreads to neighboring regions, so that the map need not be trained exhaustively on all input vectors. In the Ritter and Schulten application, dynamic information about a robot arm was stored in the map. The input vector was the configuration of the robot arm (a three dimensional vector of joint angles) while the map was a two-dimensional lattice (thus a dimensional reduction was performed). At each lattice point, configuration-specific dynamic information was stored. Simulation results showed that the calculated joint forces for each configuration converged to those which produced the desired arm movements, demonstrating that the dynamic information was efficiently learned and stored. There are two implications here for computational neuroscience. First, whereas topographical brain structures tend to be two dimensional, this shows that higher dimensional quantities may be stored, without sacrificing altogether the benefits of maintaining neighborhood relations. Second, studies such as Ritter and Schulten [24] have shown magnification effects, where training one region of the input space more than the rest causes it to occupy an expanded area in the Kohonen map (related experimental and theoretical results are due to Spinelli and Jensen [25] and Amari [26] respectively). This provides a mechanism for *adaptive resolution* in motor control, where detailed motor information for delicate tasks can be learned in *one* region of the arm's state space, without necessarily having increased neural hardware for *all* of the arm's state space.

The problem of sensorimotor association has prompted neural models such as Kuperstein [27]. In this model, a neural network monitors the tensions of the extraocular eye muscles during foveation of the tip of the hand. Using a distributed representation of muscle length information, an association is built up between the activity of the extraocular muscles and that of the muscles of the arm in its postural state. Later, when an object of interest is foveated, the correct arm muscle tensions are recalled in order to position the arm to grasp the target. The point of the computational model is that problems in stereo vision and arm kinematics can be solved through a single computational stage, without a priori knowledge of the geometry of the system. The solution is stored in an associative-like memory, built through experience. Note that issues in arm dynamics and trajectory generation are not addressed: only the terminal posture is learned.

However, as our earlier discussion has emphasized, knowing the end-state of a movement is only the beginning of determining a trajectory which will successfully reach that end-state. In earlier sections, we discussed principles or algorithms which will yield such trajectories and which are in some sense optimal. However, while such principles may *describe* well-rehearsed human or animal movements, or may be used to compute robot movements, we see in human development and skill acquisition a process of *learning* in which trajectories are generated which in some sense get ***better and better***. Such learning poses the temporal ***credit assignment problem***: How can movement error measured at any point (especially the endpoint) of a trajectory be used by the learning rule to adjust the way in which limb activation will be computed for points earlier in the trajectory, so that with repeated experience, the system comes to generate increasingly ***desirable*** overall movement (in the sense of, say, reducing some given error measure)?

But what is learned through such a process? Is it a control law for efficiently generating desirable trajectories, or is it a set of optimal trajectories per se? Temporal sequence learning networks provide an example of the latter. A desired output pattern is presented to the network which learns over time to regenerate it. Jordan [28] uses a three layer neural network to translate sensory input into sequences of motor commands (Fig. 2). The input consists of state units, based on the motor command at the previous time step, and plan units, which encode the goal of movement. Classic ***back propagation*** [29] applies to a ***feedforward*** (acyclic) network, adjusting a network's tunable parameters (its connection weights) layer-by-layer proceeding back from the output layer, to reduce the error of the network's output for each given input pattern. Jordan [28] extends this by providing recurrent connections from command (output) units to state units to allow the network to exhibit dynamic behavior, which is tuned to follow a desired trajectory. The mapping is from current plan and current state to desired next state, and thus requires that the trajectory does not twice enter the same state unless the plan changes. A second notable aspect of the network, for the application of limb configuration sequencing, is the application of a smoothness penalty. The evaluation of smoothness is via an internal network structure that allows comparison of temporally neighboring configurations. Additionally, during supervised training the desired network output need not be given in terms of the motor commands [30]. Instead, the desired observed performance of the controlled motor plant (e.g. the movement of the tip of a redundant arm) is given. The error in this *distal* performance measure is back-propagated through a neural model of the plant, to yield performance errors in terms of motor command input. These *proximal* errors are then used to train the control network.

Massone and Bizzi [31] use a Jordan style network to train minimum-jerk reaching trajectories. The training is supervised, in that the desired output (muscle activation) for a reach target is given at each time step. With ten hidden units in the layered network, they are able to store in the network's connections the information for making minimum-jerk trajectories (of six points per trajectory) to fourteen targets.

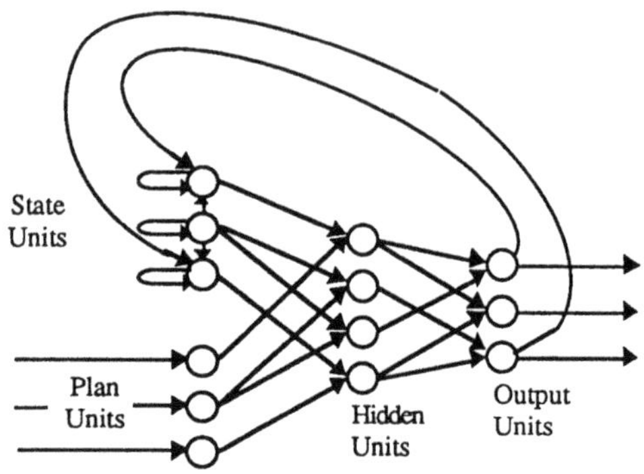

Figure 2. Jordan's neural network architecture for trajectory generation. From [30].

The Jordan-style network has internal dynamics which allows it to generate desired patterns in state space. Another variation of *back propagation*, called ***back propagation through time*** (BTT; [32]), *unravels* a cyclic network as many times as there have been network activation cycles, yielding a feedforward network of many layers: The problem of performing BTT on a recurrent neural network with m layers, run through n cycles becomes the problem of performing normal back propagation through a feedforward network of $n * m$ layers. Kawato et al. [33] use a trajectory generation network to minimize torque change in the movement of a robotic arm, solving the trajectory formation problem discussed earlier. Simultaneously, the inverse kinematics and inverse dynamics problems are solved. The learning process is reminiscent of BTT, as back propagation proceeds through the network which utilizes a spatial representation of time: For each time step there is a dedicated network to calculate the torque values to be applied to the robot's joints. Nguyen and Widrow's network [32] ***unraveling*** is expressed literally in the architecture. While such a structure is computationally useful, it is clearly biologically implausible. Further, the minimization of torque change is enforced by inhibition between torque value generation neurons in neighboring time step networks. Thus the network topography constrains the format of the optimization criterion to be based on temporal smoothness, rather than allowing an arbitrary criterion, as in reinforcement learning, which we now discuss.

It is important to note that the above learning algorithms are **supervised** in the strong sense that the pattern to be reproduced must be presented to the network at every time step for it to learn, sidestepping the need to ***reach through time*** to perform temporal credit assignment. We now turn to critic based schemes which use ***reinforcement learning*** to train networks to generate ***good*** trajectories without explicitly supervising the step-by-step generation of the trajectories.

We first review the optimization approach of dynamic programming (DP; [34]), and then describe a neural network implementation of reinforcement learning based on DP and show its applicability to learning families of trajectories defined by optimality criteria. Dynamic programming gives a recursive algorithm, which is based on optimizing the trajectory at a single time step (or *stage*), assuming all subsequent stages are optimal, and then proceeding backwards in time. Let there be n stages, and assume that the optimal *cost-to-go* $J(x,k)$, the segment of the cost function from the current stage, k say, to the end, is known for each state x. Then the cost-to-go is computed one additional stage backward in time to yield $J(x, k-1)$ by finding the control u which minimizes $c(x,u) + J(f(x,u), k)$, where $c(x,u)$ is the cost of a single step from state x under control u, and $f(x,u)$ is the state reached in this step. The procedure may then be iterated until the first stage is reached.

Werbos [35] presents a general method for tuning a plant controller in the manner of reinforcement learning. Aside from the neural network controller, a network-based critic monitors the state of the plant and produces an output $J(x,k)$ which is an estimation of the cost-to-go (cumulative future cost or benefit), as the controller-plant system proceeds from the current state. The actual cost (or benefit) at each time, $U(k)$, is given by the environment. Werbos' Heuristic Dynamic Programming (HDP) algorithm consists of iterations of two, simultaneous, adaptation procedures: The critic network learns to better predict the actual reinforcement, while the controller learns to minimize what the critic predicts. When the reinforcement is the instantaneous squared-jerk or squared torque-change, then learning to minimize reinforcement is tantamount to numerically solving the corresponding optimization problem. The neural network implementation of this algorithm is the ***Back propagated Adaptive Critic***, shown in Fig. 3.

MODELING BRAIN REGIONS WHICH CONTROL REACH AND GRASP

Neural studies have prompted models of the underlying control and learning mechanisms for limb movement generation. The overwhelmingly dominant

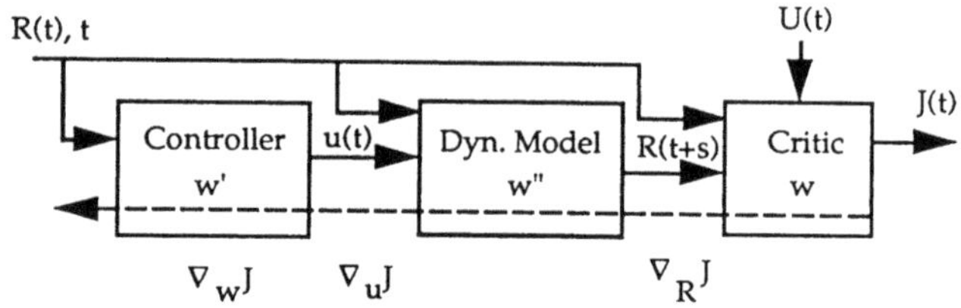

Figure 3. Werbos' neural network architecture for the **Back propagated Adaptive Critic** *(BAC; [35]), showing the three key elements: An adaptive controller, a model of the dynamics of the controlled system, and the critic, which maps external* ***punishment****, into an evaluation of the immediate state of the system. Back propagation from critic to model to controller (shown at bottom) relates the critic's evaluation to changes in the controller's behavior.*

neural area inspiring models has been the cerebellum, a part of the brain clearly involved in motor learning and coordination. We thus first review several computational models of cerebellar function.

Cerebellar modeling

The most famous model of the cerebellum is that due to Marr [36] and Albus [37] which views each Purkinje cell of the cerebellar cortex as a perceptron, with parallel fibers providing the input patterns and the climbing fiber providing the training signal.

The classical perceptron performs pattern recognition by mapping an input vector $\boldsymbol{S}$ (the pattern) to association units (cells) $\boldsymbol{A}$, the outputs of which are weighted and summed to produce the output vector $\boldsymbol{P}$. The weights of the output units (the model of the Purkinje cells) are variable, and change during the learning process to tune the response $\boldsymbol{P}$. The system has an important property of activating some common circuitry for similar inputs. This characteristic of crosstalk is the key to the perceptron's ability to generalize, that is, to apply the response learned for one input to other, similar inputs. An associated assumption about the desired function of the trained system is that for similar inputs similar outputs are desired, that is, the *control function* is a smooth, continuous one.

These models have, in particular, guided the research of Ito who, while working within a conceptual framework of adaptive control [38], has since 1970 sought evidence for the Marr-Albus theory, and sees long-term depression [39] as evidence for the Albus version of the theory, namely that conjoint activity of climbing fiber and parallel fiber input tends to weaken (rather than strengthen) the synapses from parallel fibers to the Purkinje cell.

Albus [40] has applied his ideas on the cerebellum to robot control. He views the manipulator control problem as that of determining what each joint actuator should do (e.g., how much torque each should exert) given the state of the system (a myriad of variables, including joint position, velocity, acceleration, sensor inputs, and high level commands) and the point in time. His CMAC (***Cerebellar Model Articulation Controller***) is basically a table of joint output values, whose addressing is controlled by the various inputs to the system, so that the joint output is determined by the input variables. Feedback is used to judge the performance of the system and to modify the table to improve performance. The structure corresponds to known cerebellar architecture: both cortical commands and peripheral sensory information enter the cerebellum through mossy fibers. Signals pass through several specialized layers of cells before leaving through the axons of Purkinje cells. Climbing fibers provide a second source of input, which may be involved in tuning the other pathway, thus being the pathway for training signals. Shadmehr [14] applied the CMAC to learning dynamics for control of a two-link limb during reach, showing its capability of learning how to control not only the complex dynamics of the linkage, but the muscles as well.

The Marr-Albus models concentrate on the cerebellar cortex alone, suggesting that it converts state information into motor control signals. By contrast, the Boylls-Arbib theory [41-45] has stressed that the cerebellar cortex is part of a system. Noting that the Purkinje cells are inhibitory, they suggested that lowering of this inhibition could release reverberatory activity in loops joining cerebellar and reticular nuclei, and it was this activity, rather than Purkinje outflow *per se*, which served for motor control. The theory was further refined by noting that the anatomy and physiology of Voogd [46] and Oscarsson [47] allow one to divide the cortex into microzones. They showed how activity of each microzone could provide **working memory** for the setting of various parameters for a motor schema (in the sense of Arbib [48]), and how parallel fibers could provide lateral interactions to coordinate the activity of different motor schemas. Boylls [44] modeled cerebellar function in locomotion of the high decerebrate cat as described by Shik and Orlovskii: These authors found that in the high decerebrate cat, stimulation of Deiter's nucleus during locomotion would not affect extension during the swing phase, but would yield increased extension during the support phase. Since the locomotory ***motor schema*** has been shown to be available even in the spinal cat, it seems reasonable to view the system in which the cerebellum and Deiter's nucleus are involved as providing parametric adjustment for the spinal schema. In Boylls' model of the region of the cerebellar vermis concerned with gait, the role of the cerebellum is the modulation of the motor output. In short, a standing wave of activity is set up in recurrent brainstem-cerebellar connections from which the locations of excitatory maxima choose the motor parameters (relating to stepping gait); these parameters are held constant throughout the movement.

Recently, Houk and his co-workers [49-51] have provided a partial synthesis of the above approaches, in that they combine a model of Purkinje inhibition as modulating activity in a variety of loops involving cerebellar nuclei and other brain structures with a learning model in which climbing fibers provide the training signal for modification of synapses from parallel fibers to Purkinje cells. In studying the cerebellum's role in controlling and tuning reaching movements, Houk sees the cerebellum as an array of adjustable pattern generators (APGs). (The region studied in his work is the intermediate part of the cerebellum, which is mostly concerned with the control of precise limb movements such as reach). The corticorubral tract allows cortical movement commands to send inputs to brainstem circuitry (involving the cerebellum, red nucleus, and other brainstem nuclei) believed to be involved in the execution of motor programs for reaching. Houk [49, 50] presents a computational model for the recurrent loop between the interpositus nucleus of the cerebellum, the magnocellular red nucleus (RNm), and the lateral reticular nucleus (LRN; see Fig. 4). This recurrent excitatory loop takes input from the motor cortex via RNm and through the cerebellar cortex via pontine nuclei. This brainstem system also monitors afferents, which carry information regarding the arm's state, carried to the cerebellum via the cuneocerebellar tract. The output of this brainstem circuitry is via the rubrospinal tract. The driving idea behind this model is the assumption that red nucleus output does not reflect sensory events during movement. This is supported by the fact that continuous, delayed, feedback from the periphery would generate oscillations in the system: Thus this system implements what the authors call a *quasi-feedforward mode of operation*, that is to say that the feedback is used only to end the movement.

An APG corresponds to an excitatory loop associated with a cerebellar microzone, called here a *Purkinje domain*. This domain is defined by the innervation domain of one basket cell and/or of a small cluster of cells of the inferior olive. The Purkinje cells are assumed to be bistable elements with hysteresis: They have two states, on or off. Thus persistent transition in the internal state is due to the bistable properties of Purkinje cells

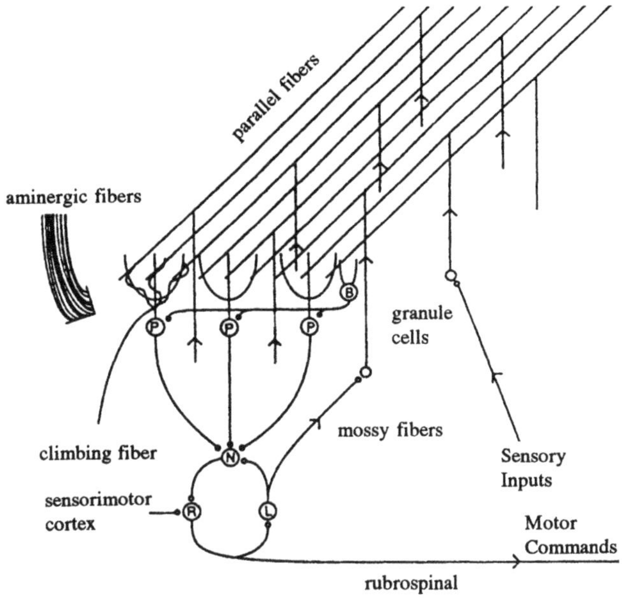

Figure 4. Houk's model for movement generation by cerebellum and brain stem. From [51].

(in reality multistable as each large dendrite would be bistable) and to the reverberatory loops.

The region of the cerebellum concerned with arm movement is represented by an array of APGs, each element of this array having a preferred direction of action. The movements are generated by linear combinations of APG activity. Preparation for movement takes place in the *selection phase*, where information about the target to reach switches off a set of Purkinje cells, allowing activity in the loop to start. A trigger from motor cortex starts the *execution phase* loop activity, which is sustained by a positive feedback loop. Proprioceptive inputs serve to inhibit loop activity, eventually stopping movement. In the *correction phase*, if the arm failed to reach its target, a crude corrective movement is generated by some extracerebellar circuit. Triggered climbing fibers cause long term depression at Purkinje cell synapses, modifying the stored motor program.

Addressing learning and control with respect to the arm's complex dynamics, and inspired by cerebro-cortical cerebellar connectivity and

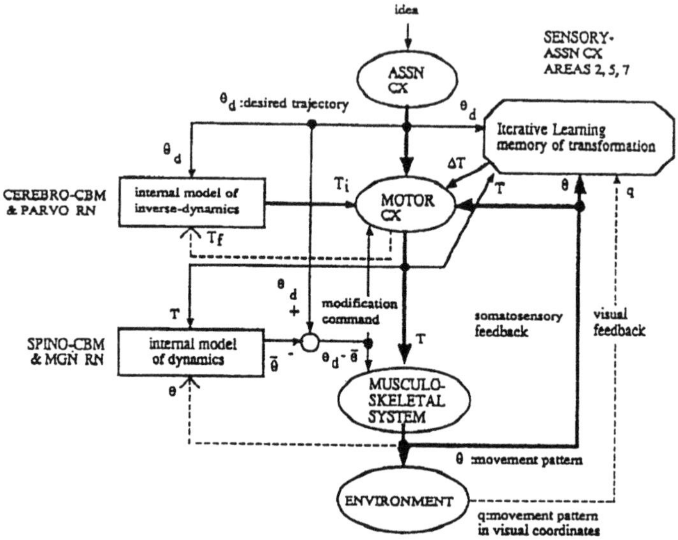

Figure 5. The model of Kawato et al. for adaptive motor control by cerebellum and related neural areas. From [5].

functionality [52], Kawato et al. [5, 6] proposed a model of motor learning with different roles for cortico-cerebellar and spino-cerebellar circuits. The biological evidence points to a kind of hierarchy of motor control involved in reaching. Here, the lateral cerebellum, along with other subcortical circuitry, compute the *inverse dynamics* of the biomechanics of the arm. That is, given a kinematic command for the arm's trajectory (from a parietal association area), this circuitry computes the command to be sent along the cortico-spinal tract, to segmental levels, which will drive the arm along the desired trajectory. In addition, a less sophisticated servomechanism uses position feedback to drive the arm along the desired trajectory (Fig. 5). As the inverse dynamics circuitry is tuned, this servo mechanism becomes inactive (although it remains in place to add robustness in case of perturbation or changes in the arm's biomechanics). It is proposed that a forward dynamics model of the arm's biomechanics is ingrained in circuitry involving the *spinocerebellum*. It performs two functions: 1) It is an internal model of the biomechanics, used by the higher level processes in lieu of afferents which, with their long delays, induce oscillations, and 2) it provides a position servo to move the arm while the more sophisticated inverse dynamics model (located in the lateral cerebellum) is being tuned, and to add stability when perturbations occur. The tuning of the inverse dynamics circuitry is via the feedback signal, which carries information about the discrepancy between desired and actual arm state. The use of this discrepancy both for low level feedback control and higher level adaptation is called feedback error learning.

Cortical modeling

Functional studies of cortical areas have inspired neural models. The seminal work of Georgopoulos et al. [53] has generated population vector models as parts of Houk et al. [51] and Bullock and Grossberg [54]. Georgopoulos et al. [55] suggest that the instantaneous direction of movement is represented in world coordinates in the cortex, and in a distributed way: They trained a rhesus monkey to reach from a center target to various peripheral targets. Motor cortical cells had discharges indicating preferred directions of hand movement, and a special vector combination of their firing rates predicted accurately the direction of hand movement. However, the interpretation of this finding is not as straightforward as first appears. While Georgopoulos suggests that motor cortex neurons code only for direction of movement, Caminiti et al. [56] showed that the firing rates of these direction sensitive cells were also modulated in relation to the initial arm configuration. Moreover, the cells were primarily in the shoulder area of the motor cortex. Thus, it is probably better to take the result in stages: within a limited area of the workspace, the direction of hand movement correlates well with the direction of shoulder movement; the firing rates of different cells in the shoulder area of primary motor cortex correlate well with different patterns of contraction of shoulder muscles and thus with different directions of shoulder movement. However (recall our earlier discussion of Feldman), a given pattern of muscle activation will cause different directions of shoulder movement for different initial positions of the shoulder. Moreover, significant displacements of elbow and shoulder will be required to position the hand. Thus, an actual trajectory of motor cortical activity in (at least) shoulder, elbow, and wrist areas will be required to move the hand to a given target. Georgopoulos et al.'s [53] population vector can, in a limited workspace, *correlate* well with the direction of movement - but it is not the neural code of the complete ensemble of neurons which *control* the movement. In any case, Burnod et al. [57] modeled the data of Caminiti et al. [56] on how initial position of the arm modulates the direction population vector. In this model of reach generation (incomplete for the reasons just given), involving parietal and motor cortices, visual target location information and kinesthetic arm position information are combined to determine how to drive the arm. A self organizing neural model develops hidden unit responses similar to the populations seen by Caminiti et al. [56], being distributed in their preferred directions of arm movement and sensitive to the initial position of the limb.

Turning from arm to hand, Rizzolatti et al. [58] have shown that in inferior premotor cortex (PMi) there are neurons which encode the visually located target of the reaching arm, while other neurons encode the shape and grasp strategy used to obtain the target. Merlo et al. [59] modeled the spatial representation of reach targets, which has been found in PMi, using a three layer neural net having as input the target location and size, and as output the arm's reach and grasp. Some hidden layer units took on the coarse coded, distributed representation found in the target sensitive PMi cells, while others were sensitive to target size.

Fagg and Arbib [60] modeled not the learning of movement per se, but rather the learning of visual-motor associations. Observations of behavior and neural activity in premotor cortex of monkeys learning to pair an arbitrary visual stimulus with one of a set of previously learned behaviors [61] were modeled with a network comprising a large number of ***motor selection columns***. Reinforcement-learning was used to recognize new visual patterns and acquire the appropriate visual-motor conditions. The architecture employs a distributed representation in which a single pattern is coded by a small subset of columns. A column is initially able to respond to many different inputs; as it learns to trigger a motor program, its responses become more narrowly defined. Each column's output is a set of votes for the various motor programs. The votes for each program are collected by ***selection units*** which drive a winner-take-all circuit to determine whether or not a particular motor program is executed.

The model is successful in reproducing the sequence of behavioral responses given by the monkeys, as well as a number of phenomena that have been observed at the single-unit level in premotor cortex.

MODELING BEHAVIORAL PHENOMENA

In contrast to the models in the first two sections of this review, which dealt with the reproduction of velocity or torque profiles for simple point-to-point movements, the models of this section address more complex behavioral conditions, more in the realm of behavioral motor control than biomechanics. Below we discuss the reproduction of reach and grasp kinematics under conditions of target perturbation and the reproduction of the effects of accuracy constraints.

Modeling perturbed reach

Are the characteristic point-to-point movement profiles discussed above predetermined and then executed, or are they developed as the movement unfolds and subject to modification by sensory input during the reaching movement? Evidence for the latter option comes from target perturbation experiments [62-64] which show that reaching movements can be modified ***on the fly***. In these experiments, as the subject began reaching for a visually determined target, the target was unexpectedly moved to a new position. With only a brief delay (100-250 ms), and without stopping, the hand modified its trajectory to begin reaching toward the new target. This implies a system which reevaluates its progress as movement proceeds, based on incoming, albeit delayed, sensory information.

Bullock and Grossberg's [54] vector integration to endpoint (VITE) model of trajectory generation is based on a continuous comparison between target location and hand location, provided by efferent copy of the motor command. A "go" signal provides the appropriate temporal scaling as well as a trigger signal to initiate movement. The repeated comparison during movement is reminiscent of the iterative correction model of reach movement generation ([65] and others discussed below). All these models consider movement paradigms in which a single target is presented. In considering the case in which a target's location is switched during the movement, Flash and Henis [66] recorded reaching kinematics for a low accuracy pointing task, under unperturbed and perturbed conditions. They argued for a *superposition* model, based on the hand-minimum-jerk model. This model fits perturbed trajectories with a trajectory that is the sum of two minimum jerk functions. The first function corresponds to the unperturbed movement to the initial target, while the second is a minimum jerk trajectory from the initial target to the perturbed target location. Given the correct onset time for the second movement and correct durations for both movements, the result is a smooth, continuous movement toward the new target.

Hoff and Arbib [67, 68] developed a minimum-jerk based, on-line control model of the control of reach which reproduces several results from low-accuracy, point-to-point reaching, and which also lays the basis for modeling the duration of movement, the coordination of reach and grasp, and the kinematics of reaching under high accuracy constraints, as we will see below. The optimization criterion (minimum jerk or minimum torque change) was used by Hoff and Arbib [67, 68] to design a feedback controller for trajectory generation which had target location as an input. By perturbing this input, the trajectory was smoothly and automatically deviated to the new target location, much as in the perturbed reaching movement studies [62-64, 69].

Modeling the speed/accuracy trade-off in reaching movements

Information theory was the inspiration for Fitts' law, a model of the speed/accuracy trade-off in goal oriented reaching movements: In 1954, Fitts [70] published results of a stylus tapping experiment, where the subject moved back and forth between two squares on a table top, the interiors of which were to be touched with the stylus. He noted a trade-off between movement time for each *tap* and the width of the target, in which the movement time (***MT***) was a linear function of the logarithm of the ratio of movement distance (***D***) to target width (***W***):

$$MT = A + B \log_2\left(\frac{2D}{W}\right)$$

where ***A*** and ***B*** are positive constants. The logarithm term was named the ***Index of Difficulty*** (ID), and was measured in bits, units of information for sensing and control. Fitts' law gave rise to an engineering based, intermittent control model [65, 71-73]. This model is based on the observation that if motor error is sampled at a constant rate, and after each sampling a corrective movement is generated which covers a constant percentage of the remaining movement distance (with the process being repeated until some threshold of accuracy is reached), then the movement's accuracy, measured by the degree of target undershoot, is exponential in the number of samplings. Put another way, the movement time is logarithmic in the required accuracy. In the VITE model [54], discussed above, with the right parameter settings the system overshoots the target by an amount exponential in the movement time, hence movement time is, again, logarithmic in accuracy, so Fitts' law is reproduced by the model. We note here that while the iterative correction model predicts a fixed degree of undershoot, the VITE Fitts' law reproduction predicts a fixed degree of overshoot. In reality, both overshoot and undershoot occur in goal directed reach, as reviewed by Jeannerod [74]. Hirayama et al. [75] present another deterministic model of the speed/accuracy trade-off. Here a neural network model learns to create arm reaching trajectories.

By putting constraints on the convergence of the neural network learning, inaccuracy in final position is created. The shortcoming of these models is in their determinism. That is, they have included no stochastic element in the computational model, essentially modeling inaccuracy without variability, which is an illogical construction.

Variability *is* considered in the impulse-variability model of Schmidt et al. [76]. The model is based on the variability of muscle generated forces, whose variations in duration and amplitude increase with their intensity. Implicit in this model, however, is the assumption of feedforward control: that variability in the driving input is uncompensated and therefore results in errors in final position. It would be preferable to have a model which takes into account the evidence for afferent influence on movement accuracy.

Meyer et al. [73] offer perhaps the most advanced model of controlling final limb position when variability is involved. They suggest that a reaching movement may be thought of as consisting of a series of independent pulses, each pulse having a standard deviation in the displacement it generates which is inversely proportional to its duration. For the two-pulse case, at the end of the first pulse, a second (smaller) pulse is generated to cover the remaining distance, correcting for the error in the first pulse but introducing errors of its own. The authors optimize the duration of each pulse to minimize the final variability, and find that total movement time (***MT***) is approximately a square-root function of the accuracy (accuracy being measured by the distance moved, ***D***, divided by the variability at the end point, ***W***). Further, they claim that if their optimization approach is applied to an ***n***-pulse movement, then ***MT*** is related to accuracy to the ***1/n*** power, and that the limit as ***n*** goes to infinity is a logarithmic relationship similar to Fitts' law. The drawback of their concept is that it does not consider the problems associated with implementation as a controller for movement, for example, delay and noise. At the end of one movement pulse, a second one is immediately generated. This implies precise and immediate feedback about limb position which is instantaneously converted into motor output, or else some method of estimating the error in a submovement before it ends.

In contrast, Hoff [77] considers an explicit model of how delayed, continuous feedback about the state of the limb is incorporated into the current efferent motor command to optimize movement: Two phenomena are captured. First, Fitts' speed/accuracy trade-off is reproduced. Second, the model captures the fact that when accuracy requirements are increased, reaching movements are modified so that more time is spent in the low velocity, deceleration portion of the movement, that is, the velocity profile exhibits skewing (see Fig. 6). (Intuitively, velocity skewing increases accuracy, since near the end of the movement and at low speed little inaccuracy is introduced, while any existing inaccuracy can be corrected). This is done by introducing the following properties to the model: First, there is a stochastic element in the mechanics of the plant, and it is responsible for variability in movement. (This is in contrast to the *deterministic* models of inaccuracy discussed above). The second property of the model is that control is based on knowledge of the plant's current state and is delayed in its arrival: To be realistic, sensorimotor delay must be included, since visual and proprioceptive information about unexpected inaccuracy cannot be gathered and applied to the trajectory instantaneously. Thirdly, to model noise it is assumed that instantaneous variability in the plant's state is proportional to its velocity. The plant, with this stochastic element, is described by the discrete time dynamics:

$$x_{k+1} = Ax_k + Bu_k, \qquad A = \bar{A} + \xi n_k, \quad \xi = \begin{bmatrix} 0 & \xi_{12} & 0 \\ 0 & \xi_{22} & 0 \\ 0 & \xi_{32} & 0 \end{bmatrix}$$

$$\bar{A} = \begin{bmatrix} 1 & s & 0 \\ 0 & 1 & s \\ 0 & 0 & 1 \end{bmatrix}, \qquad B = \begin{bmatrix} 0 \\ 0 \\ s \end{bmatrix}$$

where $\boldsymbol{x}_k$ is the hand's state (position, velocity, acceleration) at time step $\boldsymbol{k}$, $\boldsymbol{u}_k$ is the control at that step, $\boldsymbol{s}$ is the time step size, and each $\boldsymbol{n}_k$ is an independent gaussian random variable. With the above properties applied to a dynamical system, then the *best* movement trajectory is found, for a measure of goodness described by the following cost function penalizing both smoothness and accuracy:

$$\begin{aligned} J_0(x_0) &= E\{(x_N - x_G)\,|\,x_0\}^T Q_N E\{(x_N - x_G)\,|\,x_0\} \\ &\quad + E\{(x_N - x_G)^T V_N (x_N - x_G)\,|\,x_0\} + E\left\{\sum_{k=0}^{N-1} u_k^T R\, u_k \,|\, x_0\right\} \end{aligned}$$

where the first term penalizes deviation of the movement trajectory from the desired endpoint (x_G) on average (i.e., constant error), the second penalizes the variability of the endpoint, and the third penalizes lack of smoothness.

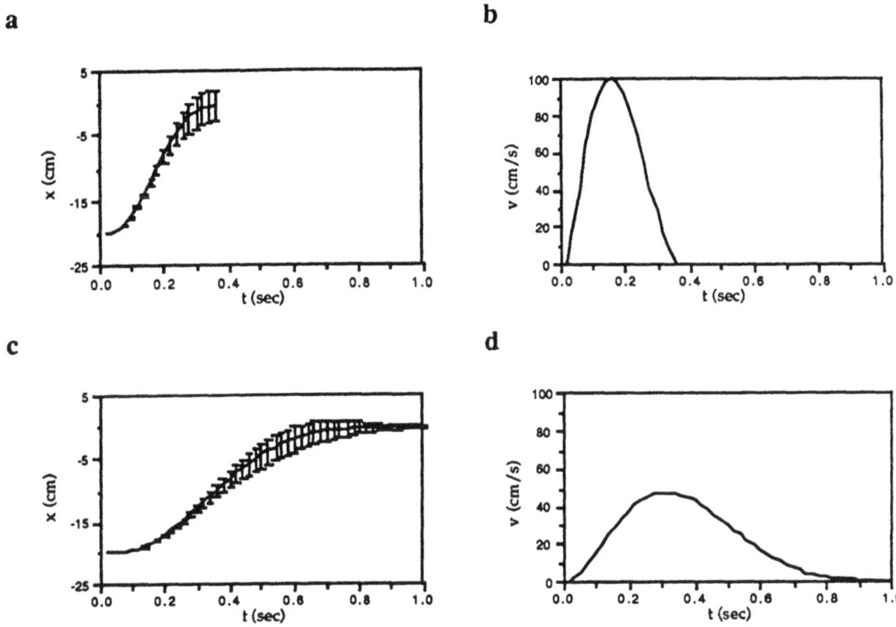

Figure 6. The skewing of the wrist's velocity profile with increasing accuracy. The top row shows a simulated quick movement which maximizes smoothness and minimizes variability. ***a*** *Position with standard deviation. Note the large terminal variability.* ***b*** *Velocity profile. Bottom row: Slow movement performing the same optimization. Terminal variability* ***(c)*** *is lower and the velocity profile* ***(d)*** *is skewed compared to the symmetrical shape in* ***b****. From [77].*

(The matrices $\boldsymbol{Q}_N$, $\boldsymbol{V}_N$, and $\boldsymbol{R}$ are constants chosen to relatively weight each of the three penalties). Without specifying a priori any particular trajectory characteristics, the emergent trajectory is shown to correspond to actual data on voluntary reaching, in terms of the variability throughout the position trajectory as well as the skewed shape of the average velocity trajectory (Fig. 6).

This model of reaching accuracy offers an alternative to the two-phase, feedforward/feedback model of reaching, showing how a single continuous control process may generate both the initial, fast, "ballistic" phase of reach and the later slow, accurate phase: The reach-to-grasp movement has often been characterized as a two-phase process with a quick feedforward phase being followed by a slow feedback phase [78, 79]. Two parallel subprocesses, reaching and prehension, occur during this two phase process, and each goes through a marked transition at about the same time. According to this view, after about 75% of the movement time the hand completes its *feedforward*, proprioception-based preshaping process and turns to a tactile-input controlled feedback process, closing until the target object has been grasped. Similarly, the reaching motion goes from a relatively quick *feedforward* phase to a slower phase which observers have characterized as a feedback phase for accurate positioning. The intuition is that feedback in the slower phase is in terms of visual perception of the hand-target discrepancy, and that the quick initial phase must lack feedback control, because of insufficient time to process such information. With the accuracy optimization model, although the *kinematics* of reach may go through a transition, the *control process* does not change. A single feedback process is responsible for both the quick initial phase and for the slow accurate phase.

Modeling the coordination of reach and grasp

Most of the models above have addressed arm reaching alone, without grasp. Arbib [48] proposed the ***Coordinated Control Program*** (CCP) to address, at a high level, the simultaneous activation of sensory guided reaching and grasping (Fig. 7). Control theorists use block diagrams in which each box represents a subsystem that is continually active, with lines linking the boxes illustrating the transfer of control signals. By contrast, computer scientists use flow diagrams to represent serial programs in which each box represents an instruction in such a way that only one box may be active at a time, with lines of the flow diagram specifying how activation is to be transferred from one instruction to another. However, since most behavior involves complex sequences of coordinated activity of a number of control systems, *coordinated control programs* (CCPs) combine control theory and computer programs into a form suited to the analysis of such coordination. In the diagrams representing such a program, there are lines representing both transfer of activation and transfer of data.

The original CCP for reach and grasp (Fig. 7) incorporates the preshape of the hand before grasping and the fast and slow phases of hand transport. The CCP has three perceptual schemas: (1) successful location of the object activates schemas for (2) recognizing the size and (3) orientation of the object. The outputs of these perceptual schemas are available for the control of the hand movement by concurrent activation of two motor schemas, one controlling the arm to transport the hand towards the object, the other **preshaping** the hand, with finger separation and orientation guided by the

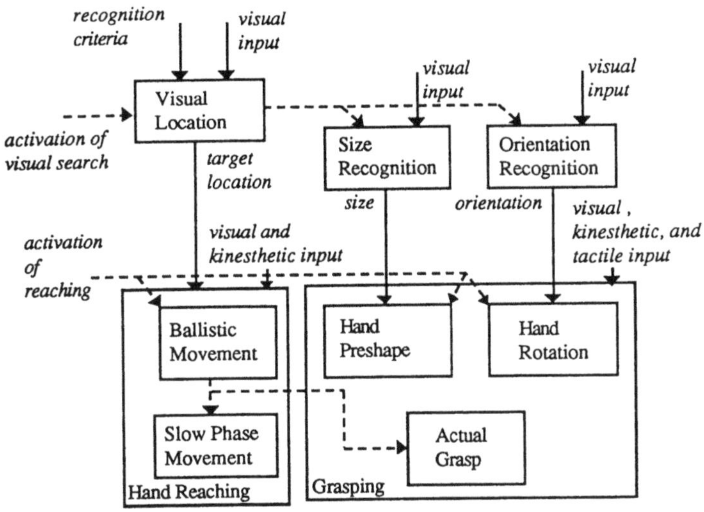

Figure 7. Arbib's Coordinated Control Program (CCP) for reach and grasp. From [48].

output of the appropriate perceptual schemas. Once the hand is preshaped, it is only the completion of the fast phase of hand transport that "wakes up" the final stage of the grasping schema to shape the fingers to **enclose** under control of tactile feedback.

However, Hoff and Arbib [67] and Hoff [77] (reviewed above) have shown that the fast and slow phases of arm movement do not, as Arbib [48] assumed, correspond to the activity of separate controllers, casting doubt on the hypothesis that it is the transition between these phases of arm transport that triggers the enclosing phase of hand movement. Hoff and Arbib [68] thus refined Arbib's original CCP model based on target perturbation data and modeled actual kinematic trajectories of hand and arm. In addition to using the minimum-jerk model for producing reaching trajectories, smoothness and magnitude constraints were used for modeling hand aperture. The key idea is that no particular phases of arm and hand movement are correlated. Rather arm and hand movements are coordinated by total duration. Thus, if a perturbation of the position of a target requires a longer transport phase, the hand movement will be optimized for this new, longer duration. Similarly, a change in object size may require time to adjust the hand's preshaping, leading to an increased duration of arm movement.

Kinematic observations of normal and perturbed movements were reproduced by the model.

In attempting to understand the schemas controlling hand movements, Arbib et al. [80] introduced the concept of the ***virtual finger*** analyzing the task of picking up a mug not directly in terms of what the five fingers do, but rather in terms of three "virtual fingers." The first (always the thumb) places itself on top of the handle. Virtual finger two goes through the handle, and can contain one, two, three or even four fingers. Whatever fingers remain constitute virtual finger three. The important point is that there are no separate programs explicitly telling the five fingers what to do in different situations. Rather, a perceptual schema takes the parameter of handle size and on that basis partitions the five actual fingers into three virtual fingers. From then on, all the control is in terms of commands being passed, and feedback being received from, the three virtual fingers.

The concept of the virtual finger tells us how to replace analysis of hand movements directly based on the mechanical degrees of freedom of individual fingers by analysis of the functional roles of the forces being applied in carrying out some task. But, having agreed to analyze the hand in terms of virtual fingers, how do we specify the movement of these units? Iberall et al. [81] argue that ***opposition space*** provides the appropriate coordinate system. For example, in *pad opposition*, the pad of the thumb opposes the pad of the virtual finger, and the opposition space is the axis along which the finger pads move to provide that opposition. *Palm opposition* is a power grip, with several fingers working together as a virtual finger to oppose the palm, which acts as a second virtual finger. The axis determines the movement of the virtual finger towards the palm (Fig. 8).

When an object is grasped, the virtual fingers moving along the opposition axis may come to rest their opposing surfaces on the object between them, rather than making direct contact with each other. This makes possible a theory of preshaping. One task of vision is to determine, from the retinal input, an opposition space embedded in the object which is to be the target for the positioning of the appropriate opposition space of the hand. There is a safety margin extending the opposition space beyond the boundaries of the object. Preshaping forms the hand so that the opposing surfaces of the virtual fingers will be correspondingly separated. An approach vector, between the origin of opposition space and the center of the opposition vector, distinguishes the orientation and distance of the hand relative to the object. As the hand preshapes to meet this specification (the ***Preshape Schema***), the arm transports it (***Move Arm Schema***) and the wrist rotates it (the ***Orient Schema***) to approximately the right position. Note that *the target position of the wrist depends on the proposed preshape of the hand* - the

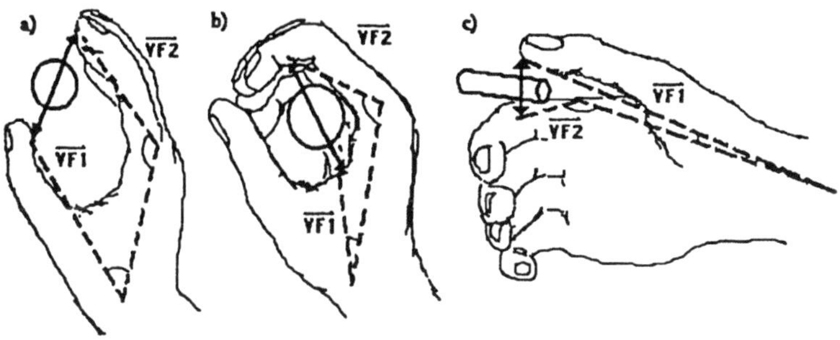

*Figure 8. Various oppositions for grasp (**a**. pad opposition, **b**. palm opposition, **c**. side opposition) of cylindrical objects, and their imbedded opposition vectors. From [81].*

arm controller must know the offset of the wrist from the center of the opposition space as embedded in the object if it is to transport the hand successfully to its goal. Thereafter, the position of the hand is adjusted to align the two opposition spaces (that in the hand with that in the object) and the virtual fingers then close along the aligned axes to firmly grasp the object under tactile control (the ***Enclose Schema***).

Iberall and Arbib [82] postulated how the various schemas of reaching and preshaping, based on opposition spaces, may be distributed in the primate brain. Currently we are revising their model to incorporate new data on areas 6 and 7, the basal ganglia, and cerebellum, to generate a global view of brain functions in this sensorimotor process, at a level of detail similar to the oculomotor model of Dominey and Arbib [83]. As we proceed, we believe that schema theory [48, 84] will provide a behavior-based functional framework from which to relate further anatomical and physiological data, so that we achieve our neural modeling goal in a structured way.

CONCLUSION

Having diverse backgrounds, modelers use different theoretical paradigms to answer questions about the nature of motor control. It seems that since this modeling field is relatively young, research to-date has centered more around exploring different theoretical paradigms than answering well known questions within a single accepted paradigm, the latter being the mark of a more mature field. To support this very opinionated statement, we ask the reader to consider Stein [85], an open discussion about what variable or variables the nervous system controls in limb movement. Discussants argued for the standard choices of position or force, but a variety of other choices were suggested including viscosity. It seems that little consensus was obtained. Stepping back, it is apparent that the classic paradigm of servo control was a starting assumption of the discussion. That is, the nervous system was being viewed as a regulator, a controller which tries to keep a variable at a specified level. It may be that this starting assumption is to be questioned, rather than that the assumption is valid and the answer to the posed question is elusive. As in the above discussion of equilibrium positions, where the endpoint servo hypothesis was rejected, it is now apparent that a different modeling paradigm is needed. Currently, optimal control is gaining support as a useful paradigm (introducing the associated question "What quantity is being minimized during movement?" with the answer that joint-torque-change is a better candidate than hand-acceleration-change), yet not all modelers have converged upon optimization as the important aspect of movement to model. For example, the model of Kuperstein [27] is more concerned with how sensory input and motor output can be related in determining target location, putting aside the questions of trajectory generation in reach. At the neural modeling level, different studies have used motor circuitry to solve different computational problems. Where Houk et al. [49-51] focus on motor pattern storage and replay in cerebellum and brain stem but do not address the complexity of dynamics computation, Kawato et al. [5, 6] focus on inverse dynamics computation in cerebellum, assuming the movement goal is stored and retrieved at higher cortical levels, and input to their model. Where learning is concerned, Houk shows how to adjust the final position in the movements generated by their neural machinery, but does attempt to optimize the entire trajectory. This problem is taken up elsewhere in more abstract models [35], but not in the framework of functional neuroanatomy. So, again, instead of answering widely considered open questions in a common framework, recent modeling efforts seem based on selecting candidate frameworks and interesting questions within those frameworks, then showing plausible (i.e., sufficient but not necessary) scenarios in which those questions are solved.

Clearly, there is no single approach being used in modeling prehension. Rather, we have a collection of complementary approaches, each having different degrees of detail and different strengths. By acknowledging the origins of these different approaches and the phenomena they seek to explain, we can better see how to use these tools to further our modeling efforts. We feel that the ultimate understanding of human motor control will incorporate:

A) a thorough understanding of the relevant functional neuroanatomy
B) the computational abilities and limitations of the component neurons, their connectivity, and distributed information representation, and
C) an understanding of the computations (as elucidated by robotics; see e.g., [86]) which must be solved in producing the observed behavior.

REFERENCES

1 N. Hogan, J. Neurosci., 4 (1984) 2745.
2 T. Flash and N. Hogan, J. Neurosci., 5 (1985) 1688.
3 B. Hoff, Biol. Cybern., submitted.
4 J.M. Hollerbach and T. Flash, Biol. Cybern., 44 (1982) 67.
5 M. Kawato, Y. Uno, M. Isobe, and R. Suzuki, Proc. IEEE First Ann. ICNN, San Diego, 1987.
6 M. Kawato, H. Miyamoto, T. Setoyama and R. Suzuki, Neural Networks, 1 (1988) 256.
7 Y. Uno, M. Kawato and R. Suzuki, Biol. Cybern., 61 (1989) 89.
8 J.M. Hollerbach and C.G. Atkeson, Biol. Cybern., 56 (1987) 279.
9 M.B. Berkinblit, I.M. Gelfand and A.G. Feldman, Biophysics, 31 (1986) 142.
10 G.E. Hinton, J. Mot. Behav., 16 (1984) 171.
11 O.I. Fukson, M.B. Berkinblit and A.G. Feldman, Science, 209 (1980) 1261.
12 S.I. Helms-Tillery, M. Flanders and J.F. Soechting, J. Neurosci., 11 (1991) 770.
13 A.G. Feldman, Biophysics, 11 (1966) 565.
14 R. Shadmehr, A Computational Theory for Control of Posture and Movement in a Mulit-joint Limb, TR USC-CNE-91-07, University of Southern California, 1991.
15 E. Bizzi, W. Chapple and N. Hogan, TINs., 5 (1982) 395.
16 T. Flash, Biol. Cybern., 57 (1987) 257.
17 M. Dornay, Y. Uno, M. Kawato and R. Suzuki, in: J.E. Moody, S.J. Hanson and R.P. Lippmann (eds.), Advances in Neural Information Processing Systems 4, Kaufmann, San Mateo, 1992, p. 627.
18 D. Bullock and S. Grossberg, in: W.A. Hershberger (ed.), Volitional Action, Elsevier, North Holland, 1989, p. 253.
19 F.A. Mussa-Ivaldi, N. Hogan and E. Bizzi, J. Neurosci., 5 (1985) 2732.

20 S.F. Giszter, E. Bizzi and F.A. Mussa-Ivaldi, in: F.H. Echman and C.D. Deno (eds.), Analysis and Modeling of Neural Systems, Kluwer Press, Moffett Field, 1991, p. 123.
21 N. Hogan, Biol. Cybern., 52 (1985) 315.
22 H. Ritter and K. Schulten, IEEE ICNN II, I (1988) 109.
23 T. Kohonen, Biol. Cybern., 43 (1982) 59.
24 H. Ritter and K. Schulten, On the Stationary State of Kohonen's Self-Organizing Sensory Mapping, Tech. Report., Dept. of Physics, Tech. University of Munich, FRG, 1987.
25 D.N. Spinelli and F.E. Jensen, Science, 203 (1979) 75.
26 S. Amari, Bull. Math. Biol., 42 (1980) 339.
27 M. Kuperstein, Science, 239 (1988) 1308.
28 M.I. Jordan, in: M. Jeannerod (ed.), Attention and Performance XIII, Motor Representation and Control, Erlbaum, Hillsdale, 1990, p. 796.
29 D.E. Rumelhart, G.E. Hinton and R.J. Williams, in: D. Rumelhart and J. McClelland (eds.), Parallel Distributed Processing: Explorations in the Microstructure of Cognition, Vol. 1, MIT Press, Cambridge, 1986, p. 318.
30 M.I. Jordan and D.E. Rumelhart, Cogn. Sci., 16 (1992) 307.
31 L. Massone and E. Bizzi, Biol. Cybern., 61 (1989) 417.
32 D. Nguyen and B. Widrow, IJCNN (1989) II-357.
33 M. Kawato, Y. Maeda, Y. Uno and R. Suzuki, Technical Report TR-A-0056, ATR Auditory and Visual Perception Research Laboratories, Kyoto, Japan, 1989.
34 R.E. Bellman, Dynamic Programming, Princeton University Press, 1957.
35 P.J. Werbos, in: W.T. Miller, R.S. Sutton and P.J. Werbos (eds.), Neural Networks for Control, MIT Press, Cambridge, 1990, p. 67.
36 D. Marr, J. Physiol., 202 (1969) 437.
37 J.S. Albus, Math. Biosci., 10 (1971) 25.
38 M. Ito, The Cerebellum and Neural Control, Raven Press, New York, 1984.
39 M. Ito, Ann. Rev. Neurosci., 12 (1989) 85.
40 J.S. Albus, Brains, Behavior, and Robotics, McGraw-Hill, Peterborough, 1981.
41 M.A. Arbib, C.C. Boylls and P. Dev, in: R. Oldenbourg (ed.), Kybernetik und Bionik/Cybernetics, 1974, p. 216.
42 C.C. Boylls, Technical Report 75C-6, Computer and Information Science Dept., University of Mass. at Amherst, 1975.
43 C.C. Boylls, Technical Report 76-1, Computer and Information Science Department, University of Mass. at Amherst, 1976.

44 C.C. Boylls, Brain Res., 159 (1978) 445.
45 J. Szentágothai and M.A. Arbib, Conceptual Models of Neural Organization, MIT Press, Cambridge, 1975.
46 J. Voogd, in: R. Llinás (ed.), The Neurobiology of Cerebellar Evolution and Development, Amer. Med. Assoc., 1969, p. 493.
47 O. Oscarsson, in: A. Iggo (ed.), Handbook of Sensory Physiology, Vol. II, Somatosensory System, Springer-Verlag, Berlin, 1973, p. 339.
48 M.A. Arbib, in: V.B. Brooks (ed.), Handbook of Physiology, Section I, The Nervous System, Vol. 2, Motor Control, Am. Physiol. Soc., Bethesda, 1981, p. 1449.
49 J.C. Houk, in: M. Glickstein, C. Yeo and J. Stein, (eds.), Cerebellum and Neuronal Plasticity, Plenum Press, 1987, p. 249.
50 J.C. Houk, in: R.M.J. Cotterill (ed.), Models of Brain Function. Cambridge University Press, Cambridge, 1989, p. 204.
51 J.C. Houk, S.P. Singh, C. Fisher and A.G. Barto, in: W.T. Miller, R.S. Sutton and P.J. Werbos (eds.), Neural Networks for Control, MIT Press, Cambridge, 1990, p. 301.
52 G.I. Allen and N. Tsukahara, Physiol. Rev., 54 (1974) 957.
53 A.P. Georgopoulos, J.F. Kalaska, R. Caminiti and J.T. Massey, J. Neurosci., 2 (1982) 1527.
54 D. Bullock and S. Grossberg, in: S. Grossberg (ed.), Neural Networks and Natural Intelligence, MIT Press, Cambridge, 1988, p. 553.
55 A.P. Georgopoulos, R.E. Kettner and A.B. Schwartz, J. Neurosci., 8 (1988) 2928.
56 R. Caminiti, P.B. Johnson and A. Urbano, J. Neurosci., 10 (1990) 2039.
57 Y. Burnod, P. Grandguillaume, I. Otto, R. Caminiti and P. Johnson, IJCNN, 2 (1990) 783.
58 G. Rizzolatti, M. Gentilucci, L. Fogassi, G. Luppino, M. Matelli and S. Ponzoni-Maggi, Exp. Brain Res., 67 (1987) 220.
59 P.E. Merlo, F. Stefano, M. Gentilucci and G. Rizzolatti, in: E.R. Causpiello (ed.), Second Italian Workshop on Parallel Architectures and Neural Networks, World Science Publishers, 1990, p. 27.
60 A.H. Fagg and M.A. Arbib, Adapt. Behav., 1 (1992) 1.
61 A.R. Mitz, M. Godshalk and S.P. Wise, J. Neurosci., 11 (1991) 1855.
62 D. Pelisson, C. Prablanc, M.A. Goodale and M. Jeannerod, Exp. Brain Res., 62 (1986) 303.
63 A.P. Georgopoulos, J.F. Kalaska and J.T. Massey, J. Neurophysiol., 46 (1981) 725.
64 Y. Paulignan, C.L. MacKenzie, R.G. Marteniuk and M. Jeannerod, Exp. Brain Res., 83 (1991) 502.

65 K.J.W. Craik, Brit. J. Psych., 38 (1947) 56.
66 T. Flash and E. Henis, J. Cog. Neurol., 3 (1992) 220.
67 B. Hoff and M.A. Arbib, in: R Caminiti (ed.), Control of Arm Movement in Space: Neurophysiological and Computational Approaches, Exp. Brain Res. Suppl., 22 (1992) 285.
68 B. Hoff and M.A. Arbib, J. Mot. Behav., in press.
69 C. Prablanc and O. Martin, J. Neurophysiol., 67 (1992) 455.
70 P.M. Fitts, J. Exp. Psychol., 47 (1954) 381.
71 L. Stark, Neurological Control Systems, Studies in Bioengineering, Plenum Press, New York, 1968.
72 E.R.F.W. Crossman and P.J. Goodeve, Quart. J. Exp. Psychol., 35A (1983) 251.
73 D.E. Meyer, R.A. Abrams, S. Kornblum, C.E. Wright and J.E.K. Smith, Psych. Rev., 95 (1988) 340.
74 M. Jeannerod, The Neural and Behavioral Organization of Goal-Directed Movements, Clarendon, Oxford, 1988.
75 M. Hirayama, M. Kawato and M.I. Jordan, J. Mot. Behav., in press.
76 R.A. Schmidt, H.N. Zelaznik and J.S. Frank, Big 10 Symposium on Information Processing in Motor Learning and Control, University of Wisconsin-Madison, 1978.
77 B. Hoff, Tech. Report USC-CS-92-523, USC-CNE-92-02, University of Southern California, Los Angeles, 1992.
78 R.S. Woodworth, Psychol Rev. Mono. Suppl., 3 (1899) 1.
79 M. Jeannerod, J. Mot. Behav., 16 (1984) 235.
80 M.A. Arbib, T. Iberall and D. Lyons, in: A.W. Goodwin and I. Darian-Smith (eds.), Hand Function and the Neocortex, Exp. Brain Res. Suppl., 10 (1985) 111.
81 T. Iberall, G. Bingham and M.A. Arbib, in: H. Heuer and C. Fromm (eds.), Generation and Modulation of Action Patterns, Springer-Verlag, Berlin, 1986, p. 158.
82 T. Iberall and M.A. Arbib, in: M. Goodale (ed.), Vision and Action: The Control of Grasping, Ablex, New Jersey, 1990, p. 204.
83 P.F. Dominey and M.A. Arbib, in: M.A. Arbib and J.P. Ewert (eds.), Visual Structures and Integrated Functions, Research Notes in Neural Computing, Springer-Verlag, Berlin, 1991.
84 M.A. Arbib, in: M. Jeannerod (ed.), Attention and Performance XIII. Motor Representation and Control, Erlbaum, Hillsdale, 1990, p. 111.
85 R.B. Stein, Behav. Brain Sci., 5 (1982) 535.
86 G. Bekey and K. Goldberg (eds.), Neural Networks in Robotics, Kluwer, Boston, 1993.

Insights into the Reach to Grasp Movement
K.M.B. Bennett and U. Castiello (Editors)

CHAPTER 16

OPTIMAL CONTROL OF REACHING MOVEMENTS

Y. UNO and M. KAWATO

ATR Human Information Processing Research Laboratories,
2-2 Hikaridai, Seika-cho, Soraku-gun, Kyoto 619-02, Japan.

SUMMARY

From computational research, it is elucidated that at least three different problems must be solved to execute visually-guided reaching movements: trajectory planning, coordinate transformation and calculation of motor commands. These problems are ill-posed in the sense that there exists an infinite number of possible solutions. However, the brain easily solves these problems by adopting certain constraints. In this chapter, we discuss optimization principles that define the unique solution for the ill-posed motor control problem; the minimum-jerk model, minimum-torque-change model and minimum-muscle-tension-change model are introduced. Several neural network models are presented to calculate the optimal trajectory and the corresponding motor command. They include Hoff and Arbib's network with an optimal feedback controller, Jordan's sequential network, our cascade network and a new model with forward and inverse models of a motor apparatus.

COMPUTATIONAL RESEARCH FOR REACHING MOVEMENTS

Recent studies of computational neuroscience and robotics have revealed computational problems in the control of voluntary movements. Saltzman [1], Hollerbach [2] and Kawato et al. [3] pointed out that the problem of controlling goal-directed limb movements can be divided into a sequence of information processing parts: trajectory planning, coordinate transformation

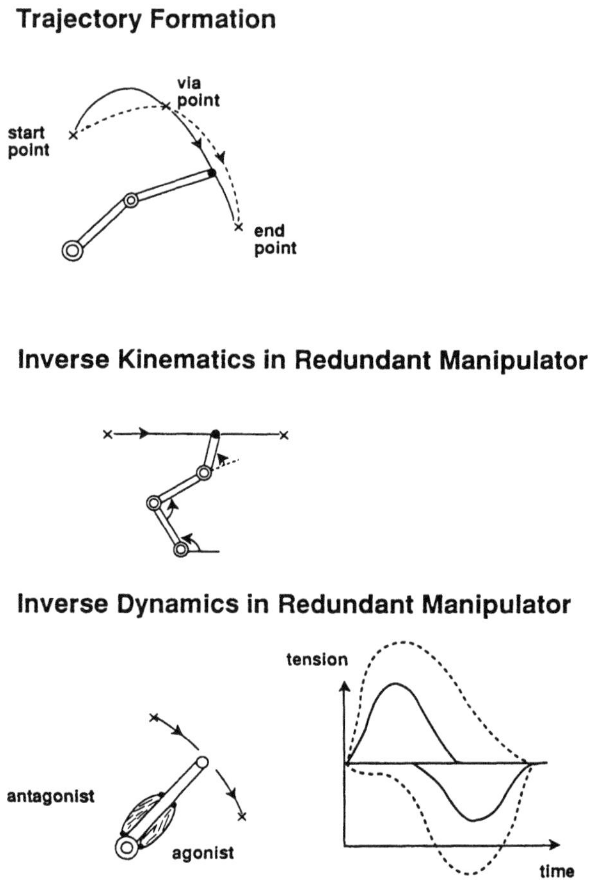

Figure 1. Three ill-posed problems in sensory-motor control. From [26].

from extracorporeal space to intrinsic body coordinates, and motor command generation, all of which are required to translate the spatial characteristics of the target or goal of the movement into an appropriate pattern of muscle activations.

In this chapter, we discuss the motor control problems from the viewpoint of computational theory. Consider a thirsty person reaching for a glass of water on a table. Here the goal of the task would be to control the movements of the arm toward the glass in order to reduce the thirst. To achieve this goal, one desirable trajectory in task-oriented coordinates should first be selected from the infinite number of possible trajectories that lead to the glass, whose spatial coordinates are provided by the visual system (determination of trajectory). Second, the spatial coordinates of the desired trajectory are transformed in terms of a corresponding set of body coordinates, such as joint angles or muscle lengths (transformation of coordinates). Finally, motor commands are generated to coordinate the activity of many muscles so that the desired trajectory is achieved (calculation of motor commands). In the robotics field, the second and the third problems are called the ***inverse-kinematics problem*** and the ***inverse-dynamics problem***, respectively.

Ill-posed motor control problems

Unfortunately, the three motor control problems are ***ill-posed*** rather than ***well-posed***. A problem is ***well-posed*** when its solution exists, is unique, and depends continuously on the parameters that define the problem. Ill-posed problems fail to satisfy one or more of these criteria. Most motor control problems are ill-posed in the sense that their solutions are not unique. Jordan [4] called this second type of ill-posed problem an indeterminacy problem. We list three ill-posed control problems in Fig. 1. For simplicity, it is assumed in this chapter that the duration of movement is already determined.

First, consider the trajectory-determination problem for a two-joint arm movement within a plane, where the starting point, the via-point, and the end point, as well as the movement time, are specified (Fig. 1, top panel). There are an infinite number of possible trajectories satisfying these conditions. Here, by the word trajectory, we mean velocity profiles as well as path shapes. Thus, the solution is not unique, and the problem is ill-posed.

A second ill-posed motor-control problem is the inverse-kinematics problem for controlling a redundant arm with excess degrees of freedom. For example, consider a three-degree-of-freedom arm in a plane (Fig. 1, middle panel). The inverse-kinematics problem is to determine the three joint angles (three degrees of freedom) when the hand position is given in Cartesian coordinates (two degrees of freedom). Because of system redundancy, even when the time course of the hand position is strictly determined, the time course of the three joint angles cannot be determined uniquely.

A third ill-posed problem is the inverse-dynamics problem for controlling an arm with agonist and antagonist muscles. For example, consider a single joint arm with a pair of muscles (Fig. 1, bottom panel). The inverse-dynamics problem is to determine the time courses of agonist and antagonist muscle tensions when the joint-angle time course has been specified. Even when the time course of the joint angle has been specified, there are an infinite number of tension waveforms in the two muscles that can realize the same joint-angle time course, as indicated by the thick and thin curves at the bottom of Fig. 1.

Approaches to resolving ill-posed problems

A number of approaches have been proposed to resolve the second type of ill-posed problem, that is, the indeterminacy problem (see e.g., [1, 5] for review).

Some of these approaches utilize a mechanism to reduce the number of degrees of freedom in the motor control system. The degrees-of-freedom of a system is the number of variables that can be changed independently. Bernstein [6] proposed that connections, physical or physiological, between muscle groups can serve to partition degrees of freedom. That is, he proposed that there are *synergies* among muscle groups that help reduce the number of degrees of freedom to be managed (see [1, 7] for related concepts of coordinative structures). The use of a feedback controller also reduces the number of degrees of freedom in the motor control network. Introducing couplings between potentially independent variables tends to decrease the number of degrees of freedom. Feedback controllers and coordinative structures are examples of such couplings. A hierarchical control strategy such as virtual trajectory control [8, 9] or the task dynamic approach [10] provides specific design principles to introduce couplings between the high-level task space, the low-level body space and the motor command space.

In engineering, an objective function (performance index) is frequently used to define a unique solution for an indeterminacy problem. This is the fundamental concept of optimal control theory [11]. This optimal control concept was also explored in biological motor control by Hogan [9] and Nelson [12]. The optimization approach is the main subject of the current chapter. That is, we introduce several optimization principles (models) that are experimentally confirmed. We also introduce optimization procedures that realize various optimization models. In particular, several neural network models that solve optimization problems are highlighted.

OPTIMIZATION PRINCIPLES FOR TRAJECTORY FORMATION

Human arm trajectory formation

In this paper, the term ***trajectory*** refers to path and speed of movement: the path is a sequence of positions that the hand follows in space, and the speed is a time sequence of movement velocity along the path. Several researchers have measured the hand trajectories in human skilled arm movements and have found common kinematic features. One elegant feature is that when moving the hand between a pair of targets, subjects tend to generate roughly straight hand paths with bell-shaped speed profiles [13-18].

In our behavioral experiments [18], subjects were asked to move their hands from one point to another using elbow and shoulder joint rotations while their wrists were braced. Arm movement was constrained in the horizontal plane at the shoulder level. Visual information about the arm location was eliminated by darkening the room. Fig. 2**B** shows hand trajectories observed in our behavioral experiments. One can see that the hand paths are roughly straight but noticeably curved for some movements.

Several mathematical models have been proposed to account for the kinematic features of human multi-joint arm movements. In the following, we introduce three different optimization models which were formulated by defining objective functions (criteria), measures of performance for any possible movements.

Minimum-jerk model

The minimum-jerk model was the first optimization model that was experimentally confirmed. Here, jerk is mathematically defined as the rate of change of the acceleration of the wrist. Considering that skilled movements are, in general, extremely smooth and graceful, Flash and Hogan [17] proposed the minimum-jerk model, which assumes that the trajectory followed by a subject's arm tends to minimize the square of the movement jerk, integrated over the entire movement:

$$C_J + \frac{1}{2}\int_0^{t_f}\left\{\left(\frac{d^3x}{dt^3}\right)^2 + \left(\frac{d^3y}{dt^3}\right)^2\right\}dt. \tag{1}$$

Here, (x, y) are Cartesian coordinates of the hand, and t_f is the movement duration. Flash and Hogan [17] showed that the unique trajectory predicted by this equation agreed well with data on movements made in front of the body. Let us explain this in a little more detail.

It can be mathematically shown that the optimal solution of the minimum-jerk model has the form of a fifth order polynomial in time using the Euler-Poisson equation. The fifth order polynomial function has six unknown parameters that should be specified to determine the solution. In the case of a discrete, point-to-point movement, the six parameters are uniquely determined from the six boundary conditions at the beginning and the end of movement. The velocity and acceleration of the hand are zero before and after the movement (4 conditions). The initial and final hand positions are given (2 conditions). Consequently, the following pair of fifth order polynomials is obtained as the optimal hand trajectory for a point-to-point movement:

$$\begin{aligned} x(t) &= x_0 + (x_0 - x_f)(15\tau^4 - 6\tau^5 - 10\tau^3) \\ y(t) &= y_0 + (y_0 - y_f)(15\tau^4 - 6\tau^5 - 10\tau^3), \end{aligned} \tag{2}$$

where $\tau = t/t_f$, and (x_0, y_0) and (x_f, y_f) are the initial hand position and the final hand position. One can easily see that the path derived from this equation **(2)** is a straight line with a perfectly symmetrical bell-shaped speed profile. Furthermore, the minimum-jerk model successfully reproduces a curved movement through a certain via-point as well as a straight movement between two points (see [17] and Figs. 5, 6 and 7). The optimal solution for the minimum-jerk model depends only on the initial, final and intermediate points on the movement trajectory; in other words, the minimum-jerk trajectory is independent of the dynamics of the musculoskeletal system.

Minimum-torque-change model

Based on the idea that movement optimization must be related to the dynamics of a controlled object (arm), Uno et al. [18] proposed the following alternative quadratic measure of performance:

$$C_\tau = \frac{1}{2}\int_0^{t_f} \sum_{i=1}^{m} \left(\frac{d\tau_i}{dt}\right)^2 dt, \tag{3}$$

where τ_i is the torque fed to the *ith* of m actuators. Here the performance measure (objective function) is the sum of the square of the rate of change of torque, integrated over the entire movement. One can see that C_τ of this last equation **(3)** is related to C_J of the first equation **(1)** because the rate of change of torque is locally proportional to the jerk. However, it must be emphasized that C_τ depends critically on the dynamics of the musculoskeletal system.

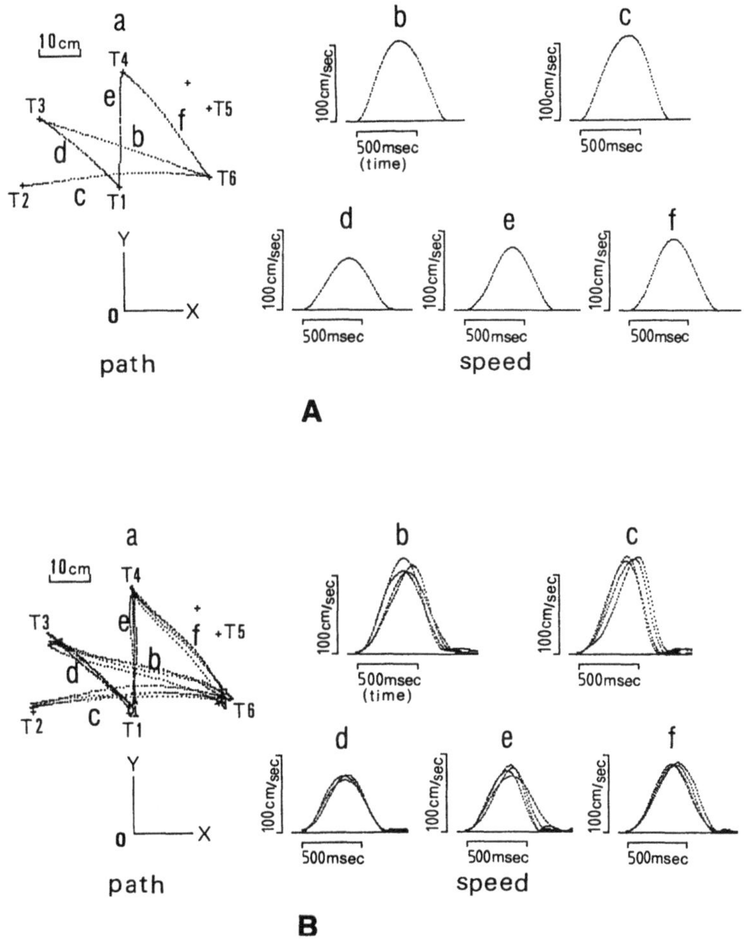

*Figure 2. Comparison of theoretical prediction by the minimum-torque-change model and experimental data for free movements between two targets located approximately in front of the body. **A**. Hand trajectories predicted by the minimum torque-change model. **T1-T6** are the targets. The origin of **X-Y** coordinates represents the location of the shoulder. **a** shows the five hand paths (b: T3 → T6, c: T2 →T6, d: T1 → T3, e: T4 → T1, f: T4 → T6). **b-f** show corresponding hand tangential speed profiles along the paths. **B**. Hand trajectories observed in human arm movements. Four trials are depicted for each movement. Again **a** shows the hand paths and **b-f** show the corresponding speed profiles. From [18].*

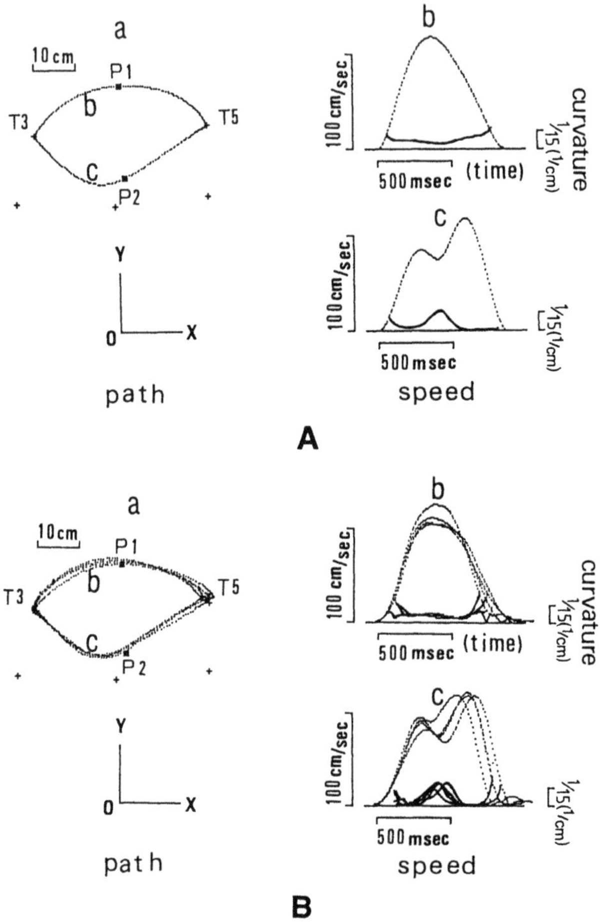

Figure 3. Comparison of theoretical prediction by the minimum-torque-change model and experimental data for via-point movements. Free movements passing through a via-point, either ***P1*** *or* ***P2*** *are considered.* ***P1*** *and* ***P2*** *are located symmetrically with respect to the line connecting* ***T3*** *and* ***T5***. ***A***. *Hand trajectories predicted by the minimum torque-change model.* ***a*** *shows the convex path (b: T3 →P1 →T5) and the concave path (c: T3 →P2 →T5).* ***b*** *and* ***c*** *show the corresponding speed (dotted curves) and curvature profiles (solid curves).* ***B***. *Hand trajectories observed in human arm movements. Four trials are depicted for each movement. The figure format is the same as that in* ***A***. *From [18].*

Fig. 2 shows minimum-torque-change trajectories (**A**) and actual human hand trajectories (**B**) for free movements between pairs of targets located approximately in front of the body. The minimum-torque-change model predicts roughly straight hand paths as shown in Fig. 2**Aa** which are always not completely straight (for example, the hand path leading from the target T_2 to T_6 is slightly convex); the corresponding speed profiles are single-peaked and bell-shaped as shown in Fig. 2**Ab-f**. For these point-to-point movements in front of the body, predictions made by both these criteria (first and third equations) agree well with experimental data. However, movement trajectories under the minimum-torque-change model are quite different from those under the minimum-jerk model in four other behavioral situations.

The first result by Uno et al. [18] concerned what happens when the starting point of an arm is to the side of the body and the endpoint is to the front. Here the movement path was curved under the minimum-torque-change model, but always straight under the minimum-jerk model. The hand paths of sixteen human subjects were all curved, supporting the minimum-torque-change model (see [18] and Fig. 4).

A second result by Uno et al. [18] concerned movements between two points while resisting a spring, one end of which was attached to the hand while the other was fixed. Here the minimum-jerk model always predicts a straight movement path regardless of external forces. The minimum-torque-change model predicts a curved path and an asymmetrical speed profile for the movement with the spring (see [18] and Fig. 6). The latter predictions again agreed closely with the data, further supporting the minimum-torque-change model.

Third, Uno et al. [18] examined vertical movement affected by gravity. The minimum-jerk model always predicts a straight hand-path between two points. The minimum-torque-change model predicts curved paths for large up and down movements, but essentially straight paths for small fore and aft movements. The speed profiles were bell-shaped for both movements (see [18] and Fig. 7). This outcome agrees closely with the data of Atkeson and Hollerbach [16], as one would expect from the minimum-torque-change model.

Finally, the most compelling evidence obtained by Uno et al. [18] concerns a pair of via-point movements, which are shown in Fig. 3. These movements involved two subcases, with identical start, ***T3***, and end, ***T5***, points but with mirror-image via-points, ***P1*** and ***P2***. Fig. 3**a** shows the convex path (b: ***T3*** →***P1***→***T5***) and the concave path (c: ***T3*** →***P2*** →***T5***). Fig. 3**b** and Fig. 3**c** show the corresponding speed profiles (dotted curves) and curvature profiles (solid curves). Because the objective function C_J is invariant under translation, rotation, and roll, the minimum-jerk model predicts identical movement paths with respect to roll as well as identical speed profiles for the two subcases. On the other hand, the minimum-torque-change model predicts two different paths as shown in Fig. 3. For the concave path, the speed profile should have two peaks. However, for the convex path, the speed profile should have only one peak. The latter prediction agrees closely with the data.

Summarizing these comparisons, we see that the trajectory derived from the minimum-jerk model is determined only by the geometric relationship of the initial, final and intermediate points in the task-oriented coordinates. The trajectory derived from the minimum-torque-change model depends not only on the relationship between these three points but also on the arm posture (in other words, the location of the shoulder relative to the three points), and on external forces. Empirical data suggest that the latter dependence is in fact the case. Wann et al. [19] also found that the minimum-jerk model fails because of its lack of information about movement dynamics.

Minimum-muscle-tension-change model

Considering that musculoskeletal systems possess muscle-tension sensors (Golgi tendon organs) and muscle length and velocity sensors (muscle spindles) but no direct joint-torque sensors, Uno et al. [20] proposed a minimum-muscle-tension-change model, in which the following objective function is minimized.

$$C_F = \frac{1}{2}\int_0^{t_f} \sum_{i=1}^{n} \left(\frac{dF_i}{dt}\right)^2 dt, \qquad \textbf{(4)}$$

where, $\boldsymbol{F_i}$ is the muscle tension generated by the $\boldsymbol{i}$th of $\boldsymbol{n}$ muscles. Here the performance measure (objective function) is the sum of the square of the rate of change of muscle tension, integrated over the entire movement. One can see that $\boldsymbol{C_F}$ of this last equation **(4)** is related to $\mathbf{C}\tau$ of the third equation **(3)** because the joint torque is the summation of muscle forces weighted by their moment arms for the joint.

Uno et al. [20] simulated the discrete point-to-point trajectories shown in Fig. 2 based on the minimum-muscle-tension-change model. We used a two-link manipulator with six muscles (elbow flexor and extensor, shoulder flexor and extensor, and double-joint flexor and extensor) as a model of the human arm. The contraction of double-joint muscles such as the biceps and the triceps feeds torque both to the elbow and to the shoulder. We [20] found that if the double-joint muscles were not included in the arm model, then the predicted trajectories were overly curved. Thus the presence of the double-joint muscles is essential to reproduce roughly straight hand paths for discrete point-to-point movements in the minimum-muscle-tension-change model. We do not fully understand the reason for this but speculate that double-joint muscles may act like linear spring actuators which effect forces on the hand in the Cartesian space.

We found that the minimum-muscle-tension-change model was better than the minimum-torque-change model in that it can reproduce human data for a wider range of inertial parameters of the arm. Recently Flash [21] criticized the minimum-torque-change model, specifically the link inertia moment value assumed in Uno et al. [18], according to her own simulation of the minimum-torque-change trajectory. The inertia moment value used in our simulation [18] of the minimum-torque-change trajectory was about double a reasonable value based on the other physical parameters of the links: mass, length and center of mass. Flash [21] pointed out that if a reasonable, smaller inertia moment value was assumed, the hand path for a point-to-point movement in front of the body was too curved compared with the human data. We also confirmed this simulation result, and concede that her criticism is valid. This is another compelling reason to extend the minimum-torque-change model to the minimum-muscle-tension-change model. Uno et al. [20] found that the minimum-muscle-tension-change model reproduced roughly straight hand paths using realistic inertia moment values, which were about half the magnitude used in our former simulation [18].

In our simulation of the minimum-muscle-tension-change model [20], the geometrical arrangement of the muscle insertion points and origins on the skeleton assumed in the arm model were not based on biological measurement. Recently, Dornay et al. [22, 23] reexamined minimum-muscle-tension-change trajectories while using a dynamic model of a monkey's arm, which is based on biological measurement, and reconfirmed the main results of Uno et al. [20].

The time course of muscle tension $\boldsymbol{F}_i$ derived from the minimum-muscle-tension-change criterion (Equation **(4)**) depends on the initial value of $\boldsymbol{F}_i$. When the muscle tension is small (its value is 0 in the extreme case) at the beginning of movement, the predicted muscle tension is always small during the movement and the stiffness of the arm is relatively low, which is another feature in skilled arm movements.

NEURAL NETWORK MODEL

We have presented several optimization models in the previous sections, but have not yet discussed how these optimization problems can actually be solved. In this section, therefore, we introduce several neural network models which can solve the optimization problems explained in the previous sections.

Simple recurrent network model for minimum-jerk trajectory

It is quite easy to analytically calculate the minimum-jerk trajectory as mentioned above. However, it is extremely difficult to assume that this algebraic procedure is actually adopted in the brain. Instead of such an algebraic procedure, Hoff and Arbib [24] proposed a network model that can generate a minimum-jerk trajectory. Their model is based on an optimal feedback controller design well known in optimal control theory. The model can be implemented as a recurrent neural network model which receives the movement duration, the target location, and the current position and velocity information, and can generate position and velocity at the next time step. The model calculates the minimum-jerk trajectory in Cartesian coordinates.

It was clearly demonstrated by this study that a simple recurrent neural network model can generate an exact minimum-jerk-trajectory in real time. Here, time in movement is represented by time in network dynamics. Furthermore, temporal interaction between reaching and grasping was simulated by introducing a new objective function for prehension. However, the model can not easily be extended to deal with dynamic optimization models like the minimum-torque-change or minimum-muscle-tension-change model. The model was prewired by a researcher, and training was not considered.

Jordan's recurrent network for optimal trajectory

In this subsection, we briefly introduce Jordan's approach to the optimal trajectory. Jordan [4] proposed a recurrent sequential network that can learn to generate a minimum-jerk virtual trajectory. In Jordan's recurrent network, first, a forward dynamics model of the motor apparatus is learned. Then, another recurrent network for motor control is attached to the motor-command input channel of the previously trained forward dynamics model. The motor control network was assumed to calculate a *virtual trajectory* [8, 9, 25]. Here, the virtual trajectory is regarded as the motor-command. Synaptic weights in the motor control recurrent network are modified so that two different objective functions are minimized. One corresponds to the minimum-jerk criterion of the virtual trajectory, and the other corresponds to the target location error at the end of movement. The learning process can be regarded as the embedding of the above two criteria into the synaptic weights of the motor control recurrent network.

Jordan's network is attractive in several respects. First, it has been shown that a network can be trained as a device for trajectory formation and control. Second, the recurrent network generates trajectories in real time. Third, movement time is naturally represented as time in the dynamics of the recurrent network. Jordan has stated that the learning scheme can readily be applied to the formation of a minimum-torque-change trajectory. However, it is not apparent whether a large number of trajectories starting from arbitrary points to arbitrary targets can simply be stored as synaptic weights in the recurrent motor control network. It can mathematically be shown that his original sequential network is too small to store a method to calculate the minimum-jerk or minimum-torque-change trajectories rigorously.

Cascade neural network model for a dynamically optimized trajectory

Since the dynamics of the human arm is nonlinear, finding the unique trajectory that minimizes the objective functions of the minimum-torque-change or minimum-muscle-tension-change model is a nonlinear optimization problem. Kawato et al. [26] proposed the cascade neural-network model to solve the nonlinear optimization problem. In this subsection, we present the cascade neural-network model to find the optimal trajectory that minimizes the torque-change criterion $\mathbf{C}_\tau$. This network model can also readily be applied to trajectory formation based on the minimum-muscle-tension-change model.

For simplicity, we explain the neural network model in the case of a single-degree-of-freedom controlled object. There is no need to change the basic structure and algorithm of the model for extension to a multiple-degree-of-freedom case. We need only to increase the number of neurons according to the degrees of freedom in the motor command, the body space and the task space.

The model is a repetitively-structured, time-invariant, cascade neural network as shown in Fig. 4. The state of the controlled object is described by the joint angles of the arm, θ, and angular velocities, $d\theta/dt$. The motor command is denoted by τ; joint torque is a special case of the motor command. Generally, the state of the controlled object is described as the following differential equation:

$$d^2\theta / dt^2 = f(\theta, d\theta / dt, \tau) . \tag{5}$$

Although this equation is expressed in continuous time, the network adopts a discrete-time representation. That is, the model consists of many identical four-layer network units; the jth network unit corresponds to time $j\Delta t$. Here Δt denotes a time step (unit of time). If there are N network units, the model can generate a movement of duration up to $N\Delta t$. Each network unit calculates the change of the trajectory within a unit of time, $\Delta t \bullet f(\theta, d\theta/dt, \tau)$, between the first and third layers. The fourth layers of the network units (i.e., the output line on the right side of Fig. 4) represent the estimated time course of the trajectory.

Network operation is divided into a learning phase and a trajectory-formation phase. In the learning phase, the common input torque is fed to both the controlled object and the neural-network model. The realized trajectory of the controlled object (arm) is used as a teaching signal to acquire the forward dynamics model between the first and third layers of the network unit. The back-propagation learning algorithm [27] can be applied to the modification of the synaptic weights. The back-propagation learning is biologically implausible, and hence it can be replaced by other learning algorithms such as *associative reward-penalty learning* proposed by Barto and Anandan [28].

Once this learning is completed, the cascade neural network provides a forward dynamics model of the controlled object; that is, each network unit can calculate $\Delta t \bullet f(\theta , d\theta/dt , \tau)$ corresponding to Equation (5). Fig. 4 shows the trajectory-formation phase in which the cascade neural network calculates the optimal trajectory and the corresponding necessary motor command using relaxation computation. The higher motor center gives information to the fourth layer of the network about locations of the desired target point, the desired via-point, and the locations of obstacles to be avoided. The minimum-torque-change criterion is embedded as hardware in the model (see electrical resistance in Fig. 4). That is, neurons representing motor commands at a neighboring time are connected by an electrical resistance so that their states can be brought to a common value. It can be shown mathematically that the cascade neural network is executing a steepest descent motion with respect to the objective function of the optimal control and will reach a stable equilibrium point where the summation of the smoothness criterion multiplied by the electrical conductance of the gap junction and the error in movement conditions is minimum. Consequently, the torque time course required to generate the minimum-torque-change trajectory can be calculated by relaxation. In our computer simulation, the

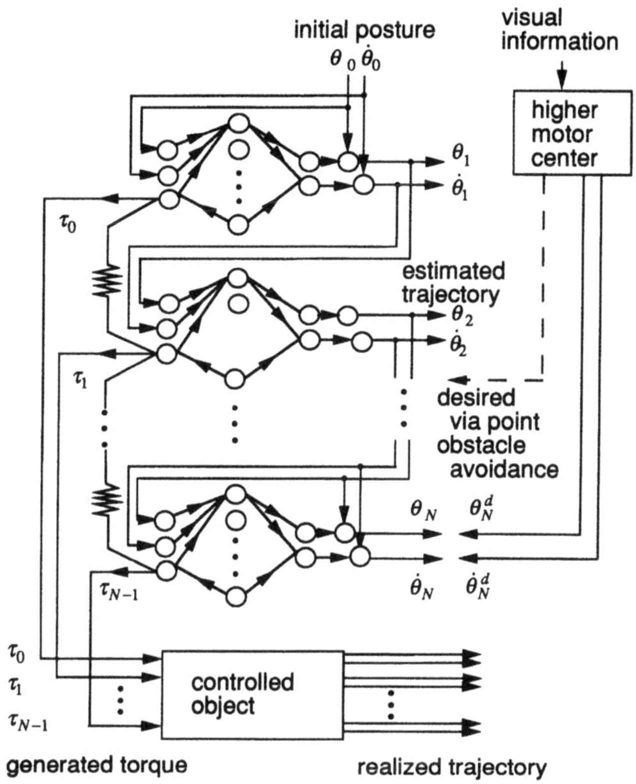

Figure 4. A cascade neural network model for trajectory formation based on the minimum torque-change criterion. From [26].

cascade network generated point-to-point discrete movements, via-point movements, and movements that avoid obstacles, based on the minimum-torque-change criterion. In contrast to Jordan's scheme, learning is used only to acquire a forward dynamics model of the motor apparatus, and the smoothness constraint and task constraint are satisfied by relaxation calculation. Thus, a trajectory can be generated and controlled only if the forward dynamics model has been acquired. This is both the advantage and disadvantage of the cascade model. A computationally intensive relaxation process requires a few biologically implausible mechanisms. The first is that a large number of iterations are required for relaxation calculations. A second weak point is the necessity of back-propagation during the relaxation calculation. Back-propagation is biologically implausible. Thirdly, time is represented by a cascade array of motor-command neurons in the cascade neural network model. We can not totally deny spatial representation of time in the brain, but it seems rather awkward.

Trajectory formation based on forward and inverse models

To resolve the disadvantages of the cascade network, Wada and Kawato [29] had the novel idea of using both forward and inverse models of the motor apparatus during optimization calculation. As far as we know, this is a totally new method in the field of optimal control. Fig. 5 shows the structure of the new model, which consists of four subnetworks: a feedforward network for the inverse dynamics and kinematics model of the controlled object, a recurrent network for the forward dynamics and kinematics model of the controlled object, a network that executes smoothing of the motor commands, and a network that calculates an approximate trajectory of the nonlinear optimization problem. Let us explain the fundamental relaxation procedure in this model. First, the approximate trajectory formation mechanism generates a suboptimal trajectory. Although this suboptimal trajectory does not attain optimal performance, it satisfies the multiple-point boundary conditions. Second, the inverse model of the controlled object calculates the motor command necessary to realize the suboptimal trajectory. Then, the pair (the trajectory and the motor command) satisfies the dynamic relationship for the controlled object as well as the task specifications for the target point and via-points. Third, the motor-command smoothing circuit smooths out the motor command waveform. After this smoothing operation, however, neither the boundary conditions nor the dynamic relationship is satisfied. Fourth, the forward model of the motor apparatus operates on the smoothed motor command and calculates the resulting trajectory. The new trajectory and the smoothed torque give a valid pair, which satisfies the dynamic relationship. However, the boundary conditions are still violated. Finally, the approximate trajectory formation mechanism generates a corrective suboptimal trajectory which compensates for the error in the boundary conditions. This corrective trajectory is summed with the previous trajectory. At this point the boundary conditions are satisfied. However, the dynamic relationship is violated. Returning to step two above and using the inverse model, the correct motor command is calculated. After this, the four operations, inverse dynamics, motor-command smoothing, forward dynamics, corrective trajectory formation and summation, are repeated in this order

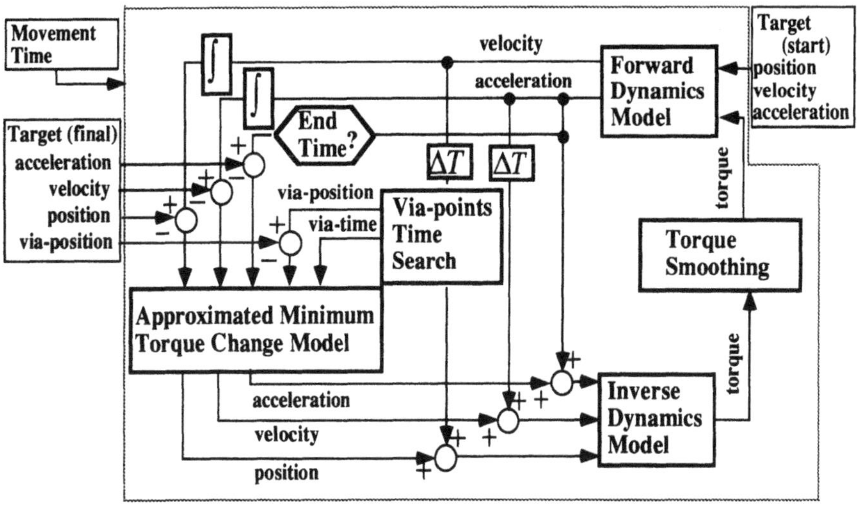

Figure 5. The neural network structure proposed for arm trajectory formation based on minimum-torque-change model using forward and inverse models of the controlled object. From [29].

until a satisfactory solution is obtained.

As an approximate optimal solution, Wada and Kawato [29] investigated a linear optimal control problem associated with the original nonlinear optimization problem. The nonlinear dynamics of the controlled object can be locally linearized around a specific trajectory and the motor command. Then, it is rather easy to obtain an exact optimal solution for this linearized dynamics while using the same objective function. Wada and Kawato [29] mathematically proved that the necessary and sufficient condition for the convergence of the above iterative procedure is that the optimal solution to the nonlinear problem is obtained. Furthermore, under a reasonable assumption, it can be shown that the objective function decreases for every step of the algorithm. Thus, under this circumstance, the optimal solution can be obtained with the new neural network model.

The new neural network scheme presented here plays a major role in trajectory formation and feedforward control. In addition, if a feedback component is added to the inverse dynamics model and/or the forward dynamics model, this scheme can correct trajectories based on sensory information as well as the Hoff-Arbib model.

Let us intuitively explain why the new model resolves the three disadvantages of the cascade neural network model. First, back-propagation is not needed because both the forward dynamics and inverse dynamics models are used in the new model. In the cascade or Jordan's model only the forward dynamics model was used, and hence back-propagation through it was essential to transform the terminal error condition into the motor-command space. In the new model, the inverse dynamics model executes this operation. Second, the number of iterations is dramatically reduced because of the overall structure of the model. Wada and Kawato [29] confirmed by computer simulations that only one to five iterations were required to find the approximately optimal solution for a point-to-point movement. Finally, because the motor-command smoothing can be executed by a simple temporal filter, spatial representation of time is not needed; that is, the movement time can be represented as time in the network dynamics.

DEVELOPMENT OF OPTIMAL CONTROL RESEARCH

Any computational theory or neural network model with biological or psychological relevance should possess the capability to resolve the ill-posed motor control problems. Some models introduced in this chapter have been discussed from this viewpoint. Any optimization process to resolve the ill-posed motor control problems is exceedingly complicated. Do living things really possess optimization models like those presented in this chapter? Lower animals such as insects would not solve the ill-posed motor control problems explicitly, because their movements are stereotyped. However, we believe that the primate brain resolves the ill-posed motor control problems using some optimization procedure. The architecture of the optimization models presented here is considerably artificial, and hence it is not clear whether these models are biologically plausible. We have put emphasis on the computational theory for motor control rather than on biological aspects. At the end of this chapter, we consider future tasks on the optimal control research of human reaching movements.

In the early sections, we introduced three different optimization models in which a ***smoothness constraint*** is imposed on the different spaces; the minimum-jerk model is defined at the task space, the minimum-torque-change model is defined at the joint-torque coordinates, and the minimum-muscle-tension-change model is defined at the muscle tension level. The minimum-jerk model can only determine the desired trajectory in the task-oriented coordinates, and hence cannot resolve the ill-posed inverse-kinematics or inverse-dynamics problems for redundant arms. To resolve all three ill-posed problems on motor control, Flash [25], Mussa-Ivaldi et al. [30], and Massone and Bizzi [31] combined the minimum-jerk model with other approaches (e.g., *virtual trajectory hypothesis*). On the other hand, the minimum-muscle-tension-change model can resolve the three different ill-posed problems. However, even if the muscle tensions are all specified, the firing patterns of motor neurons can not be uniquely determined, because there exists a vast number of neurons that are involved in motor control. Therefore, we should further examine the optimization principle at the motor command level of the central nervous system. In addition, it seems more plausible that the origin of the smoothness resides in the central nervous system rather than in the periphery.

What is the optimization principle at the motor command level? If muscle tensions are proportional to motor commands, it is quite easy to extend the minimum-muscle-tension-change model to a minimum-motor-command-change model. It is, however, well known that muscle tensions depend on muscle length, shortening velocity, fatigue and external loads. In particular, Mannard and Stein [32] showed that the relation between isometric force change and neural stimulus input can be expressed as a linear second-order system. In this case, the smoothness of muscle tensions would be derived from minimizing the activities of motor neurons (i.e., motor commands). In addition, the minimum-motor-command criterion might serve to economize energy and effect a compatibility with the low stiffness of muscles which is another feature in skilled arm movements.

In this chapter, we have introduced several neural network models for trajectory formation. These neural network models can be applied to other optimization problems. For example, let us consider human grasping movements. Determining a prehensile hand shape is an ill-posed problem, because there are many ways to grasp any one object. Uno et al. [33] proposed a neural network model that determines the optimal hand shape in grasping movements. As shown in Fig. 6, the network model consists of five layers of neurons, and the input layer (1st layer) and the output layer (5th layer) have the same structure. The operation of this network is divided into a learning phase and an optimization phase as in the cascade neural network model. In the learning phase, first, many grasping movements are performed by trial and error using feedback control, etc., and then the visual images of grasped objects and prehensile hand shapes are perceived by several sensory organs.

The neural network shown in Fig. 6 learns the relation between the visual images and hand shapes. Both a signal $\boldsymbol{x}$ representing the visual image of an object and a signal $\boldsymbol{y}$ representing the hand shape to grasp it are fed to the

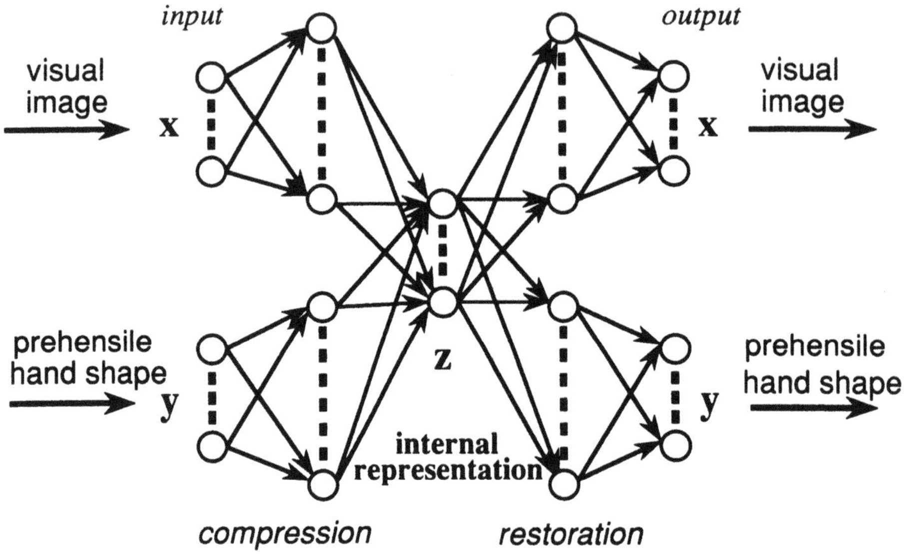

Figure 6. A neural network model for recognizing objects and planning hand shapes in grasping movements.

1st layer. The synaptic weights of the network are modified so that the 5th layer outputs the same signals x and y as are fed to the 1st layer. As fewer neurons are set in the 3rd layer than in the 1st layer, signals x and y are compressed and integrated. After network learning, an internal model that represents the relation between visual and motor information is embedded in the network. Based on the internal model and certain criteria, the network can design a hand shape to grasp any given object in the optimization phase. When a visual image x of an object is given, the network calculates the optimal prehensile hand shape y using relaxation computation.

The presented network model treats only a static configuration of the human fingers. The model should be developed so that it can learn the dynamics of human hand manipulation. The coordination of reaching and grasping movements can be regarded as an ill-posed problem. However, it is not clear at the present stage what criterion defines the optimal grasping movement. A coordinated control scheme for reach and grasp is discussed in the previous chapter.

ACKNOWLEDGEMENTS

We thank Dr. Yo'ichi Tohkura of ATR Human Information Processing Research Laboratories for his continuing encouragement. We are also grateful to Drs. Menashe Dornay and Yasuhiro Wada for allowing us to duplicate their results and for their discussion on subjects treated in this chapter. Preparation of this manuscript was supported by a Human Frontier Science Project Grant to Mitsuo Kawato.

REFERENCES

1 E.L. Saltzman, J. Math. Psychol., 20 (1979) 91.
2 J.M. Hollerbach, TINs, 5 (1982) 189.
3 M. Kawato, K. Furukawa and R. Suzuki, Biol. Cybern., 57 (1987) 169.
4 M.I. Jordan, in: M. Jeannerod (ed.), Attention and Performance XIII, Erlbaum, Hillsdale, 1990, p. 796.
5 M.I. Jordan and D.A. Rosenbaum, in: M.I. Posner (ed.), Foundations of Cognitive Science, MIT Press, Massachusetts, 1989, p. 727.
6 N. Bernstein, The Coordination and Regulation of Movements, Pergamon, London, 1967.
7 M.T. Turvey, R.E. Shaw and W. Mace, in: J. Requin (ed.), Attention and Performance VII, Erlbaum, Hillsdale, 1978, p. 557.
8 E. Bizzi, N. Accornero, W. Chapple and N. Hogan, J. Neurosci., 4 (1984) 2738.
9 N. Hogan, J. Neurosci., 4 (1984) 2745.
10 E.L. Saltzman and J.A.S. Kelso, Psychol. Rev., 94 (1987) 84.
11 A.E. Bryson and Y.C. Ho, Applied Optimal Control, Blaisdell, Waltham, 1969.
12 W.L. Nelson, Biol. Cybern., 46 (1983) 135.
13 J.A.S. Kelso, D.L. Southard and D. Goodman, Science, 203 (1979) 1029.
14 P. Morasso, Exp. Brain Res., 42 (1981) 223.
15 W. Abend, E. Bizzi and P. Morasso, Brain, 105 (1982) 331.
16 C.G. Atkeson and J.M. Hollerbach, J. Neurosci., 5 (1985) 2318.
17 T. Flash and N. Hogan, J. Neurosci., 5 (1985) 1688.
18 Y. Uno, M. Kawato and R. Suzuki, Biol. Cybern., 61 (1989) 89.
19 J. Wann, I. Nimmo-Smith and A.M. Wing, J. Exp. Psychol., Hum. Percep. Perf., 14 (1988) 622.
20 Y. Uno, M. Kawato and R. Suzuki, Proc. 4th Symp. Biol. Physiol. Eng., 1989, p. 299 (in Japanese).

21 T. Flash, in: J.M. Winters and S.L.Y. Woo (eds.), Multiple Muscle Systems: Biomechanics and Movement Organization, Springer, New York, 1990, p. 282.

22 M. Dornay, Y. Uno, M. Kawato and R. Suzuki, in: J.E. Moody, S.J. Hanson and R.P. Lippmann (eds.), Advances in Neural Information Processing Systems 4, Kaufmann, San Mateo, 1992, p. 627.

23 M. Dornay, Y. Uno, M. Kawato and R. Suzuki, in: Proc. 31st SICE Annual Conference, 1992, p. 1131.

24 B. Hoff and M.A. Arbib, submitted.

25 T. Flash, Biol. Cybern., 57 (1987) 257.

26 M. Kawato, Y. Maeda, Y. Uno and R. Suzuki, Biol. Cybern., 62 (1990) 275.

27 D.E. Rumelhart, G.E. Hinton and R.J. Williams, Nature, 323 (1986) 533.

28 A.G. Barto and P. Anandan, IEEE Trans., SMC-15 (1985) 360.

29 Y. Wada and M. Kawato, Neural Networks, in press.

30 F.A. Mussa-Ivaldi, P. Morasso and R. Zaccaria, Biol. Cybern., 60 (1988) 1.

31 L. Massone and E. Bizzi, Biol. Cybern., 61 (1989) 417.

32 A. Mannard and R.B. Stein, J. Physiol., 229 (1973) 275.

33 Y. Uno, N. Fukumura, R. Suzuki and M. Kawato, Adv. Neural Inform. Processing Systems, 5 (1993) 311.

Author index

Subject index

www.ingramcontent.com/pod-product-compliance
Ingram Content Group UK Ltd.
Pitfield, Milton Keynes, MK11 3LW, UK
UKHW020230290926
15470UKWH00002B/457

9 780444 899316